IMPERIAL ENGINEERS

RICHARD HORNSEY

IMPERIAL ENGINEERS

The Royal Indian Engineering College, Coopers Hill

UNIVERSITY OF TORONTO PRESS
Toronto Buffalo London

ISBN 978-1-4875-0686-5 (cloth)
ISBN 978-1-4875-3505-6 (EPUB)
ISBN 978-1-4875-3504-9 (PDF)

Library and Archives Canada Cataloguing in Publication

Title: Imperial engineers : the Royal Indian Engineering College, Coopers Hill /
Richard Hornsey.
Names: Hornsey, Richard, 1964– author.
Description: Includes bibliographical references and index.
Identifiers: Canadiana (print) 20210377194 | Canadiana (ebook) 20210377372 |
ISBN 9781487506865 (cloth) | ISBN 9781487535056 (EPUB) |
ISBN 9781487535049 (PDF)
Subjects: LCSH: Royal Indian Engineering College – History. | LCSH: Engineering
schools – England – London – History. | LCSH: Engineering students – England –
London – History. | LCSH: Engineers – India – History.
Classification: LCC T108.L66 H67 2022 | DDC 620.0071/14212 – dc23

We wish to acknowledge the land on which the University of Toronto Press
operates. This land is the traditional territory of the Wendat, the Anishnaabeg, the
Haudenosaunee, the Métis, and the Mississaugas of the Credit First Nation.

University of Toronto Press acknowledges the financial support of the Government of
Canada, the Canada Council for the Arts, and the Ontario Arts Council, an agency of
the Government of Ontario, for its publishing activities.

Canada Council
for the Arts

Conseil des Arts
du Canada

ONTARIO ARTS COUNCIL
CONSEIL DES ARTS DE L'ONTARIO

an Ontario government agency
un organisme du gouvernement de l'Ontario

Funded by the   Financé par le
Government   gouvernement
of Canada   du Canada

Canadä

Contents

Figures and Tables

Figures

Tables

Abbreviations

Note on place names: Many of India's cities have been renamed since the end of British rule. However, this book uses the former British names to maintain consistency with historical sources.

CH (years)	years of study at Coopers Hill
CHM	*Coopers Hill Magazine*
CHS	Coopers Hill Society
CSAS	Centre of South Asian Studies, University of Cambridge
DFO	divisional forest officer
DS	divisional superintendent, Telegraph Department
ICE	Institution of Civil Engineers
IOR	India Office Records and Private Papers
KCIE	Knight Commander of the Most Eminent Order of the Indian Empire
OCH	Old Coopers Hill man
PP Cd.	parliamentary command paper, UK
PWD	Indian Public Works Department
RIEC	Royal Indian Engineering College, Coopers Hill
SAS	sub-assistant supervisor, Telegraph Department
UCL	University College London

Acknowledgments

For their hospitality and patient assistance at the archives of their respective institutions, I am indebted to Kevin Greenbank and the University of Cambridge Centre of South Asian Studies, Anne Barrett at Imperial College, and Carol Morgan from the Institution of Civil Engineers. I am also grateful to the staff of the Asian and African Studies Reading Room of the British Library for their help.

I thank the vicar and staff of St. John's Church, Egham, for enabling me to locate and view the Coopers Hill memorials in their lovely church.

Part of this work was performed under the auspices of the York-Massey Fellowship at Massey College, Toronto. I am grateful to the principal, members, and staff of the college for their hospitality.

Paul Thomas and David Strong read early versions of this manuscript and offered their valuable comments and suggestions.

Stephen Shapiro at the University of Toronto Press has been a great source of insight and support throughout the preparation of this book.

Lastly, thank you to my family for everything else.

IMPERIAL ENGINEERS

1

Dastardly Murder

Introduction

On the morning of 30 August 1893, between Moradabad and Rampur in northern India, a young man named Edmund Elliot was the victim of a "cold-blooded and dastardly murder"[1] when Robert Hawkins shot him twice in the back. As Elliot lay writhing on the ground, Hawkins fired a third bullet into the man's stomach and "Mr. Elliot ceased to struggle."[2]

People were affronted. In the relatively small British-Indian community the murder was front-page news and the subsequent trial was followed with interest. This was not the "done thing"; cowardly shots in the back, insubordination against a superior, a hint of insanity. Very quickly Hawkins was vilified as "a wrong-headed, fiery-tempered man, who seemed to limit his sense of duty to obstructing and insulting his superior officer in every possible way." In contrast, Elliot was a man of "uncommon professional abilities," "unsparing of himself," and together with his wife a "genial and open" host.[3] But when arrested, Hawkins said, "I don't call it murder. If you only knew half of what I have had to put up with, you would say I was justified."[4]

Far from being hot-headed rivals in love, these two men were supposedly level-headed engineers, working on the Bareilly-Rampur-Moradabad Railway. But what did that work entail and how were they trained and selected for it? What pressures of the job and life in India drove Hawkins to thoughts of murder?

Elliot had been educated at the Royal Indian Engineering College in England expressly so he could join the engineers building the vast transportation, irrigation, and communication infrastructure of British India. As educational institutions go, the Royal Indian Engineering College (RIEC) – known informally as Coopers Hill because of its location – had a relatively brief existence.[5] While it was open for just thirty-five years, between 1871 and 1906, its 1623 graduates[6] contributed to the world in an increasingly wide variety of ways until the

Second World War. At the time it was a high-profile institution, governed in a rather military style but with the aspirations of an Oxbridge college. This book traces the history and impact of Coopers Hill through the lives and careers of its students, professors, and staff.

Because engineering works were such a major source of British national pride during the late nineteenth century, technical education became a focus of attention. This led to an important period of transition in engineering education, marking the switch from the training of engineers by apprenticeships to university programs. In 1871 "engineering" was either civil or military – the subject matter was mostly the same with a strong emphasis on construction, but the applications were different. By 1906, however, civil and mechanical engineering were becoming distinct fields, and electrical engineering was emerging as a new discipline. Moreover, the working lives of college graduates spanned the tremendous social and political transitions of the early twentieth century in Europe and India.

Royal Indian Engineering College

From the outset, the college's existence was political. It was created as a government initiative to increase the supply of high-quality civil engineers for the Indian Public Works Department (PWD). At the time, "high quality" meant "educated in England," and the college was built at the Coopers Hill estate, near Egham on the outskirts of London. Its creation was opposed by the Government of India because they saw no need for a new college, especially one funded from their revenues. The feathers were also ruffled of the PWD establishment in India, which was then staffed largely by the military Royal Engineers. Later, the authorities transferred the training of telegraph officers to Coopers Hill and added Britain's first school of forestry.[7]

The college's eventual closure was similarly political and controversial. By 1906, British universities were offering enough engineering degrees for the needs of India to be met without Coopers Hill. The British government was also fed up with the expense and inconvenience of operating an engineering school. This time, in an ironic twist, the Government of India now opposed the closure.

This political contradiction was summarized in "The History of Coopers Hill College in Two Chapters":

> Chapter I, 1870
> Secretary of State [for India] – "I propose to establish Coopers Hill College."
> The Government of India – "We do not like the idea."
> Secretary of State – "Oh! don't you? Then consider the College established."

Chapter II, 1903
Secretary of State – "I propose to disestablish Coopers Hill College."
The Government of India – "Don't do that, the College has been a great success."
Secretary of State – "Oh! has it? Then consider it disestablished."[8]

But Coopers Hill was more than a political creation, it was a vigorous intellectual and social community. To the "Coopers Hill man,"[9] the three intense, formative years spent at the college embodied all the honour and traditions of an Oxbridge college. Its esprit de corps led to a social and professional network that spanned the world and supported men like Edmund Elliot in their frequently isolated and arduous workplaces. To the professors it was a prestigious institution for furthering their research or educational career. For the staff it was the source of great pride and loyalty. This community was driven by the principles, achievements, and prejudices of its members, and it is to the lives of these people we must turn in order to understand the human story behind the college's place in the world.

Coopers Hill officers were in the field for nearly seventy years, counting from the first appointee to the last retiree. It is useful at the start to reflect briefly on the huge scale of public works in India during this time.

When Coopers Hill was established, revenues and expenses from India were roughly balanced, at a little over £50 million per year. Of this, the Indian Public Works Department's budget took a sizeable £7.5 million, "more than the entire sum raised for the Government from its opium monopoly."[10]

Coopers Hill engineers were particularly involved in the construction of railways and irrigation systems. During the college's existence, an astonishing 22,980 miles of railways were built in India.[11] Another 12,000 miles or more were added before the last Coopers Hill man retired.[12] It is estimated that these railways were carried over 175,000 bridges[13] and through numerous tunnels. At the same time, the area of Indian farmland under irrigation increased by 33 per cent, mainly as a result of canal construction. By 1902, some 44 million acres (18 million hectares) were being watered, representing nearly one-fifth of the country's agricultural land.[14]

From 1878, Coopers Hill would provide training for the Indian Telegraph Department. By 1904, just before the college's closure, there were 2127 combined postal and telegraph offices in India. They transmitted 7.3 million paid messages per year along nearly 60,000 miles of line.[15]

Starting in 1885, the college also offered a forestry program. Thirty years later, when Coopers Hill graduates were in senior positions in the Indian Forest Service, more than 20 per cent of the total area of British India was under the control of the Forest Department, totalling a staggering 250,000 square miles (65 million hectares), roughly the area of France.[16]

Taken together, these numbers reflect a truly remarkable technical achievement, one that was led by the graduates of Coopers Hill. Since their construction, these public works have affected the lives of hundreds of millions of people and continue to underpin the infrastructure of one of the world's most populous and significant countries.

Scope

This is primarily a book about an engineering college and its students. Much of the story is set in England and concerns the establishment and history of the college, life as a student, and the Coopers Hill Society that was formed after the college's closure. Its focus is particularly on the lives and work of the graduates, and how their education at Coopers Hill prepared them to be engineer, telegraph, and forest officers in India. Information for this comes from official documents (college calendars, parliamentary papers, etc.), personal recollections and diaries of Coopers Hill students and staff, and articles in college magazines and technical publications.

Although most do not appear here, this research has traced the careers of more than a thousand men from the time they joined Coopers Hill as eighteen-year-olds through their work in India until their retirements (or perhaps until early deaths). As a result of this familiarity, the perspective here is generally sympathetic to the college and its graduates. It also tends to be a view from the inside – how *they* regarded the technical issues, what context was important to *them*, how *they* experienced life and work in India. These men were socially and financially relatively privileged and shared the contemporary attitudes of their background. But most were ordinary people who struggled with their technical studies and fulfilled their duties in India and other countries to the best of their abilities.

That said, Coopers Hill was a creature of the colonial system, and the majority of its graduates served the British Government in India. They typically saw their roles in terms of duty to the empire, and for the most part, their class and upbringing led them to support the ideologies and goals of that government. Although very few Coopers Hill graduates played significant roles in shaping government policies, by building infrastructure and administering forests in India they were an important part of the British colonial machinery. They were therefore complicit in the broader issues of colonialism, environmental impact, and imperial legacy. This book addresses these issues where they relate to the college and its graduates, but the purpose here is not to assess the broader legacy of the British Raj.

The railway, irrigation, and road infrastructure built in India by the British, in many cases by Coopers Hill engineers, has often been considered one of the beneficial legacies of colonial rule. The Coopers Hill men certainly thought this way. Colonial authors such as Maud Diver in *The Unsung* and Col. E.W.C. Sandes in *The Military Engineer in India* naturally agreed. Historian Ian Hay wrote that the Public Works Department men ensured "the dams hold, the canals irrigate, the grass grows, and the British Raj endures."[17] Some more recent authors such as Kartar Lalvani in *The Making of India* adopt a more balanced but similarly positive view of the legacy left by British engineers.

Passionate nationalists, of whom Shashi Tharoor is a prominent recent example, are opposed to this perspective. He wrote about the railways, "In its very conception and construction, the Indian railway system was a big colonial scam." In this view, the infrastructural legacy of the Raj was an incidental by-product of the imperial objectives of maintaining military supremacy and increasing agricultural and resource revenues – a by-product, moreover, that came at great cost to the Indian people. Of Coopers Hill specifically, Tharoor stated that the entrance examination was "designed to exclude the majority of Indian candidates."[18] The nature of the entrance exam and the admission of Indians to both Coopers Hill and the Indian Public Works Department are addressed later in the present work.

Between these contrasting views lies a wealth of historical enquiry into the motivations, implementations, and implications of British public works in India. Detailed references will be given in the relevant sections, but works such as *Technology and the Raj* (Macleod and Kumar) and *Science, Technology, and Medicine in Colonial India* (Arnold), explored the broad social and economic effects of the combination of technological change and imperial policies.[19]

In specific disciplines related to Coopers Hill, Ian Kerr has examined the construction and history of India's railways, including an account of the Indian workforce that actually constructed them.[20] The multifaceted role of the telegraph in British India has been explored by Chaudhuri in *Telegraphic Imperialism*. British policies concerning forest management and irrigation had lasting social and environmental consequences that have gained great modern interest and have been addressed in major works such as *Nature, Culture, Imperialism* (Arnold and Guha), *Environment and Empire* (Beinart and Hughes), *The British Empire and the Natural World* (Kumar, Damodaran, and D'Souza), and *This Fissured Land* (Gadgil and Guha).[21]

In part, the present work can be regarded as an engineering counterpart to David Gilmour's *The Ruling Caste* and *The British in India,* insofar as it relates the experiences of individuals living and working in India.[22] But whereas those works (and many others) concentrate on members of the army or civil service, the focus here is on the training and experiences of the British Empire's technical services, which have received considerably less attention. By focusing on the period from 1871 to 1906, it also serves as a prequel to Aparajith Ramnath's *The Birth of an Indian Profession,* which considered the emergence of the Indian engineering profession after the closure of Coopers Hill.[23] In addition, it complements and extends the previous partial histories of the college. In 1960, J.G.P. Cameron, a former student, wrote *A Short History of the Royal Indian*

Engineering College – a brief overview for circulation to Coopers Hill alumni[24] – whereas a more detailed account of the governmental and institutional aspects of the college was given by Brendan Cuddy in his doctoral dissertation, "The Royal Indian Engineering College, Coopers Hill (1871–1906): A Case Study of State Involvement in Professional Civil Engineering Education."[25]

The present volume is therefore the first complete history of the Royal Indian Engineering College, a vision originally conceived more than a century ago by its graduates.[26]

Goals

In this context, this book has two overall goals: to understand the culture, education, and achievements of its students, and to assess the significance of Coopers Hill in the picture of the British Empire and its engineering education.

The college's brief existence and present obscurity[27] belie its contemporary prominence. With its government backing, a prominent founder, outstanding professors, and the lure of a guaranteed job in India, Coopers Hill made a big impact in its early years. The entrance examinations were widely advertised, the annual prize day was attended by dignitaries and reported in national newspapers, and its sporting achievements were legendary – all of which drew attention to the fledgling college.

In progressing from Cheltenham College to Coopers Hill, and to India, Edmund Elliot was following an established pathway. Studying at Coopers Hill was expensive.[28] It was designed to attract relatively wealthy families by offering their sons a career in India comparable, but never quite equal, in prestige to the army or civil service.[29] By combining a technical education with traditional "manly" pursuits (Elliot, for example, was captain of boats), Coopers Hill was a natural destination for young men with an English public school education. And as a college that was considered comparable with Oxbridge and Trinity College Dublin,[30] it contributed to the increased respectability and wider appeal of engineering to the upper middle classes.[31]

Although not the only institution at the time to offer engineering training, Coopers Hill at its opening provided a systematic, comprehensive technical education that was ahead of its time in Britain, particularly in the combination of theory and practice. It brought engineering education into the spotlight and arguably stimulated other institutions to expand and enhance their own programs. Ironically, their success in this regard contributed to Coopers Hill's eventual closure.

The capital costs of Coopers Hill and any annual operating deficits were financed by the state out of revenues from India. The public universities were unhappy about the government subsidizing a competitor, particularly in the 1880s when the college was opened to students not going to India.[32] It was also a source of anxiety for successive administrators of Coopers Hill because of repeated demands to make the college self-sustaining. This experiment in state-run education effectively concluded when the government declared that Coopers Hill was not an institution of higher learning at all, but a training establishment more akin to a military academy.

In 2018, another new technical university was established in England with initial support from the government. Like Coopers Hill nearly 150 years earlier, the New Model in Technology and Engineering (NMiTE) intends to offer a new engineering education designed to meet the needs of its time. Indeed, the parallels between opening an engineering school in 1871 and today are surprisingly strong. The issues of financial sustainability, start-up funding for facilities, student recruitment, and building reputation are essentially the same as those faced by Coopers Hill. So too are the challenges of including practical experience in the curriculum and meeting the expectations of external bodies (the Public Works Department or accreditation bodies). For the same reasons, student tuition fees in engineering are still relatively high. A study of Coopers Hill, one of the first engineering schools, therefore reveals the origins of dilemmas in engineering education that are still relevant.

A Legacy of Monumental Works

Other dilemmas result from the nature of the engineering work itself. At the last Coopers Hill Prize Day in July 1906, Sir John Ottley stated, "The RIEC may be abolished but the name of 'Coopers Hill' will ever live in the monumental works erected in India by her sons."[33] The Coopers Hill men viewed their legacy in the wholly positive light of their Indian accomplishments. In view of the number of their works still in use today, this belief has justification: William Johns and Charles Cole's Mushkaf-Bolan Railway is still in use, likewise Herbert Harington's Kalka-Shimla Line and George Rose's Khojak Tunnel; Robert Gales's Hardinge Bridge still stands; the descendants of Eustace Kenyon and Ernest Hudson's telegraph system faithfully transmitted messages throughout India until 2013; the mile-long Mettur Dam constructed in part by Vincent Hart still holds back almost 100 billion cubic feet of water; and John Benton's Swat River Canal still carries water through the lengthy tunnel constructed by William Sangster.

But that is not the whole story, for even perfectly sound technical engineering can have far-reaching and unanticipated consequences. Eloquent arguments have been made that British rule left no positive legacy in India.[34] In addition, much has been written specifically on the social and environmental impacts of railway construction, irrigation projects, and forestry practices in British India, implemented by engineers, with the conclusion that there were undoubtedly negative effects.[35] Knowledge and policies varied with time but, broadly, social and environmental consequences were either known, but subordinate to British imperial interests, or unexpected.

The first category would include such issues as agricultural policies, enabled by irrigation, that moved people away from well-understood farming practices and left them susceptible to famine;[36] deforestation due in part to the vast numbers of wooden sleepers (ties) used in railway construction, as well as being used to power locomotives;[37] and a transportation and communication infrastructure more suited for maintaining British authority and trade than for serving the Indian population.[38]

Into the second category would fall creation by irrigation of damp breeding grounds for mosquitoes that transmit malaria;[39] the long-term effects of coppicing in wet climates;[40] silting of waterways;[41] and the increased urbanization and spread of disease facilitated by the railways.[42] In many cases, the engineers of the time were aware of the issues, but solutions were either not known or not implemented because of cost.[43]

Engineering at the time of Coopers Hill, especially on the scales possible in India and Africa, was perhaps the first instance of a paradoxical situation that is still with us. They had the technology, wealth, organization, and determination to undertake engineering projects with massive implications. However, they frequently lacked the understanding of what those implications might be, or the motivation to develop such understanding. Facing today's global climate change, a plastic waste crisis, and the imminent extinction of one million species, it seems we have learned little from experiences in India.

Overview

With the foregoing as motivation and context for this study, four broad questions guide the discussion: (1) Why was Coopers Hill established and how did it go about its fulfilling its mandate? (2) What did the college graduates do in India and how did their education equip them for this work? (3) How did increasing diversity and progress in engineering education contribute to the college's closure? (4) What does the extraordinarily long-lived Coopers Hill Society tell us about the men who attended the college?

Chapters 2 through 4 address question 1. The government's motivation for establishing Coopers Hill is examined with reference to the education and supply of engineers for India prior to 1871. George Chesney's brilliant leadership made Coopers Hill a reality and shaped the college's ethos, while the Duke of Argyll's political support was instrumental in obtaining parliamentary approval. The admission process and academic curriculum during the first decade of 1871 to 1880 were designed to build the professional and personal characteristics that Chesney believed were needed in India. Chapter 3 examines how the college survived and maintained its relevance in the face of fiscal challenges and external changes during the years from 1871 to 1896, the period of the first two presidents. College magazines, personal diaries, and later recollections are used in chapter 4 to recreate a picture of student life that shows how successfully Coopers Hill managed to educate engineers and build their esprit de corps.

Chapters 5 through 7 approach the second question by using a combination of personal accounts and official sources to build a picture of the graduates' experiences in India. Most of our personal accounts of life in British India come from government officials or military men. In contrast, first-hand accounts of the work of technical officers in the field are rare. Unlike their civil service counterparts, telegraph officers and engineers were involved in construction projects, sometimes on a very large scale. Ernest Hudson's diaries, written between 1889 and 1908 (chapter 5), are one of the very few accounts that survive of the logistics and organization of telegraph line construction in India. His descriptions of inspecting and maintaining the lines reveal a wealth of information about how this fundamental component of India's infrastructure worked in practice.

While forest officers were not involved in construction works, their technical work underpinned the vast Indian forestry system, the results of which have been highly controversial. The improved understanding of the work and perspectives of a forest officer obtained from the 1892–1911 Osmaston diaries in chapter 6 therefore offer a valuable insight into these issues.

Because engineers in the PWD constructed many types of infrastructure, chapter 7 examines the work of Coopers Hill engineers primarily through selected technical publications and secondarily from personal accounts. It highlights the natural and logistical challenges faced by the engineers and their innovative solutions, as well as some of the unforeseen issues that resulted.

These chapters therefore provide a valuable addition to the understanding of the roles and practices of technical officers in the field in India. Additionally, information in these three chapters is used to analyse how their education at Coopers Hill prepared them for their duties, thereby addressing question 2.

The tenures of the college's third and fourth presidents were marked by crisis, as described in chapter 8. The thirty-five years of Coopers Hill's existence coincide with a period of significant change in the technical, educational, and societal landscapes of British and Indian engineering. Chapter 8 therefore approaches question 3 by considering the response of Coopers Hill to emerging engineering disciplines, such as electrical engineering, and the increasing student diversity as it approached the turn of the twentieth century. During the same period, the numbers of engineering students at British colleges and universities had also grown rapidly. The influence of these factors on the decision to close Coopers Hill is studied.

Chapter 9 then looks at events after Coopers Hill was closed in order to address question 4. It traces the history of the Coopers Hill Society over its nearly sixty years and discusses the attitude of its members towards Indian independence, the progress of professional ethics, and issues related to retirement and pensions, including their life expectancy.

Finally, the strands of the book's narrative are drawn together into concluding remarks and reflections that gauge the long-term impact of Coopers Hill and its graduates.

2

"This College Has Been Established at Cooper's-Hill"

Backdrop

The *Pioneer* newspaper first broke the news of Edmund Elliot's murder on the Bareilly-Rampur-Moradabad Railway two days after the crime itself – first the bare facts, and over the next few days with increasing (but not wholly consistent) detail. During the trial of Robert Fenelon Hawkins, which opened on 13 November 1893,[1] more details of the tragedy came to light. Duties of the accused "included the measuring of work done by contractors and superintending the construction of a bridge over the river Kosi."[2] The trial is informative because it represents one source of contemporary information about the lives and duties of engineers in India.

The events on the morning of Hawkins's dismissal for insubordination were described as follows:

Mr. Elliot was killed on the 30th. On 28th August an order was passed by Mr. Elliot and communicated to [Mr. Hawkins]. It was kept in a correspondence book containing official orders about the work. Orders were written on the left hand side by Mr. Elliot, and prisoner made any note he wished to make on the right hand side. This order of the 28th called on prisoner to leave and proceed to Moradabad, where he would be settled up with. [After his dismissal]

Prisoner was directed to meet Mr. Elliot on the morning of the 30th at chain 190 on the railway line, where they would go over the measurements, and he was asked to have the contractors present. Chain 190 was about half way between Kosi and Mundha. About 7 o'clock on the 30th Mr. Elliot drove to the river and crossed it. A tum-tum was waiting for him and he drove towards Mundha. Several khalasis,[3] whose presence was required to make measurements, met him and kept close to the tum-tum till they reached chain 190, where Mr. Elliot left the tum-tum and went towards the embankment.

Prisoner was standing at the foot of the embankment, which is from five to ten feet high. Jhinguri, one of the khalasis, followed Mr. Elliot closely, the others were a little farther off, but could see and hear what followed. When Mr. Elliot went up to prisoner, something passed between them, described by the witnesses as a salutation. That was all that was said. At this time the prisoner was holding two books in his left hand. Mr. Elliot went past the prisoner to go up the mound, and as soon as he passed, prisoner took out a pistol and fired at him. He was so close to prisoner that his clothes were singed. He turned round and the prisoner fired again, and Mr. Elliot fell, and after an interval prisoner fired a third time.

Prisoner then went along the side of the embankment to his hut on the river bank, and told his Madrasi clerk, Ramayar, to muster the men and to tell them all that he had shot Mr. Elliot. He then returned to the hut, and sent off a duplicate telegram to Mrs. R.F. Hawkins [his wife] and Mrs. V.E. Hawkins, Madras [his mother] – "Battle of Kosi over. I have shot Mr. Elliot. You will therefore never see me again. Good-bye."[4]

Elliot's body was taken to Moradabad that same evening for burial, accompanied by Mrs. Elliot and some friends. The newspaper report commented, "Socially Mr. and Mrs. Elliot's genial and open hospitality has been well known much beyond the small circle at Rampur."[5]

It seems that the root of the conflict was the "dissatisfaction on the part of Mr. Elliot with [Hawkins's] work, and the belief on the part of prisoner that he was unjustly accused; and that proper deference was not paid to his social position." No doubt this refers in part to the different backgrounds of the two men; Elliot was a "British-Indian," whereas Hawkins was an "Eurasian." At the time of Coopers Hill, "Anglo-Indian" could refer to the children of British parents serving in India,[6] but later came to be used in place of Eurasian, people of mixed heritage. The term "country born" was used at the time to describe "non-transients, who could include domiciled Europeans or Anglo-Indians."[7] Writing in 1935, James W. Best commented wryly, "During my time in India 'Natives' became Indians, Eurasians became Anglo-Indians, the old-time Anglo-Indians became Europeans, and I suppose that the old-time English should become Anglo-Anglians."[8]

G.A. Campbell, the executive engineer in charge of the Bareilly-Rampur-Moradabad Railway gave evidence of his knowledge of the conflict between Elliot and Hawkins, to the effect that he had never observed Elliot treating Hawkins with anything other than "politeness and consideration." The court also satisfied itself that the long, drawn-out nature of the disagreement precluded a defence of temporary insanity. The defence countered that the evidence of a doctor who was asked by the police to determine the matter and who "spoke with the prisoner for two or three minutes, and from that he judged that he was not insane," was inadequate, and that the jury should take that into account during sentencing.

The defence's appeal was in vain, and the judge, in imposing the death penalty, commented that he had "never known a more scandalously unprovoked, indefensible and inexcusable murder."[9] Hawkins was hanged in January 1894.[10] It is apparent that the establishment was, consciously or not, stepping in to protect its lines of authority, with little sympathy for Hawkins's personal demons, or the fate of his wife and young children.[11] Nevertheless, we do learn something about being an engineer in India. We see the supervising engineer being responsible for his section of the railway line, with European and Indian subordinates responsible for individual tasks, and contractors doing some or all of the implementation, such as embankments and bridges. He might live in a small community near the worksite, with a boss at a distant town. Orders were communicated through a correspondence book, to allow for periods when the engineers did not often meet each other. The case offers a glimpse of the misery that a poisonous atmosphere in a small community could cause, where people and their families both worked and socialized together, contrasting ideas of "genial and open hospitality" with issues of social position. We also find that the thirty-six-year-old Elliot was in a position of some authority on local engineering projects, but that Hawkins, just two years younger, was only a temporary engineer. These matters will be discussed further in chapter 7.

Public Works in India

The Royal Indian Engineering College was designed to provide a direct pipeline of young men like Edmund Elliot from the relatively well-off families of Britain and Ireland to the railways, canals, and telegraphs of India. A big attraction of the college in its early days was that a position in the Indian PWD was guaranteed for successful graduates. In later years this ceased to be the case as the demand in India slowed, but this was the very reason for the formation and funding of Coopers Hill.

"Public works" comprised "the construction and repairs of all state buildings, civil and military, as well as the prosecution of roads, railways, and irrigation works." Public works was under the auspices of the Military Board until 1854, when Lord Dalhousie separated it into a standalone department. Chesney describes the typical administrative structure:[12] "A Chief Engineer is placed at the head of the Public Works Department in each province, who is also secretary to the provincial government. Under him are the Superintending Engineers of Circles, while the actual execution of work is conducted by the next grade of officers, styled Executive Engineers, aided by Assistant Engineers, with a staff of subordinates. As a rule, the same engineer carries out all the works, whether of roads or buildings, within his district or division."[13]

As usual, Elliot had been appointed as an assistant engineer and had been promoted to executive engineer by the time of his death. When Coopers Hill opened in 1871, the demand for railway engineers (and public works engineers) was on the increase. A particular exception to this trend occurred in the early 1890s when there was a major construction slump, coincident with Elliot's murder.[14] This may account for Hawkins's anxiety about not getting proper credit for his work; it was a tough time to be a temporary engineer looking for railway work.

Engineering Education in England

In the late 1860s, during the recovery from the stagnation following the violence of 1857, the budget for Indian public works had reached £7.5 million.[15] The shortage of qualified engineers available to construct these works had become pressing.

The two key words here are "qualified" and "engineer," neither of which was especially well defined in the 1860s. The idea of engineering as a profession had been taking hold in Britain since the formation of the Institution of Civil Engineers (ICE) in 1818. Nevertheless, there remained the uncertainty – and one that unfortunately still persists – about whether the engineer drives the train or designs and builds the train. In *Daily Life in Victorian England,* Mitchell notes that "someone called simply an *engineer* was likely to be a skilled worker [metalsmiths, shipbuilders, engine drivers, maintainers of machines] rather than a professional," while the more educated and scientific men called themselves "civil engineers."[16] Originally, the word "civil" distinguished its practitioners from "military" engineers, who were responsible for infrastructure and logistics in the army. The ICE thus claimed to represent all non-military engineers, even after the formation in 1847 of the Institution of Mechanical Engineers and other organizations representing the specialized disciplines of mining engineers, electrical engineers, and naval architects. Edmund Elliot, like many of his colleagues, joined the ICE.

Until the formation of Coopers Hill, there was arguably no comprehensive university-level education available for civil engineers in England. The training of engineers had traditionally been by the pupilage system, whereby a man would join the workplace of an established engineer for several years to learn the profession, essentially as an apprentice. This idea remained firmly rooted in the psyche of the profession until well into the twentieth century. Roughly speaking, only one-third of applicants to the ICE in 1885–6 had a technical education; this had increased to two-thirds by 1913–14.[17] However, the first cracks started to appear in the pupilage system in around 1870.

There was a growing sense that British engineering was falling behind competitor countries, a view crystallized by the dismal showing of British technology at the Paris Great Exhibition in 1867, and perhaps even earlier.[18] "It was the Exhibition of 1867 in Paris which gave the nations, and especially England, a final lesson. By that Exhibition we were rudely awakened and thoroughly alarmed. We then learnt, not that we were equalled, but that we were beaten – not on some points, but by some nation or other on nearly all those points on which we had prided ourselves."[19]

The issue was first brought to public attention by Lyon Playfair, an Indian-born Scot, professor of chemistry, and Liberal politician, who had served as juror in the great exhibitions of 1851, 1862, and 1867. In a letter to the chairman of the commission enquiring into school reform at the time, he reported the opinions of "many eminent men of different nations" and was "sorry to say that, with very few exceptions, a singular accordance of opinion prevailed that our country had shown little inventiveness and made but little progress in the peaceful arts of industry since 1862."[20] His letter appeared in the 19 July 1867 edition of the *Engineer*, along with support from other eminent scientists. One correspondent had "to admit that our British portion of the display was generally meagre and defective."[21]

Playfair concluded with a recommendation that the government should conduct a more formal enquiry into the issue. Instead, it was the ICE itself that "felt it their duty to interest themselves in that part of the inquiry which bore on their own profession." *The Education and Status of Civil Engineers in the United Kingdom and in Foreign Countries*[22] consisted of a compilation of reports from groups examining different aspects of the issue. For a report that stemmed from an anxiety that British technical education was neglected, the tone of the overview of the status of engineering education is remarkably complacent. The overview is reproduced below to show succinctly the establishment view of the matter on the cusp of the Royal Indian Engineering College's creation. To the modern eye, the deficiencies of this approach are clear:

> There is ... in England no public provision for engineering education. Every candidate for the profession must get his technical, like his general education, as best he can....
>
> The education of an Engineer is, in fact, effected by a process analogous to that followed generally in trades, namely, by a simple course of apprenticeship, usually with a premium,[23] to a practising Engineer; during which the pupil is supposed, by taking part in the ordinary business routine, to become gradually familiar with the practical duties of the profession, so as at last to acquire competency to perform them alone, or, at least, after some further practical experience in a subordinate capacity.
>
> It is not the custom in England to consider *theoretical* knowledge as absolutely essential. It is true that most considerate masters recommend that such knowledge should be acquired, and prefer such pupils as have in some degree attained it, and

it is also true that intelligent and earnest-minded pupils often spontaneously de-vote themselves, both before and during their pupilage, to theoretical studies; but these cases, though happily much more frequent now than formerly, really amount only to voluntary departures from the general rule.[24]

Questions of practice versus theory and work experience versus higher ed-ucation have resonance to this day. The crux is that engineering requires both theory and practice, and educational systems continue to grapple with achiev-ing the appropriate balance of the two in what is often, unlike medicine and law, a direct-entry degree. In the early days of engineering when scientific under-standing of materials was primitive, "rules of thumb" and other empirical tools passed down from master to apprentice in the "shop" served the profession well. But as projects became more ambitious, diverse, complex, and expensive, it became necessary to predict with greater certainty how new designs would perform *before* making them.[25] This demanded a knowledge of mathematics and theory. Increasingly engineering advances would come directly from mak-ing and applying scientific breakthroughs, and a "school culture" developed.[26] This new reality also demanded a higher ethical standard from engineers, lead-ing to the legal regulation of the profession and pledges such as the Ritual of the Calling of an Engineer in Canada.

In the 1860s, however, the main issue with the pupilage system was the com-plete absence of any quality control. The ICE did not conduct exams or enforce any technical expectations of those who completed their training. Many en-gineers received excellent training through the pupilage system, but this was far from guaranteed. Although Fleeming Jenkin was exaggerating to make his point, his inaugural address as professor of engineering at the University of Edinburgh in 1868 revealed something of the situation: "In England, the path to employment as an engineer lies through the office or workshop of a civil or mechanical engineer, and therefore we find that young Englishmen and their parents crowd the doors of the offices and workshops, offering premiums of £300 or £500 for the mere permission to pass three years unheeded inside the magic gates, which must be passed to gain an entrance into the profession."[27] A more pragmatic version noted, "The office of an Engineer is not, and probably cannot be made, an educational establishment. In the first place, he has too much personal occupation to be able to attend much to pupils; and in the next place, where work has to be got through, a pupil will inevitably be more or less kept to what is most pressing, or what he can do best."[28]

In addition to specialist establishments such as the Royal School of Mines and the Royal College of Naval Architecture, the 1870 ICE report listed a num-ber of institutions that offered elements of an engineering education and had been doing so since the early decades of the century. Typically, an institution offered full or part programs leading to certificates or diplomas in engineering.

This group included University College London, Edinburgh, Glasgow, Trinity College Dublin, and the recently formed Owens College, Manchester (precursor to the University of Manchester). King's College London was at the forefront of post-secondary education in engineering, offering a three-year program to thirty to fifty students, amongst the largest enrolments in the country. The total number of people taking a technical education in Britain was therefore modest, largely because the profession did not see this as an acceptable route to becoming an engineer.[29]

The prospectus for Owens College was typical in stating that "the course will furnish a thorough scientific groundwork for … the higher branches of the Engineering profession, but it is not intended to supersede the practical training which can only be obtained in the office of a Civil, or the workshop of a Mechanical, Engineer." The exception was Trinity College Dublin, whose students, on completing the program, were "supposed to be at once competent, without pupilage, to undertake professional practice." But the pattern was consistent – scientific background at the institution, practical engineering in the workplace. The curriculum at Coopers Hill would be the first to offer a combined education in theoretical and practical engineering.

Subjects covered by these courses included a strong emphasis on natural sciences and math, some coverage of civil and mechanical engineering, drawing, and surveying. The three-session course at UCL was typical: pure and applied mathematics, applied mechanics, physics, physical laboratory, chemistry, chemical laboratory, civil and mechanical engineering, mechanical drawing and designing, surveying and levelling, and geology.

After 1870 the number of universities offering technical education grew swiftly. Arguably the first recognizable engineering degree was offered at the University of Glasgow in 1872 as a result of the pioneering work of W.J. Macquorn Rankine. Oxford and Cambridge Universities ventured into the applied sciences with the creation of their great Clarendon and Cavendish labs in early 1870s, and the appointment of A.B.W. Kennedy to the chair of Civil Engineering at UCL in 1874 led to a revival of that institution's programs. Universities at Newcastle, Leeds, Bristol, and Sheffield established engineering studies in the 1870s, followed by Birmingham, Nottingham, and Liverpool in the next decade.[30] By 1925 half of the applicants to the ICE held a technical degree, and by the Second World War the normal route to membership was a three-year degree and two years of practical training.[31]

So, at the time Coopers Hill was being contemplated, the pupilage system was strongly entrenched in the profession and would remain so for some decades. University programs were seen as optional supplements to the core apprenticeship. However, there were a few champions of a more systematic approach to educating engineers who were gradually having an impact. By the early 1900s the notion of a university education in engineering had achieved more widespread acceptance. The relatively short existence of Coopers Hill therefore precisely bridged these two worlds. The question of the extent to which Coopers Hill blazed the trail or merely reflected the changing times will be addressed later.

The 1870 ICE report contained a section devoted to suggestions from eminent men for the optimal education of engineers.[32] Many were modifications of the pupilage system, such as introducing two years of technical education, starting at an age of fifteen or sixteen, at a university to prepare young men for pupilage. There was also talk of introducing examinations, conducted either by the universities or by the institution, to determine technical competency to join the profession. Eventually, the ICE introduced entrance examinations in 1896. Nowadays, many countries have a legal framework requiring practising engineers to hold a licence. To become licensed the engineer must demonstrate technical competence through a series of examinations or by graduating from an accredited college or university engineering program.

A Man of Unquestionable Talent

While the engineering profession was working on its response to Playfair's concerns about British competitiveness, another set of reforms was being proposed, this time in India. The remarkable General (later Sir) George Tomkyns Chesney (1830–95)[33] published two works: a history of British rule in India with suggestions for its future development entitled *Indian Polity* (in 1870), and an exploration of the role of the Royal Engineers in Indian public works projects, the rather less catchy *Memorandum on the Employment of the Corps of Royal Engineers in India* (in 1868).[34] Partly on the strength of this work, Chesney became the champion and first principal of the Royal Indian Engineering College.

Chesney (see figure 2.1) entered the East India Company's Addiscombe College in 1847, shortly before his seventeenth birthday.[35] The curriculum there focused on mathematics and sciences with an application to military engineering. While professing military ethos and discipline, the reality under some principals seems to have been rather different – fisticuffs with the townspeople, forgery of letters of leave, and spying on the cadets by the sergeants.[36] George Chesney finished the standard two-year program at Addiscombe in 1848, one of five engineering cadets. He does not appear to have partaken of the cadets' hijinks (or was too smart to get caught), for on graduation he was awarded the prize of a sword by the directors of the East India Company "as a testimony of their approbation of your exemplary conduct at this institution."[37]

Chesney proceeded to India in 1850 to join the Bengal Engineers, where he was wounded in action at Delhi during the Indian uprising. Later, he joined the Public Works Department and for two periods (1856, 1858–9) served as an assistant principal at the Thomason Civil Engineering College in Roorkee,

Figure 2.1. General Sir George T. Chesney, KCB CSI CIE, in later life (*Illustrated London News*, 6 April 1895, 406)

India.[38] Between 1860 and 1867 Chesney was head of the new Public Works Account and Finance Department, where he devoted his energies to reorganizing public works and financial administration, advocating for reforms to increase their efficiency. Field-Marshal Lord Roberts of Kandahar later referred to Chesney as "a man of unquestionable talent and sound judgement."[39] *Indian Polity* and the *Memorandum* were both informed by these experiences.

Battle of Dorking

Strangely, Chesney would become a household name not for any of these achievements but for a work of fiction. *The Battle of Dorking*, originally published anonymously in *Blackwood's Magazine* in May 1871, proved to be an immediate and immense success; within two months over 100,000 issues had been sold,[40] Blackwood's were printing more, and translations into other languages were underway. Speculation about the author's identity was rampant – the *Pall Mall Gazette* was the first to get the correct answer in its 18 July edition.

In *The Battle of Dorking*, an old army volunteer tells his grandchildren the story of the successful invasion of Britain fifty years previously by a German-speaking foreign power. As he was writing the story, Chesney summarized the plot to the publisher, John Blackwood: "We have the quarrel with America and Russia, dispersion of all our forces, followed by rising in India. Sudden appearance of Germany on the scene. Sentimental platitudes of Messrs Gladstone & Co., triming [sic] leaders in the 'Times.' Destruction of our 'Field Line' by new torpedoes. Arrival of 100,000 Sanscrit-speaking Junkers brimming over with 'Geist' and strategy. Hurried defence of the chalk-range by the volunteers and militia, no commissariat, line turned, total defeat, retreat on London, occupation of that place, and general smash up."[41]

The story caused a furore, which further fuelled the huge surge in readers. To some it was a grave insult to the honour and expertise of the British forces. To others it was a wake-up call. For most it undermined the nation's sense of superiority. This had already been shaken by the decisiveness of the very recent Prussian victory over the French using modern technology, and the resulting shift in the balance of European power. The story's repercussions were felt in the country's highest levels.

In superbly dramatic fashion, Chesney had drawn attention to his concerns about the cost-cutting agenda of Prime Minister William Gladstone's Liberal government. As Chesney put it in March 1871, "If the Government [Gladstone's] had brought in a really good Army Bill, the words would have been taken out of my mouth, but as it turns out, they have left me with virgin soil to work upon."[42]

Although it was not the first story of its kind, the timing, good storytelling, and compelling military authenticity of *The Battle of Dorking* have established it as "the first major example in what would become a long line of popular pre-1914 invasion-scare narratives."[43] The sales and impact of *Dorking* are all the more remarkable for being the work of a first-time author of fiction, written in the midst of all the preparations for Coopers Hill.[44]

Chesney was later described as "an officer of statesman-like intelligence, untiring energy, and great administrative power,"[45] but equally telling is John Blackwood's contemporary summary: "He is a little chap, very bright and fresh-looking."[46]

Engineers for India

Chesney's thoughts in the *Memorandum* about the shortage of engineers for the Indian PWD therefore carried weight. The primary cause of the shortage was the swiftly increasing investment in Indian public works, combined with a reduced supplement of engineers from the army (because of better opportunities elsewhere). The Indian government responded with a request for more engineers, while desiring the current proportion of military and civilian personnel to remain unchanged. In 1868 the proportions were roughly equal, a big change from a decade earlier when almost all engineers were military. Because much of the recent recruitment was of civilians, "the majority of the upper ranks belong to the [military] class, and by far the greater part of their juniors to the [civil] class."[47] The equal proportion could not be maintained in the short term because it took time to train Royal Engineers, first through the Royal Military Academy, Woolwich, and then at the School of Military Engineering at Chatham. Moreover "the authorities at the Horse Guards" refused any immediate transfer of engineers from other duties. So the effective situation was that civil engineers from a variety of sources were being recruited to fill the gap.

Chesney noted first that thirty engineers "of standing and professional experience" had been sent to India in February 1867 on five-year covenants. Shortly afterwards, another fifty were sent on similar terms, although these were "on the whole younger men than the first batch."[48] Chesney observed that this recruiting of qualified engineers to India was unlikely to be sustainable on the terms the government was offering,[49] especially as it necessitated convincing established engineers in Britain to move to India. Salary was a persistent source of tension between the civilian and military engineers because the latter continued to draw their military pay in addition to a salary from the PWD. The sporadic influx into the PWD of engineers inexperienced in Indian affairs also caused its difficulties: "The admission of men of standing in large numbers to [the railway] branch has been resented by those already belonging to it, and whose advancement has, of course, been retarded in consequence."[50]

Second, engineers were recruited through the annual competitive examinations for civil engineer appointments held at the India Office. The idea was that young men who had completed at least three years as a pupil with an established engineer and who passed the exam would be sent to India at state expense to receive language and further technical education at one of the Indian civil engineering colleges. This scheme was started in 1859 while Lord Stanley was secretary of state for India, and these men were accordingly dubbed "Stanley Engineers." Unfortunately, the scheme struggled from the first, with few candidates achieving the required passing grade in the examination, and then those men performing poorly at the Indian colleges. Indeed, such was the government's desperation for engineers that the college part of the scheme was dropped, and the probationary engineers were deployed as soon as they arrived. The PWD was also concerned about "maintaining two quite different standards of qualification for admission into the service, a high one in India, a low one in England."[51]

Even as early as 1862, the PWD was "rather doubtful to what extent any mere competitive examination can be trusted in the selection of persons for this sort of occupation."[52] Writing in 1868 after some years of the Stanley Scheme, Chesney observed that the PWD was hardly likely to obtain *more* engineers when the scheme had rarely reached its target in the past, especially when on the most recent occasion only twenty of forty-four applicants had passed the examination. Chesney pinpointed the difficulty and foreshadowed what was to come:

> Even if the competition were severe, the system appears so far defective that, in a brief examination, the candidate's acquaintance with some of the most important parts of an engineer's education – his practical knowledge of work and surveying – cannot be tested properly, while in theoretical engineering there is not much scope for examination, since this is a subject in which, under the present system of professional training, the pupils of civil engineers receive ordinarily no methodical instruction, but are left to pick up their knowledge as best they may.... The difficulty of the case suggests the idea whether it would not be worth while to establish a regular school for the young civil engineers intended for India, where they could receive a sound professional education, and a thorough course of surveying gone through under supervision.... It would be easy to arrange also that the students should get a practical acquaintance with works. An invaluable result of such an establishment would be the esprit de corps it would tend to foster, and an identification of its alumni with the interests of Government, a thing very much wanted in the department at present.[53]

Despite the continued addition of some military engineers into the PWD by various means, the make-up of the department was therefore inevitably moving from the military to the civil. In 1861, civil engineers made up 39 per cent of the total Public Works Department engineers (201 of 518), increasing to 59 per cent (533 of 896) by 1869.[54]

Chesney advocated approaching the Royal Engineers again: "Is it impossible, then, I would respectfully ask, that the authorities at the Horse Guards should be asked to reconsider their first opinion, and under the emergency of the case, to lend at least a portion of the officers now wanted for service in India?"[55] To make it palatable to Horse Guards, he suggested that the mix of men lent by the Royal Engineers should be in the usual proportion of field officers, captains, and subalterns, rather than just the young men the PWD originally requested, to avoid unbalancing the composition of the force in Britain. He also proposed that men being trained in Chatham could forgo part of that course in favour of an earlier start to practical training in India.

The response to these suggestions would prove inadequate to meet India's needs. Nevertheless, George Chesney's impact through *Indian Polity* and the *Memorandum* would lead to his critical participation in the establishment of the Royal Indian Engineering College, which would meet that demand.

Thomason Civil Engineering College, Roorkee

As mentioned above, the Stanley Engineers completed their training at one of the Indian civil engineering colleges. While several universities and colleges in India offered some elements of engineering training (including the short-lived Fort William Civil Engineering College in Calcutta, where Chesney served briefly as principal),[56] the only college that took Stanley Engineers was the Thomason Civil Engineering College at Roorkee.[57] Established in 1847, Roorkee was the country's principal engineering school[58] "designed to give theoretical and practical instruction in Civil Engineering to Europeans and Natives, with a view to their employment on the public works of the country, according to their several qualifications and the requirements of the Service."[59] Unlike Coopers Hill, Thomason College thrives today as part of the Indian Institutes of Technology system, maintaining a strong reputation in a wide range of engineering and applied technology fields.

As early as 1845, the progressive intent of the college's founder, James Thomason, lieutenant-governor of the North-Western Provinces, seems to have been to bring Europeans and Indians under the same roof for an engineering education. In particular, theoretically educated students from the British-style schools and colleges in India (mostly, but not necessarily, European) and the experienced practical men supervising the actual PWD construction (wholly Indian) should learn from each other: "It is evident that, if we can bring the former and the latter of these together … the one class might impart to the other the knowledge either of theory or of practice, which when found united, go to produce really a valuable and useful officer."[60]

Unfortunately, the departmental structure of the college seems to have encouraged the separation of the different groups of students. Given the political situation of the time, with Britain still expanding and consolidating its rule of India, it would have been impossible to contemplate European and Indian students participating on an equal basis. Nevertheless, Roorkee was still exceptional in that it admitted both groups of students to the same institution.[61]

Of the three departments, the first was the so-called Engineer Class, designed to prepare engineers from the military or with a civil background (both Europeans and Indians) directly for PWD or other employment. Eight appointments per year were guaranteed to graduates of this class.[62] The second was the Upper Subordinate Class going to serve as overseers of public works under the authority of the engineers. These were non-commissioned officers from the army or Indians. Lastly, the third department was exclusively for Indian students seeking qualifications to work on public projects under the overseers.

An Engineer Class student at Roorkee would engage in a curriculum consisting of mathematics and civil engineering, together representing more than 60 per cent of the total marks, and surveying, drawing, physical sciences, and technical "Oordoo" (Urdu).[63] In the major block called Civil Engineering, students were instructed in construction techniques for buildings, bridges, roads railroads, and irrigation works. They also learned building materials, mechanics, and ground tracing, and visited engineering works and constructions sites.

The Civil-English category was for graduates from British schools in India or people specifically educated in England for Roorkee. Candidates had to be between eighteen and twenty-two, submit to the principal an application, testimonials to good moral character, and a medical certificate to certify perfect vision and a "sound constitution." There was also an entrance examination comprising languages (225 points), physical science (75), history (50), mathematics (310), and drawing (50). It was also possible to complete an approved equivalent exam through the University of Calcutta or elsewhere. Notably absent was the classical education that was especially valued by the upper and middle classes in England and would be required for entry into Coopers Hill.

Indian students in the Engineer Class had to meet the same entrance requirements, but there were up to six positions carrying a stipend of Rs50 per month. The actual number of Indian students in the Engineer Class was small. Reviewing the "Yearly Lists of Passed Students" in the 1871–2 Roorkee calendar, it is noticeable that more Indian names appear in Engineering Class during the early years of the college, with none appearing between 1862 and 1870.[64]

On graduation, each of the eight or so engineers taken on by the PWD could expect to earn Rs100 per month as an engineer apprentice, with the expectation of promotion after a year to assistant engineer third grade on a salary of Rs250 per month. The Roorkee program lasted nineteen months (from 1 January to August of the following year), for a cost of roughly Rs1900, a sum that could be earned back in fifteen months of working for the PWD.

In his 1868 *Memorandum*, Chesney was quick to acknowledge that "there can hardly be any doubt as to the propriety, not to say obligation, of throwing open the department to the natives of the country; but for some time to come at any rate the number of those qualifying for employment is likely to be but small."[65] Indeed, in the previous two decades only in 1859 did the number of Engineer Class students at Roorkee break into double digits; more typically it hovered between four and six. This number was limited mainly by the supply of suitably qualified recruits from local schools.[66] This was clearly insufficient to meet Chesney's predicted demand for PWD engineers.

Moreover the newly arrived principal, Capt. A.M. Lang RE (Royal Engineers), in 1871 did not believe that capacity could be increased without additional investment in staff and facilities, and he presented a long list of additions and repairs needed for the college buildings.[67] This did not prevent him from urging the PWD to increase its quota of eight positions for graduates of Roorkee: "Unless then more appointments are guaranteed to the passed Civil Engineering Students, I cannot see with unmixed satisfaction any further expansion of this particular class; nor can I hope to see many Natives join it although I consider that they have perhaps the first claims upon the College, and should be more encouraged to enter the higher grades of the P.W. Department."[68] To achieve this, he believed that *all* Indian students who passed should be offered positions in the PWD.

But more than the cost and difficulty of increasing the number of Engineer Class graduates in India, it is clear that the authorities had a strong desire to educate future engineers for India in England, where they could keep a close eye on the "character" of the students and the quality of their training. In the following clever paragraph from his defence to Parliament of the proposal for Coopers Hill, Chesney simultaneously praises Roorkee's success, uses that success as evidence that the state can operate a good engineering college, draws attention to Roorkee's remoteness from "civilization" (both technical and social), and points out the expense of supporting Roorkee, while downplaying the costs of the new college (because it would attempt to cover its costs by charging tuition).

It may not be out of place to refer to the success of Roorkee as a proof of what may be done by Government in the way of affording systematic training for its own servants. Roorkee labours under the drawbacks of climate and of distance from the great centres of scientific and professional movement, and those of its pupils who have been born and educated in India necessarily start at a disadvantage compared with young men brought up under the influences of European civilization. Yet Roorkee turns out so useful a body of servants, that, although their education (which is wholly gratuitous) is given at great cost to the State, the expenditure has always been ungrudgingly bestowed, and while there is hardly any branch of the India Education Department as to the utility of which public opinion is not often strongly divided, no doubt has ever been expressed about the propriety of maintaining Roorkee, and while the annual charge for it is almost continually increasing, no item in the Estimates is more certain to pass unchallenged. If so much can be done in a remote corner of India, we may reasonably expect a high degree of success here [England], without the cost.[69]

As we now know, the proposal to establish the Royal Indian Engineering College was successful; however, the outcome was in significant doubt up until the last moment.

A New Indian Civil Engineering College

A Political Champion

In 1870, it was generally agreed there was a pressing need for engineers in India because of the increasing numbers of public works being carried out there. There were, however, some powerful obstacles to the creation of a new college. In India, the Royal Engineers who had so far been responsible for the bulk of Indian public works, and whose officers still comprised the upper echelons of the PWD, wanted to maintain their position of dominance. Moreover, Roorkee saw itself and the other Indian colleges as the natural suppliers of the needed engineers.

In England, the bulk of the civil engineering profession, and the establishment interests they represented, strongly favoured a continuation of the pupilage system. Some vocal members of the profession argued that engineers should receive higher education in both theoretical and practical training, but most still saw little need for engineering colleges. Within the education profession, institutions such as King's College, University College, Glasgow, and others who already had engineering activities, understandably believed that it would be more cost-effective for the government to support an expansion of their activities rather than to create a new institution.

In the face of these numerous and entrenched objections, George Chesney needed a powerful ally in government to advance the cause of a new English engineering college. He found this ally in the form of the Duke of Argyll, secretary of state for India. Fortunately, there *was* no more powerful player in the post-1858 world of Indian governance than the secretary of state.[70] To understand why, a brief overview is necessary of the main elements of Britain's governance of India in 1871.

In India, the viceroy/governor-general served as head of the government and local representative of the British monarch. He was advised by a council of six appointees, together grandly titled the Government of India, which was responsible for all matters of civilian and military administration. Along with its substantial retinue of administrative functionaries, it was located in Calcutta during the cool season (November to April) and at the cooler mountain resort of Simla in the summer.[71] While the Government of India wielded strong administrative powers that flowed down to provincial governments and then the district officials, there remained a strong reporting and approvals line back to London, where the strategic policies and finances were determined. The Government of India maintained five departments: Home, Foreign, Financial, Military, and Public Works.

In London, the India Office oversaw the Government of India and was headed by the secretary of state for India. He was the constitutional advisor to the Crown on the British governance of India, as well as a member of the Cabinet, accountable to Parliament for the administration of India. In reality the oversight by Parliament was not strongly exercised because funding for the India Office came from India itself and the Treasury had no control. The Government of India Act of 1858 gave the secretary of state power to "superintend, direct and control all acts, operations, and concerns which relate to the government and revenues of India."

The Council of India, also in London, was also established in 1858 to advise and moderate the actions of the secretary of state. This group consisted nominally of fifteen members, nine of whom must have had at least ten years' recent service in India. In 1870, membership of the council was by appointment of the secretary of state for a term of ten years. This collective authority in London was often referred to as the Secretary of State in Council.

In general, the secretary of state was required to present all non-urgent, non-secret matters to the council for a vote. On many matters he could overrule the council, provided he gave his reasons for doing so. One of the matters on which he could *not* overrule the council concerned "the grant or appropriation of any part of the revenues of India." Since Coopers Hill was to be paid for by the Government of India, this clause meant that the establishment of the college needed the approval of the Council of India. However, it will prove relevant that meetings of the council were called by the secretary of state, and that the attendance of only five members (including himself) was necessary for quorum.

Council of India Approval

George Douglas Campbell, eighth Duke of Argyll, served as secretary of state for India in the government of William Gladstone from 1868 until 1874.[72] Argyll had imperialist leanings, supporting the governance of India from London, but was content to consolidate the achievements of his predecessors in defining the office. One modern author described him unkindly as "gouty [and] lackadaisical."[73] While the former was certainly the case, Argyll arguably did not live up to his intellectual, political, and social potential, but his interests in India and education made him a natural and influential ally for George Chesney.[74]

Argyll's later recollection of events hints at Chesney's indirect approach to Argyll, as well as his view of his own authority:

> I did not then know Major Chesney at all and had no communication whatever with him. The first suggestion of a plan to found a new college at home, at the cost of the Indian revenue, for the special purpose of training civil engineers, came from my late friend Mr. Thornton, then at the head of the Public Works Department in the India Office.... After much consideration I adopted the suggestion

with full conviction of its great importance.... The idea was absolutely new. It needed a large immediate outlay, and a permanent annual cost not easily to be estimated, because it would be an experiment and might fail.[75]

Argyll went on to say, "Accordingly I found some leading members [of council] strongly opposed to the plan, whilst all were most doubtful and reluctant." In this he is not exaggerating. When he first circulated Chesney's draft proposal for the new college for comments in March 1869, only five of the fifteen Council of India members were in support; two were neutral and eight were opposed.[76] The reasons for the opposition closely followed the obstacles described above, according to the predilections of the individual members: use Royal Engineers, or other military personnel from Woolwich and Sandhurst; revive Addiscombe; or expand existing universities and colleges. So when Chesney returned to India in May 1869, prospects for a new college did not look good.

Although few council members raised the issue of cost, Argyll was concerned about the state of Indian revenues and asked for a rough budget for the college to be prepared. It estimated an operating expense of £17,000 per annum, assuming one principal and a teaching staff of eighteen, and including accommodation for the students. Offset against this was a proposed annual tuition fee of 100 guineas each for 150 students, bringing the annual cost to the Indian revenue to £1,850, or £37 per specially trained engineer.[77]

The secretary of state was satisfied with this estimate and took the next step towards formal approval of the scheme. This was for the Council of India in London to approve a dispatch to the Government of India informing them that he had decided there would be a new college, that entry would be by competitive examination, and that students passing the program would be offered positions in the Indian PWD. In November 1869, by a vote of eight to three, the council approved the communication to India.

Predictably, the Government of India was unhappy to be informed that the decision had already been made in London to change completely how they recruited civil engineers, and at their own expense. According to a much later account of the establishment of the college, Chesney was present at the meeting in Simla where the Government of India discussed how to respond to Argyll "and marshalled the arguments in favour of the scheme at some length, but, he feared, with little effect." The account went on to quote from Chesney's diary: "I expected criticism, or only cold support, but not the determined opposition I encountered. It is hardly likely that the India Office will persist in the teeth of the Government of India, especially as I hear that the Council was not unanimous."[78]

When it responded in March 1870, the Government of India expressed its belief that by resolving the discrepancy between military and civilian salaries, as they had just done, India could recruit enough civil engineers through existing channels. They also drew attention to the many assumptions and uncertainties in the India Office's calculation that the college would be self-supporting. But ultimately they had little choice but to accept the decision already made in London, even though they had "great doubts that any real necessity exists for its immediate establishment."[79]

The final step in the formal process for approving the new college was for the Council of India to reply to the communication from India. Argyll noted that he did not have "the full concurrence" of the Government of India but retained his "original opinion as to the expediency, and indeed, necessity, for establishing a college."[80] He rejected the claim that equal pay would resolve the situation and reiterated that the new college would pay for itself. The council met on 19 July 1870 for the final ratification of the proposal. The motion carried by six votes to five, a majority of just one at a meeting where four known dissenters were absent.[81] With understatement, Argyll summed up the process: "It was only after much discussion both in Council and out of it, that I succeeded at last in securing the consent of that body."[82]

The exchanges between London and India also included the proposal that George Chesney should be relieved of his duties in India and appointed as the founding president of the new college. Again, there was little option but to agree, and Chesney was duly appointed, with his start backdated to 24 April 1870.

Questions in Parliament

While this marks the official approval of the new civil engineering college, enabling Chesney to return to England to seek a suitable site for the college, Argyll was not yet out of the woods. "And when [the consent of the Council of India] had been secured a new danger arose. The House of Commons as a whole has the best disposition towards our Indian Empire, and very rarely interferes with its government. But when it does, the India Office is too often at the mercy of many personal or sectional interests at home."[83] The college would be funded from Indian revenues, and although Parliament might not have to approve the creation of the college itself, it did have oversight of the India Office finances. But even before Parliament had been officially informed of the India Office's decision, news of the proposed college had leaked out from sources in India to the journal *Engineering*, which wrote accurately in May 1870 that the secretary of state's view was "either that the system of education in this country is defective, or that a competitive examination is not the best means for testing the capabilities of an engineer."[84]

The first mention in Parliament of the new scheme was on 5 August 1870 during a "Statement on Indian Finance" by the undersecretary of state for India, M.E. Grant Duff. When questioned about a passing comment concerning a new college, he responded, "The creation of an engineering College, into which young men should be drafted by competitive examination, had become a matter of paramount necessity."[85]

David Plunket, MP for the University of Dublin, questioned Grant Duff again on the matter a few days later[86] but the House then rose for recess, allowing plenty of time for the "strong home interests" feared by Argyll to register their objections to the new college. Henry Fawcett, MP for Brighton, nicknamed "the Member for India" on account of his active engagement in Indian affairs, immediately signalled his intention to raise the matter in the House. Argyll, perhaps belatedly, canvassed the opinions of his government colleagues and found them, including Prime Minister Gladstone, to be lukewarm to the project. Over the next four months, "memorials" of protest were received from University College London (who also took their complaint to the prestigious journal, *Nature*),[87] King's College London, Owens College, and several other institutions.[88]

Chesney was dismayed that the institution of which he was already president was running into political difficulty but reacted with a detailed justification for the college that addressed numerous anticipated criticisms. He also responded thoroughly to objections raised by the Royal Engineers. Meanwhile Grant Duff replied to University College London that, having supplied only three engineers to India, it could hardly claim that the new college would cause them significant hardship (an argument apparently suggested by Chesney),[89] and coolly dismissed their concerns: "The objections which you put forward had, indeed, already suggested themselves to [the secretary of state]; and a reconsideration of them in the form in which you now express them, has not altered the conclusions at which he had previously arrived."[90]

It was clear that two issues were becoming paramount: the college's proposed monopoly on supplying civil engineers to India, and the political necessity for the college to be self-supporting.

Battle was rejoined on 24 February 1871 when Sir Francis Goldsmid (MP for Reading) raised several concerns. These were largely taken from the May 1870 *Engineering* article[91] and focused on whether the new college was in opposition to the general trend towards competitive appointments. He also raised the matter of the PWD's low rates of pay. Argyll had anticipated the need for a revised and more detailed projection of the college's costs, so Grant Duff was ready with his answer when Sebastian Dickinson (MP for Stroud) raised the issue on 3 March:

> There will be no charge on the revenues of India on account of the Engineering College; the fees will be slightly in excess of the charges, including interest on the buildings and plant, say on £90,000. There will be 11 professors and instructors on salaries varying from £700 to £300 per annum. Of these, nine will be entitled to

pensions under the provisions of the Superannuation Act, and two will not be entitled to pensions. If my hon. Friend would like the figures here they are – Annual sanctioned charge for College, as per regulations of Secretary of State in Council, £18,350; interest on buildings, &c, say £90,000 at 4 per cent, £3,600; total, £21,950. Fees, 150 students at £150, £22,500; difference, £550.[92]

These numbers would prove to be wildly optimistic, and efforts to achieve the promised self-sustainability would plague the college throughout its history.

Having received no satisfactory response to his point on 24 February, Goldsmid then rose to move a resolution that no "young men qualified by character and attainments for admission into the service of the Government of India as Civil Engineers" should be excluded because they had not attended the new college. Goldsmid's main concern was that young men, however brilliant, would under the new scheme be prevented from joining the PWD if they were unable to pay the £150 per annum tuition. Grant Duff's response was that paying £150 for three years with a guaranteed job was a good return compared with the expensive and ineffective pupilage system of the past.

Despite this logic, and after lengthy speeches (including support for Argyll from the chancellor of the exchequer) the resolution was narrowly adopted. Finally, on 18 May 1871, Goldsmid received a commitment from Grant Duff that students who passed the entrance examination but chose to study elsewhere would be appointed to the PWD, "provided they satisfy independent examiners that they come up in all respects to the standard of qualification which will be required from the College students at their final examination."[93]

This was formalized in the college calendar, which specified that candidates who "preferred to pursue their studies elsewhere" needed to provide evidence of good moral conduct in the intervening three years, be in good health, and have spent at least eight months under a civil or mechanical engineer. In the first year, one candidate, Henry W.V. Colebrook, took advantage of this option. He qualified for admission in the 1871 entrance examination but chose not to take up his place. He sat and passed his final exam in 1874 and entered the PWD.[94]

Taking Shape

Response to the New College

Reactions in India to the establishment of the Indian Civil Engineering College followed lines similar to those in England. Additional concerns were raised about the readiness of engineers fresh from England for the challenges of Indian conditions, both climatic and technical.[95] The *Amrita Bazar Patrika* newspaper caustically observed, "This college is to be established not here as might be supposed but in England for the benefit of India. Indistinct rumours of such a proposal now and then reached the Indian tax-payers who are to provide the necessary funds." It further noted that the college was open only to British-born subjects, and thus not to "Natives of the country from which the fund is to be derived.... Why was not the college established here or Roorkee improved? The Natives then might have served their own country."[96]

This concern was echoed by engineer Frederick Tyrrell in his essay criticizing the PWD of the 1860s. Tyrrell observed that the new college would be placed in the hands of the same people (namely the Royal Engineers) who had been responsible for the present inefficiencies and corruption in the PWD. Tyrrell continued, "An Engineering College for India – if there must be a College, – should be in India. The best Engineers for India must be, and will be at some future time, the sons of our Indian fellow-subjects.... We have excellent stuff among the Natives for Engineers, – great intelligence; a singular aptitude for figures; the patient eye and hand for drawing; above all the faculty and habit of minute attention to details that elude our observation."[97]

And of course it was Roorkee that felt the sting of the new college most deeply. Captain A.M. Lang, principal of Roorkee, was cautious in his 1871 annual report, merely noting that "the future effect of the out-turn of the Coopers Hill College in no way affects the supply from this College in the *current* year, whatever it may have on the fortunes of this College hereafter."[98] A year later, however, he was much more forceful:

> That the regular annual supply from [Coopers Hill] of a large number of assistant engineers, to whom appointments on more favourable terms are guaranteed, must affect the engineer class of this college there is no doubt whatever: in fact the English civil engineer class ... has already been affected by the uncertain prospects held out to all beyond the small "guarantee" number.
>
> While Cooper's Hill supplies a want (long felt in England) of a first class college, devoted exclusively to educating civil engineers, Roorkee must continue to perform a similar part in regard to the European, Anglo-Indian and native communities in this country.[99]

Lang's frustration is understandable as he saw investment and opportunities diverted away from his college towards "fostering its later rival." To add insult to injury, his public comments in the Thomason College calendar earned him a rebuke from the Government of India, which felt he should focus on reporting the internal affairs of the college "without committing the Principal of the College to views which may not meet with the approval of higher authorities."[100] Despite the efforts of Lang and his successors, little changed while Coopers Hill existed, and the debate about the relative qualities of the men from each college was recurrent.

Coopers Hill

Lang referred to "Coopers Hill" in the last quote. Since Chesney's appointment as president in April 1870, and in the face of the ongoing political challenges, the secretary of state had taken the major step of buying the Coopers Hill estate. Indeed, the *Engineer* had reported this as early as 2 December 1870.[101]

In his 1960 *A Short History of the Royal Indian Engineering College*, Cameron related how Chesney supposedly learned about the Coopers Hill estate from the publican of the "Bells of Ouseley" while taking a boating trip on the Thames.[102] Whether or not this was true, Chesney was unable to visit Coopers Hill until October 1870. Earlier that summer, he and the appointed architect, Sir Digby Wyatt, had reported on five possible locations.[103] The list included Coopers Hill (see figure 2.2), but they had been unable to visit and were reluctant to make an official appointment through the India Office for fear of escalating the price. Chesney and Wyatt were authorized to open discussions on two of the other possibilities, the Oatlands Park Hotel in Weybridge, just south-east of Coopers Hill, and the Imperial Hotel, Malvern. However, both negotiations fell through – the price of Oatlands Park was nearly double what the India Office was willing to pay.[104]

In the meantime, however, Chesney was able to visit Coopers Hill and wrote a report on 6 October strongly recommending its purchase.[105] He argued that Baron Grant, the current owner, was keen to sell quickly because he could not afford the upkeep and the price would therefore be good. And even if the college did not ultimately go ahead, it would be straightforward to resell the estate without financial loss. Chesney thought that it was "impossible to speak too highly of the general suitability of the place for the proposed purpose."[106] It was near London, had plenty of space for sports fields and practising surveying (and, later, forestry), and it was within easy reach of the Thames for rowing. At this point the river runs past the meadow of Runnymede, where King John signed the *Magna Carta* in 1215. To the south-west the gentle, wooded Coopers Hill rises from the meadow, with the college buildings just over the brow. It is a quintessentially beautiful English country scene, "one of the prettiest parts of the country."[107] Nearby is the village of Englefield Green with its Barley Mow pub, which would feature strongly in student life. Windsor Great Park is also close by, where students would walk at weekends, compete in military drill competitions, and learn horse riding. In Chesney's opinion it was "difficult to overrate the difference there would be in the character given to the place by locality, between a College in such a spot or one set up in a dingy suburb of London."[108]

Over the years, speechmakers at the college repeatedly succumbed to the temptation of quoting poetry about Coopers Hill. Usually, it was from Alexander Pope's "Windsor Forest," but John Denham's "Coopers Hill," first published in 1642, is also appropriate. The poet is standing on Coopers Hill, reflecting on

Figure 2.2. Coopers Hill buildings at the college opening (*Illustrated London News*, 25 November 1871, 501)

the overview his vantage point provided of the scenery and the political issues of the time.[109] In keeping with this tradition, several lines from the first version of Denham's poem express the attraction of Coopers Hill as a site for the college – near enough to the centre of the action in London, but sufficiently removed to create a self-contained academic community:

> So rais'd above the tumult and the crowd
> I see the city, in a thicker cloud
> Of business, than of smoke, where men like ants
> Toil to prevent imaginary wants[110]

Wyatt, the architect, concurred with Chesney's assessment of the estate's value and suitability.[111] So on 9 November 1870, the Council of India gave its approval for the purchase of Coopers Hill for £55,000, and the transaction was completed at the end of the following January.[112] Around £20,000 would be needed to make the buildings suitable for the college, including a new wing designed to bring the available student accommodation up to 100, as illustrated in figure 2.3. Although the eventual occupation was expected to be 150, it was decided not to

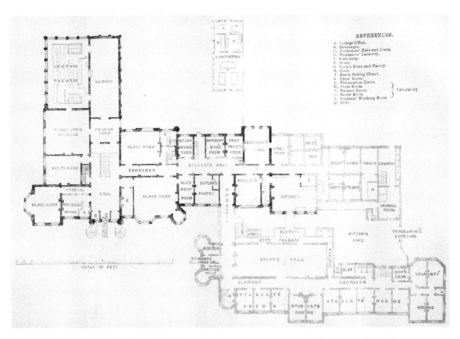

Figure 2.3. Plan of college buildings, 1871. Lighter sections indicate modifications made specifically for the college (*Builder*, 5 August 1871, 606)

incur the additional expense of the extra rooms until it was clear how many students would take advantage of the option afforded by Goldsmid's resolution to just sit the final examination. By the end of 1872, only 2 of the 100 entrants had done this, so Chesney felt secure in asking the India Office for the funds to construct the remaining accommodation (later called the New Block).[113]

Wyatt's plans for the first phase of the renovation appeared in the *Builder* in August 1871 with a description of the project:

> The principal staircase, with which the architect was loath to interfere, but which of course could not be common to the professorial, domestic, and students' departments, has been left intact, and the difficulty of providing separate access to those departments has been surmounted by the addition of the two new staircase turrets shown on the plan. These lead to the president's and professors' apartments, for whom fitting accommodation has been provided on the first and second floors. The offices for the resident officers are in the basement. In addition to the president's and married professor's quarters, accommodation is provided for three unmarried professors in the mansion.[114]

It is interesting that the *Builder* refers to the *Royal* Indian Civil Engineering College, as this was incorrect in 1871. Queen Victoria's assent for the use of the title did not come until 1875,[115] and the official document of the college, the calendar, changed from the Indian Civil Engineering College in 1874–5 to the Royal Indian Engineering College in 1875–6.

Gentlemen, Character, and Esprit de Corps

Esprit de Corps

As early as 1868, Chesney believed that a new engineering college should bring increased cohesion to the Indian PWD, along with the associated integrity and commitment to a "service": "An invaluable result of such an establishment would be the *esprit de corps* it would tend to foster, and an identification of its alumni with the interests of Government, a thing very much wanted in the department at present."[116] This concept reappeared in his proposal for the establishment of the Royal Indian Engineering College:

> Nothing has been said here of the advantage to be derived from the healthful esprit de corps which such an institution is calculated to engender ... the maintenance of this high standard of integrity of public spirit among all members of the department, civil as well as military, will be of incalculable advantage to the public service ... I believe that in no way can the sort of sentiment which it is desired to engender be better cultivated than by bringing the future members of the service together while under education, in the enthusiasm of youth, to cultivate a spirit of honourable pride in their calling, looking forward to service in India as their common goal.[117]

Four of the members of the Council of India, including some of the college's strongest supporters, agreed that esprit de corps would be one of its main contributions.[118] The *Spectator*, musing over the pros and cons of the new college in late 1870, commented more specifically on how the PWD at that time suffered from a lack of esprit de corps: "The scrap service they have got together, though very much better than it ought to have been under the circumstances, is insufficient, and wanting as a service in some highly needed qualities. The 250 Royal Engineers employed are, first of all, soldiers, and look to rewards other than reputations for the cheap building of dykes."[119] In essence, the idea was to create an institution that combined the traditions of the military academies and the Oxbridge colleges. Chesney believed that the PWD needed an injection of such high-minded spirit to improve its performance. In 1906, when the college was closed and most of the positions of authority in India were occupied by Old Coopers Hill men, their engineering accomplishments and sterling character were seen – by themselves – as among the major legacies of the college.

As Argyll prepared the ground in Parliament, he commented that a specialized college could imbue the engineers with "something which would make them serve the Government, from the feeling that they belonged specially to its service." In response, Sir Francis Goldsmid pointed out the dangers inherent in Argyll's contention: "If by esprit de corps you meant pride founded on the consciousness of real superiority of the body to which you belonged, it might be beneficial. But if you meant a vain belief in fancied superiority, it was worse than useless, and might be more fitly designated esprit de clique."[120]

Although Chesney was trying to solve the immediate problem of fragmentation in the PWD, a consequence of such an esprit de corps is the potential exclusion of talented and qualified outsiders and a reinforcement of social and racial hierarchies. It represented, in effect, a microcosm of British rule in India, and these tensions are part of what makes Coopers Hill interesting. Later chapters will explore how the college tried to achieve this, the extent to which it succeeded, and some of the consequences.[121]

Character

If building esprit de corps was the means of achieving the ideal collective behaviour of Coopers Hill men in India, then the two distinct but overlapping concepts of character and honour represented the desired characteristics of individuals.

Character was an umbrella term for a set of personal attributes that included zeal (reflecting energy and industry), prudence, perseverance, virtue, and self-sacrifice. For Britain's ardent imperialists, to which group the four ex-military presidents of Coopers Hill must surely have belonged, the concept of character was essential to the prosperity of the British Empire. These "ultra-imperialists" believed that, even as the empire had been founded on British character, the continuance and expansion of the empire was essential for maintaining the strength of that national character in the future.[122]

The importance of character and the necessity of avoiding stagnation were central to the education in the English public schools from which Coopers Hill drew its students (see chapter 3).[123] Afterwards, a career in India was seen as a way of continuing to build character, as befitting "men of strong purpose and firm will, and high ruling and organising powers, men accustomed to deal with facts rather than with words."[124]

In 1907, Lord Curzon of Kedleston, former viceroy of India, addressed Oxford University on the importance of frontiers in the development of the nation's character. With deference to his audience, he stated, "Outside of the English Universities, no school of character exists to compare with the Frontier; and character is there moulded ... in the furnace of responsibility and on the anvil of self-reliance."[125] One frontier Curzon had in mind was the "gaunt highlands of the Indian border." Coopers Hill telegraph officer Ernest Hudson, for example, relished the sense of being at the cutting edge of the empire when he served on the frontiers of Baluchistan and Burma. Several other examples of Coopers Hill graduates' work on these frontiers will be discussed later.

To these imperial leaders, then, the establishment of Coopers Hill was an additional tool in the struggle to keep the evils of stagnation at bay. Their college would join Oxford and Cambridge as a vehicle for preparing public school men for service in India. But unlike the ancient universities, Coopers Hill would train a new breed of technically minded young gentlemen to build the infrastructure that not only brought prosperity and security of the empire but contributed to its magnificence and "ornamentalism."[126]

Honesty was an element of character especially valued in Indian civil engineers because of the relative autonomy of engineers in the field, the large sums of money at stake in major construction projects, and the opportunities for corruption in such complex contractual environments.

High Principle

Chesney's and Argyll's emphasis on integrity at Coopers Hill was not simply a platitude but stemmed from an engineering controversy that occurred in India as the proposals for Coopers Hill took shape during 1869. For reasons known only internally,[127] the government of the Earl of Mayo took it upon itself to issue a notification that effectively accused the entire civil engineering profession of corruption. Dubbed by *Engineering* "The Simla Calumny," Public Works Department notification 242 (approved in Simla on 31 August 1869) noted, "In the Civil Engineering profession in England, it is a recognized practice for Civil Engineers employed by public companies and otherwise, to receive, in addition to the salaries paid them by their employers, commission on contracts given out, or stores and materials ordered or inspected by them."[128] In other words, kickbacks. Accordingly, the Government of India decreed that engineers should consider their Public Works Department salary to be their "sole legal remuneration" and that "every infringement" of this rule should be reported to the authorities. It was signed by Richard Strachey, R.E., later a member of the Coopers Hill Board of Visitors.

Civil engineers in India and Britain were outraged at the supposition that they routinely accepted "commissions" from contractors, and letters flooded newspapers and trade journals. Following an urgent meeting between the ICE and Argyll on 27 October, the government swiftly retracted its comments. On 19 November, under the title "The Strachey Manifesto," *Engineering* commented, "We were hardly prepared for so prompt a retraction by the Indian Government of the slanders thrown upon members of the civil engineering profession." Claiming that its notification had been "misunderstood," circular no. 84 issued at Simla attempted to explain away its initial position before including the all-important statement: "The Government of India unreservedly declares its complete confidence in the body of civil engineers in its service."[129]

The professionalism of civil engineers was clearly in the mind of those involved with the new Indian civil engineering college. In the published parliamentary papers on the establishment of Coopers Hill was a note by William Thornton, 28 February 1869, that spoke of the value of esprit de corps: "The most valuable element in esprit de corps is a corporate sense of honour, or, what is part of the same thing, a dread of doing anything to disgrace the corps."[130] However, an apparently earlier version of this note in the India Office Records contained an additional clause that was removed before publication. In that version, Thornton's sentence continued, "… and the engineering profession generally, unless greatly belied, is sadly deficient in the honesty and upright dealing by which such an enlightened sense of honour could not fail to be accompanied."[131] These incendiary words, seemingly written several months before the Simla Calumny document and removed on account of the controversy, are highly revealing of the India Office's prevailing notions about civil engineers.

The idea of professional integrity was therefore foundational to the establishment of the college as a response to both real and perceived shortcomings of the profession in India. At their speeches on opening day in 1871, the Duke of Argyll, for example, referred to the students' "duty as gentlemen," while George Chesney spoke of the need for a "pure and honourable tone of public feeling." Commenting on the opening, the *Engineer* wrote that the college was "exceedingly well calculated to foster a gentlemanly, liberal tone among its residents, and encourage feelings of self-respect, manliness, and integrity."[132] In this the college apparently achieved this objective. Much later, looking back on thirty years' worth of graduates, James Glass, the former chief engineer of Bengal, commented, "In high principle the majority of Coopers Hill men are very excellent."[133]

Gentlemen

By some alchemy, the collective esprit de corps and the individual traits of character and honour were unified and given substance by the nebulous concept of the English gentleman. In warning of the dangers of being ruled by such gentlemen, Harold Laski facetiously described the characteristics of the species:

> He must not concern himself with the sordid business of earning his living; and he must be able to show that, at least back to his grand- father, none of his near relations has ever been engaged in trade. It is desirable that he should have attended one of a limited number of schools, preferably Eton or Harrow; and it is practically

essential that he should have been to Oxford or Cambridge. He must know how to
ride and shoot and cast a fly. He should have relatives in the army and navy, and at
least one connection in the diplomatic service. It is vital that he should belong to a
club, urgent that he be a member of the Conservative Party, and desirable that his
ideas should coincide with those of the *Morning Post*.[134]

However, the concept of the Victorian gentleman in the last quarter of the
nineteenth century was complex and changing.[135] Being a gentleman was no
longer mainly a matter of birth but one of upbringing and education, which
in turn depended on wealth.[136] In 1903, then president of Coopers Hill, John
Ottley, commented that a few students came from Eton, Harrow, and Win-
chester, with more from newer schools such as Clifton and Cheltenham, and
many good students from Oundle, Repton, and Bradfield.[137] Recruitment to
the Indian Civil Service followed a similar pattern of being from second-tier
public schools. This was consistent with the social backgrounds of those students – most were the sons of professional men (67 per cent), businessmen (21
per cent), and minor landholders (12–13 per cent).[138] Although quantitative
data are not available for Coopers Hill students, information from obituaries
and other sources indicates that their backgrounds were similarly middle and
upper-middle class.[139]

According to the late Victorian definition, Coopers Hill students could therefore claim the necessary background to be gentlemen and demonstrated most
of Laski's characteristics. But clearly their notion of what it meant to be a gentleman and an engineer extended further. Laski continued, "The gentleman's
characteristics are a public danger in all matters where quantitative knowledge,
unremitting effort, vivid imagination, organized planning are concerned. How
can the English gentleman govern India when he starts with the assumption that
the Indian is permanently his inferior?"[140] With respect to the first sentence,
the following chapters will demonstrate that, contrary to Laski's assertion, the
gentlemen from Coopers Hill men as a whole displayed the knowledge, effort,
imagination, and organization expected of them as engineers. A gentleman
who was also an engineer encompassed a strong sense of duty, technical competence and efficiency, and both personal and professional honour.[141] They may
not have emerged from Coopers Hill fully prepared in all these aspects, but the
groundwork was done. Speaking of the early batches of college graduates, a
"highly placed source in India" commented on this reality: "Cooper's Hill men
are gentlemen, and nice fellows, but they are not engineers."[142]

In an earlier article, the *Engineer* felt similarly: "Cooper's Hill turns out educated gentlemen, trained mathematicians, good linguists, admirable draughts-
men, but it does not turn out engineers in the full sense of the term."[143] These
double-edged remarks referred mainly to the graduates' lack of practical experience when arriving in India and were hotly contested by the men concerned.

Nevertheless, they do suggest that the college was making the desired impact on personal conduct.

Unfortunately, the Coopers Hill men did not generally contradict Laski's second observation, on the equality of Indians. From opening day, Argyll urged students to "cultivate kindly feelings" towards Indians: "The students of this College should remember that the first part of their duty as gentlemen, as Christians, and as faithful servants of the Indian Government, was to cherish with the utmost respect and affection the natives of that great country."[144]

But Argyll did not advocate equality. The first-hand accounts of the graduates' lives in India discussed later in this book do indeed show that they were quite willing to respect the technical skills of their Indian subordinates and contractors on which the major construction projects relied. But they were still subordinate. It would not be until many years after the closure of Coopers Hill that Indian engineers achieved anything like equality.

After that closure, the Government of India spoke of the "high tone and standard" it sought in its engineers, prompting a Coopers Hill man to write to the *Spectator* of the continued importance to India of recruiting men "who are gentlemen by character and training."[145] Alicia Cameron Taylor perhaps best summarized the contemporary ideal of the Coopers Hill graduates (along with prevailing colonial attitudes): "a body of men – gentlemen of God's making and man's – whose lives were dedicated to service in a great Eastern Empire, the inhabitants of which look to Englishmen not only for intelligence, but for character – integrity, chivalry, and a delicate sense of personal honour."[146] The engineers were not alone in subscribing to this "cult of imperial honour." Especially in the Indian Civil Service, honour provided the moral superiority that self-justified British rule in India, as expressed by Cameron Taylor above. Honour was also the mechanism they used for keeping power in their own hands, and hence for maintaining the structure of the Raj.[147]

By sharing the social environments, belief structures, and many of the professional duties as other Britons in India, engineers also fully endorsed the ideologies of the Raj. However, the nature of their profession was materially different. They did not "dispense justice," maintain order, or collect taxes. Instead, they constructed major works of infrastructure that were themselves symbols of imperial honour and ornamentalism. Indeed, much of the engineers' sense of professional honour was tied to that of their edifices.

Professional Codes

In their jobs, engineers were responsible for substantial budgets, supervised large workforces, and interacted extensively with Indian subordinates, contractors, and suppliers. There is little indication that the Coopers Hill men did anything other than take their responsibilities seriously. Inefficiency, incompetence, or worse, corruption, were genuinely anathema to them. Rather, their weakness was a tendency towards technical hubris, a belief in their superiority over nature.

In this light, the concept of a gentleman professional engineer emerges as an unwritten but well-understood (to insiders) code of professional behaviour, which was fostered at Coopers Hill. The gentlemen's code permitted quietly dealing with transgressions without loss of face to the establishment. But there was genuine shock in the Coopers Hill Society at the trial of a corrupt engineer in India in 1908. It was "almost an unheard of thing to find a man amongst P.W. Engineers taking bribes on commission."[148]

After the closure of Coopers Hill, engineers for India were still recruited mainly from England. Professor William Unwin reassured the Coopers Hill Society that his graduates from the Central Technical Institution were "quite the same type as those who entered Coopers Hill."[149] This continuation perpetuated the concepts of gentlemanly honour used by the British to maintain power by excluding those who did not fit and never could. In the present context, this particularly affected Indian engineers, who were perceived as deficient in the qualities of the gentleman, regardless of their technical competence.[150]

After a spate of high-profile engineering disasters in the 1920s, the engineering societies in several countries began developing formal codes of professional practice. These codifications of expectations and practices were needed for another reason too. With the increasing cultural, technical, and educational diversity of engineers, the old concept of the gentlemen's code was no longer widely shared or even understood. In its place, however, the early phases in the development of engineering ethics extracted and formalized two concepts that reflected the ideals of the Coopers Hill gentleman engineer. The first was loyalty to fellow engineers and the profession, echoing the ideals of honour and esprit de corps. The second was loyalty to the interests of the employer or client, which Coopers Hill engineers would have termed efficiency.[151] The Coopers Hill "high standard of honour and efficiency"[152] was therefore a precursor of modern codes of engineering ethics.[153]

Opening Day

The preliminary Coopers Hill prospectus appeared in the *Engineer* on 2 December 1870 (later revised to comply with the 3 March 1871 House of Commons resolution), and notices advertising the first competitive examination for entry into the college appeared in February 1871. In the first year of operations, the exam and the start of term would be one month earlier than usual, in June and August, respectively to allow the incoming class of students to be present at the opening.

The grand opening day of the Indian Civil Engineering College took place at Coopers Hill on Saturday, 5 August 1871.[154] A special train was laid on to transport the 200 dignitaries and guests between Waterloo and Egham stations, with carriages added for the final three miles to Coopers Hill. Assembled in the large lecture hall with Chesney, Argyll, and the students were seven members of the Council of India (none of whom had voted against the college's establishment).[155] They were joined by members of Parliament, representatives of the civil and military engineering professions, prominent military and civil servants in India, and "a large party of ladies."

Herbert McLeod, one of the first professors to be appointed, described the day from within the college:

> After breakfast the students were sent into the Salisbury room from whence we sent for them one by one to speak to us. There are some very nice fellows among my lot. Coles was the only one who did not appear. After having left us, they walked in the grounds. The students assembled in the Halifax room at 12. Saw them to their places in the theatre....
>
> When the Duke of Argyll came we were all introduced to him at the door. We went into the theatre and the professors sat in the second row in the middle with the students behind. The Duke gave a capital opening speech describing the reasons of founding the College. The Colonel [Chesney] followed with a capital speech.
>
> We then adjourned to the dining hall to lunch. It was nicely managed.[156]

The Duke of Argyll used his speech to restate the reasons for the establishment of the college and making a strong point about the inadequacy of the British system of training engineers. Indeed, so strong was his political point that the *Engineer* regretfully reported, "It was impossible for any one to listen to the Duke's speech without feeling that it reflected gravely upon the educational status of the profession.... The impression left on the mind by a portion of the speech, was that the usual course pursued by engineers in their professional and technical training, was utterly inadequate to turn out competent men."[157] Argyll went on to encourage the students to embrace both the technical and cultural aspects of their education. He regretted that the 1857 rebellion had generated a "spirit of hostility and contempt" towards Indians by British civil servants, and therefore urged students to "cultivate kindly feelings towards the natives of India."[158]

For his part, Chesney took the opportunity to describe the course of study for students at Coopers Hill. He explained and justified the balance between subjects, the combination of theory and practice, and the inclusion of compulsory instruction in Hindustani and Indian history. He received a generally favourable review from the pro-education, pro–Coopers Hill *Engineer*.

"There is no necessity of concealing the fact that the college course will be a severe one, and will thoroughly test the abilities of the competitors at the final examination for appointments.... Those who graduate from Cooper's Hill will have the great advantage of being trained in this desirable manner, and will consequently add to the practical information of the engineer – wherever they may be obtained – that scientific knowledge which is gradually becoming indispensable."[159]

This chapter will close here, with Chesney standing in the lecture room of his new college, in a beautiful country estate, launching his new enterprise. How pleased and excited he must have felt looking out at the fifty fresh engineering students, the dignitaries in their stiff, formal attire, the accompanying ladies with their large skirts, and members of the staff and press.

Chesney's closing words are particularly apt:

We have here no traditionary influences to watch over us, no cloistered shades, with their hallowed relics, invoking us to guard the sacred genius of the place, no accumulated honour handed down to our safe keeping, and, what is more, no historic prestige to condone or extenuate folly. We shall be judged by what we are, not by any conventional standard confusing right and wrong.... Gentlemen, if it be a great thing to belong to a family descended from a distinguished ancestor, it is a still greater to be the distinguished founder of it. You, gentlemen, will be the ancestors of this college; and, if you accept this trust in the needful spirit, if you determine to maintain an etiquette of manners and a standard of morals which shall banish anything like immorality or "bad form" from the precincts of the place, its future success in this respect will be certain. Everything in such case depends on making a good beginning.... As we begin so shall we go on; and if you accept this responsibility, as I am sure you will do, you will be amply rewarded when, during [and] after life in India, you may be able to reflect that any reputation achieved by Cooper's Hill this respect is due to the character established by you, its earliest inmates.[160]

3

The Prime Years: 1871–1896

Unknown Country

In September 1871, after the opening ceremony, the president, the four professors, instructors, the fifty students, and the college staff came together to begin the work of the college in earnest. Imagine what a hustle and bustle there must have been, with new facilities to get used to, professors getting classes and laboratories ready, laying in the food and other supplies, and dealing with the remaining contractors (finishing the third block of fifty student rooms). And above all, the arrival of the students and their families, converging by train and carriage on the new college in its country village. The buildings, old and new, previously used to the decorum of a country house, now subjected to the raucous, excited voices of fifty eighteen-year-olds.

It seems that most of the students had no previous connection with India; it was "a country, unknown to most of us."[1] Indeed it was later lamented that the costs of sending a son to the college – tuition of £150 plus living expenses, totalling £280 per year – was beyond the means of most Indian officials.[2] This is consistent with what we have already heard about the relatively low rates of pay for civil engineers in India. What would induce young men, the "sons of persons of the middle class in easy circumstances,"[3] to commit to an engineering career so far from home? Some of them were not even especially interested in engineering at all. And, in the words of Alfred Newcombe, one of the first batch of Coopers Hill graduates to go to India in 1874, such a career had the disadvantage that "during the quarter of a century or more that an English official serves in India he is more or less an exile from his own country."[4]

While there were doubtless many individual reasons for going to Coopers Hill, the first few years probably attracted the pioneers and adventurers for whom joining a new college and a launching career in faraway places was an exciting prospect. Also, a highly competitive entrance exam would itself have been tempting to some ambitious applicants.[5] Yet others, several of whom had already attended university, saw it as an opportunity to further their engineering education.

On the positive side, there was a clear and well-publicized demand for engineers in India, instilling confidence in students and parents alike that the career would provide steady work and a good return on the investment. Over the history of the college. roughly two-thirds of the graduates took up appointments in India; presumably the remaining third considered the engineering education offered by Coopers Hill to be a good preparation for other careers. This pattern is still seen today; in Canada engineering is a popular choice of degree, despite the high grades necessary for admission, and the relatively high workloads and tuition fees. But less than half of the graduates become licensed engineers.[6]

There was also the sense – very strong in the first Coopers Hill men when they came to retirement at the turn of the century – that by building the transportation, communication, and irrigations infrastructures of India and other countries in Africa and the Americas, they were engaged in a great and noble enterprise making the world a better place (from their perspective). This flowery example from a non–Coopers Hill engineer is typical: "In the whole history of governments … no alien ruling nation has ever stamped on the face of a country more enduring material monuments of its activity than England has done, and is doing, in her great Indian dependency.… [T]he total number of individual works of exceptional magnitude and importance comprised in the whole, probably surpasses that to be found in any equal continuous area in any other part of the world."[7] In 1872, J.G. Medley, lately principal of Roorkee, explained the traditional appeal of a career in India: "A young Englishman, very little past thirty, … finds himself governor of a district as large as three English counties, with a population of 300,000 souls, to whom he is the embodiment of the Government, and who look up to him for advice or direction on all possible and impossible subjects."[8]

Competitive Entrance Examination

Whatever their motivation, 220 "British-born subjects of good character and sound constitution," between the ages of seventeen and twenty-one, presented themselves for the college entrance examination. Each had responded to one of the advertisements that had appeared repeatedly in a wide range of British national and regional newspapers and engineering journals between December 1870 and the end of April 1871.[9] Of these, 108 did not pass a qualifying examination and were not permitted to continue with the full process.[10]

Sixteen separate examinations in a variety of subjects were offered by the Civil Service Commission. They were held at the London University Building in Burlington Gardens on the nine days from Tuesday, 13 June, to Thursday,

22 June 1871 (with Sunday off). Each day consisted of two three-hour exam periods.[11] In the first offering of the Coopers Hill examination, 9500 marks were available: 4500 for languages (English composition, history and literature, and translation to and from Latin, French, German, and Greek), 4000 in science and mathematics, and 1000 in drawing.[12] Set texts were assigned for the English literature component, including in 1871, for example, works by Shakespeare, Bacon, Milton, Pope, and Scott. History was limited to the period from 1650 to 1714, although other periods were specified in later years.

Passing grades in each of English composition and mathematics were compulsory. Beyond that, however, there was no minimum total required for entry into the college; the fifty highest-scoring candidates would be offered admission. Of the top twenty-five placed students, twenty-three received marks in optional mathematics, twenty-one in history and literature, twenty in French, eighteen in sciences, sixteen in freehand drawing, and fifteen in mechanical drawing.[13]

The *Times* published the list of the top fifty candidates and their marks on 11 July 1871. First place went to John Watkin with 3838 marks,[14] while number fifty was Robert Way with 1354 marks. Five of the top candidates withdrew for various reasons, so men ranked fifty-one to fifty-five were admitted in their places.

It is noteworthy that even the top-placed Watkin was awarded well under half the marks available, while the lowest admissible candidates were receiving less than 15 per cent. In fact, it was not expected that any candidate would score highly on all components. Given the structure of the grading, any man could qualify for the engineering school knowing just the bare minimum of mathematics and no science at all, if he was reasonably proficient in literature or languages. Even more surprising is that when the *Engineer* published its early version of the prospectus on 2 December 1870, the original number of marks assigned to science was a mere 750 but later increased to 2000 to encourage applicants who were better prepared to be engineers.[15] Eventually, the number of marks for mathematics was increased further to 2500 and for science to 2100.

The college prospectus included a statement that appears odd at first sight: "A minimum of one fifth of the total number of marks allowed for each subject, except mathematics, will be deducted from all marks gained by a candidate." This meant what it said – that any applicants scoring less than 20 per cent in a particular subject would have their marks reduced to zero (later, the numbers were adjusted but the concept remained). The reason was that "a candidate ought to be allowed no credit at all for taking up a subject in which he is a mere smatterer," and the rule was intended to discourage candidates from scraping together a few marks from each of many exams.[16]

The consequences of the entrance examination structure became clear during the first year as some students struggled at the college. One, Colin Edward

Boyd, deemed by Chesney to be an "egregious failure" midway through his first year, was reprieved thanks to a petition of support from his fellow students.[17] But he was required to leave at the end of the year anyway when he failed his year examinations.[18] Boyd placed fortieth in the entrance exam with 1485 marks that came from compulsory English (180/500), compulsory mathematics (408/1000), Latin (548/1000), and French (349/750).

Anecdotal information comes from a piece entitled "Tales of Old Coopers Hill" published in the May 1899 edition of the *Coopers Hill Magazine*. The article told the semi-humorous story of four students who entered Coopers Hill in 1872 and who founded the *Oracle*, the college's first student newspaper: Edmund du Cane Smithe, Morice Leslie, Arthur Macleane, and Alexander Russell.[19] The four were given soubriquets:

- The "Head Boy," for being head of his year.
- "My Lord" received his title from his "lordly and striking appearance" and "splendid patronizing air."
- The "Man of Fashion," so called on account of his "coats of a specially refined cut," his skill at waltzing, and the piano in his room.

The fourth student, nicknamed "the Classical Scholar," was described as follows:

> The "Classical Scholar" from the marks he scored getting into the College … was supposed to possess marvelous stores of erudition in dead languages and useless literature generally; whether this was so or not he certainly possessed a most complete ignorance of anything connected with Civil Engineering and was popularly supposed from his entry as a first year, to his last day as a third year student, to be wondering what all the lectures were about, and what he was there to learn. He was nearly always "gated" and owed it to the kindly coaching of "My Lord" and the "Man of Fashion" that he pulled through his finals.[20]

Given the witty tone of the article, we should probably not treat this description too literally, but it does indicate that the effect of the entrance examination was well known among the students. That is was also well known by the faculty was shown by comments from George Minchin, professor of applied mathematics: "All of those 50 men were not very well qualified for the work of the College … because a great number of them got in from the public schools where Classics had been extensively taught, and Classics formed a part of the entrance examination."[21] This was not an accident but a strategic decision by George Chesney.

Because of the relatively high fees for the college, it was attended by the sons of relatively affluent families. In this, the college was consistent with the trend of applicants for associate membership of the Institution of Civil Engineers, which showed even as late as 1906 that almost 70 per cent were from at least the upper middle class.[22] The difficulty for Chesney was that the British public schools typically attended by those students before Coopers Hill valued and taught a traditional classical syllabus, as befitted the education of "gentlemen." The college entrance examination was therefore the result of a deliberate decision by Chesney, who wrote, "It appears very desirable that the test for admission should contain *nothing technical*, the certain effect of such a requirement in the present state of English education would be to narrow injuriously the field of selection.... [T]he test is of a kind which a boy from the upper forms of any good school should be able to undergo without any special cramming, and this is the main point to be arrived at. Having got the best material available, the technical part of education should begin after the college is entered."[23]

In this light, the examination is better regarded as a measure of potential, or possibly simply of work ethic, rather than preparedness for a technical education. In counterpoint to the Classical Scholar, there were many examples of students whose place at Coopers Hill was based on good classics marks and who went on to successful engineering careers. One such was Frederick Hebbert, who placed twenty-second with 1816 marks, of which 1466 came from English, history and literature, Latin, Greek, and French. Edmund Elliot, who entered Coopers Hill in 1875, was not one of the top fifty candidates listed in the *Times* and was presumably admitted when a higher-placed candidate withdrew; the fiftieth-placed candidate that year received a mark of 957.[24]

Despite Chesney's intention, many applicants were unsure that their education equipped them adequately for the competitive entrance examinations then in vogue not only for Coopers Hill but also for the British and Indian Civil Services. This uncertainty in turn led to the rise in the 1860s and 1870s of a big industry in private tutoring organizations, colloquially known as "crammers."

Looking back on his time as one of the first students at Coopers Hill, Sir Stephen Finney (CH 1871–4), commented that of the fifty students in his year "13 came from a 'crammer' named Johnstone at Croydon; 9 came from Ashton's and 7 from Wren's." The remainder came from various universities and "a few, I think 5, of whom I was one, came from public schools direct."[25] A notice in the *Coopers Hill Magazine* of 1902 reported the death of another crammer: "In the seventies, when there was keen competition in the Entrance Examination of the College, several of those who were successful were coached by Mr. Croome, at his establishment in Slough."

Although cramming for the Stanley examination was one of the things Chesney was trying to eliminate with the entrance examination, the number of successful applicants to Coopers Hill in the first decade who passed through a cramming establishment or private tutor was more than 60 per cent.[26]

Walter Wren was the acknowledged master of tutoring for the Indian Civil Service, some years filling half the available positions with his students. His organization was so well established that they even fielded a cricket team, playing a match at Coopers Hill at least once, on 12 March 1887, which the college won.[27]

Academic Structure

Competition

Wren's and the other reputable cramming schools owed their success to a combination of their focus, intensity, and motivation.[28] Or in the words of one pupil, "We are working like so many horses now. It is just six weeks to the exam."[29] It was this kind of energy that George Chesney wanted to bring to Coopers Hill by limiting the time that students had for idleness. But it was quickly apparent that the knowledge a guaranteed position awaited even those who merely scraped through the final examinations for some reduced the motivation to work hard during the year. This sole reliance on final examinations was an unforeseen consequence of Goldsmid's parliamentary resolution that allowed students to qualify for the PWD without attending Coopers Hill. It meant that course work or other assessments at the college could not contribute to the final qualification because the external students were not able to participate. Some students therefore chose to do little during term-time and cram just enough to pass the exams.

Chesney tried to remedy the situation by instituting prizes, awards, and fellowships. In the 1875–6 session, students were also offered the choice (insofar as it was possible) of province in India and branch of the profession (irrigation, railways, etc.) based on their ranking in the examination.[30] Recalling these early years after his very successful career, Sir Trevredyn Wynne (CH 1871–4) noted, "All the men in college in those days were certain of getting appointments in India provided they qualified, and there was therefore no keen competition. The men who worked helped those who did not but, in order that the honest workers might not be supplanted by those whom they helped, it was decided at a private meeting that the workers should have first choice of the appointments offered whatever the order, based on the examination results, might be."[31] In this way, the students assisted Chesney to ensure that hard work and collegial behaviour paid off.[32]

In later years, as Wynne noted, competition became much tougher as PWD positions became scarcer, and the incentive of offering choice positions to higher ranked students did increase diligence.[33] Chesney would reinforce his message by delivering speeches to the assembled students, about which one remarked sarcastically, "Colonel Chesney occasionally gave us a lecture on discipline, although the syllabus did not lead us to expect such treats. Among those we still remember with relish was one on 'Gambling and Sunday card-playing,' … and the last his valedictory address about 'Regularity in All Things.'"[34]

An Enormous Amount to Do

And there it is the rule at that Engineering School,
That the jack of all trades you must learn to be;
So they cram your wretched brain, till no more it can contain,
At the place called the R.I.E.C.[35]

W.E. Bagot (CH 1885–8)

The schedule of classes and other activities at Coopers Hill was, in theory at least, quite heavy. Students were tested "by periodical examinations and by assigning values to the drawings, surveys, notes, &c., executed by them while at the College."[36] A minimum mark in each of the required subjects and in overall average was required at the end of the first and second years to continue the program. In the last term of the third year, "special examiners not connected with the College" oversaw final examinations, project work, and *viva voce* questioning.

After first year, subjects were divided into four branches: (1) Engineering, including Drawing, Architecture, Surveying, and Accounts, (2) Mathematics, including the Mechanics of Engineering, (3) Natural Science, and (4) Hindustani, and History and Geography of India.

At the military academies, a single seniority list based on aggregate marks determined the order of appointments. Chesney argued that such a list was not needed at Coopers Hill and that division into separate branches allowed a fairer assessment of each student's abilities in the different disciplines.[37] Until the introduction of the Professional Course in 1877 (see below), Branches 3 and 4 were completed by the end of second year. These marks were combined with results from Branches 1 and 2 achieved at the end of third year to provide the final result. Optional languages, such as French, German, Latin, and Greek, could be counted towards the final result, but the college offered no instruction.

The day-to-day workload was heavy, as it was at Roorkee and is in every university engineering program nowadays. Modern students and faculty would echo George Minchin's observation: "The students at the College have an enormous amount to do.... They have a large number of subjects which are compulsory."[38]

As table 3.1 shows, the college day was formidably long, starting early and running into the evenings on weekdays and with classes on Saturdays.[39] Indeed, the most frequent minor infraction of college rules was tardiness for the early morning class. Professor Minchin later recalled that students and instructor alike were frequently sleepy at evening classes, and they were eventually stopped.[40] To discourage wild Saturday nights and sleeping in on Sundays, attendance at worship in Sundays was obligatory, even though this

Table 3.1. Coopers Hill Daily Schedule, Autumn 1872

7:00 a.m.	Physical exercise
7:45 a.m.	Classes (six days)
9:00 a.m.	Daily prayers in lecture rooms, followed by breakfast
10:00 a.m.	Classes (six days)
1:00 p.m.	Lunch
1:30 p.m.	Classes (five days)
3:00 p.m.	Flexible time
6:30 p.m.	Dinner and coffee
8:15–9:30 p.m.	Classes (five days)
Sunday 9:00 a.m.	Church

Source: "Regulations, Indian Civil Engineering College, Coopers Hill" (BL IOR L/ PWD/8/7, 313–17)

was not Chesney's original preference.[41] Initially this was at a special service at the nearby St. Jude's Church, but from 1874 onwards the college had its own chapel.[42]

Each student was assigned a tutor from the academic staff who was "responsible for exercising a certain degree of personal supervision over each student in his division." It was to the tutor that a student applied to obtain permission to, for example, leave the college on Saturday night or obtain exemption from chapel or other mandatory activities. Otherwise, "the sole responsibility for the discipline and management of the College, and for the general superintendence of the studies, is vested in the President."[43] In public rooms, at lectures, and for chapel, "academical dress" was required.

The academic topics covered at Coopers Hill were broadly similar to those included in the programs at University College London, King's, Edinburgh, and other institutions in Britain, and at Roorkee in India. Most taught surveying and levelling, but the coverage at Coopers Hill was more extensive and more practical, given its paramount importance for the canals and railways of India. This was important too because young engineers in India were often given surveying duties in their early years. Robert "Snippy"[44] Egerton thought that this set the college's education apart from others': "Surveying ... was ground into us. I do not know anywhere else where it is so taught. The whole of our three years' course we were always at surveying; we were taught on projects of different kinds. We had to get up railway projects, canal projects and things of that sort. I do not know any other place where that is done."[45]

To these more-or-less standard topics was added the course in Hindustani and the history and geography of India (with obvious logic, although Chesney felt the need to defend the decision). Members of the Indian Public Works Department were expected to speak one or more of the Indian languages, and language examinations were an important part of the promotion process, so this preparatory work at the college provided its graduates with a useful head start.

With an eye to blending practice and theory, in which Coopers Hill was unusual, Chesney stated, "Engineering is before everything a practical business, and a knowledge of the principles can in no way obviate the need for practical experience in this any more than any other active profession. But what may be insisted on is that not only is there no opposition between the two things, but that the one is the greatest possible aid to the other."[46] Practical training was obtained by students by spending two terms of their third year under the supervision of a practising civil or mechanical engineer. Students typically worked in foundries, water works, locomotive works, and navy dockyards.

Another unique element of the curriculum was accounting. As the former head of the Public Works Account and Finance Department, Chesney was well aware of the vital importance of accounting to the practising engineer in India and was determined to prepare the Coopers Hill engineers appropriately, even to the extent of teaching the course himself.[47]

Early Professors

The teaching staff of the college was initially quite small, consisting of just four professors (Construction, Mathematics, Surveying, Physical Science) and three instructors (Landscape Drawing, Mechanical Drawing, Hindustani, and Indian History). An additional professor of Applied Mathematics was added midway through the academic year and another in Hydraulics and Mechanical Engineering at the start of the 1872 session, along with two additional instructors (Surveying and Geometrical and Architectural Drawing). Of these eleven, only two did not reside in the college or villas built in the grounds.

Several of these gentlemen can be seen in figure 3.1, dated 1882,[48] although some staff changes had taken place by that time, particularly the replacement of Chesney by Sir Alex Taylor. With due attention to status, the academics are wearing gowns and mortar boards, unlike Mr. Pasco and Mr. Whiffin, the bursar and secretary respectively.

These staff were hand-picked by George Chesney to implement his vision of an engineering school that combined practical experience with theoretical learning, so it is interesting to review what each brought to Coopers Hill.

Callcott Reilly (1829–1900) – A Lovable Man

First to be appointed and most senior was (Joseph) Callcott Reilly, professor of construction, for a salary of £700 per year, along with an unfurnished residence. Reilly was a self-taught engineer who, after running away to sea as a

Figure 3.1. Coopers Hill senior staff, 1882 or 1883 (Cameron Taylor, *General Sir Alex Taylor*, facing 260)

teenager, later worked in a variety of practical engineering positions, including work foreman and draftsman in companies in the north of England, making steam engines and masts for iron ships, and as a consulting engineer in London.[49] "It was thus by dauntless energy and sheer hard work that the man, who had begun life as an apprentice in the merchant service, gradually made himself a name as a designer of bridges and as a practical engineer imbued with scientific principles."[50]

In three lengthy, detailed papers between 1860 and 1870, Reilly set out careful theoretical analyses of the stresses in wrought iron girders used in bridge construction that showed how appropriate design using the "modern" theory of materials could achieve the "utmost economy of material."[51] Moreover, he used examples of bridges actually constructed to demonstrate the validity of his approach. The second of these works won Reilly the Institution of Civil Engineers' Telford Medal and premium (financial reward) in 1866 and clearly appealed to George Chesney.

Reilly was superb at the job, and twenty-six generations of students loved "Pomph," as he was known.[52] "His house was the centre of hospitality and of kindness,"[53] and students would call socially on "Mrs. Pomph" at Coopers Hill

Villas.[54] At the Coopers Hill Winter Dinner in 1911, William Cameron spoke for many of his colleagues: "To the majority of those present to-night the name of Reilly brings a flood of affectionate memories. We think of his kindly nature and of his keen appreciation of a joke."[55]

There were disadvantages, however. In excuse for not having contributed anything to the first edition of the *Oracle*, the "Classical Scholar" pleaded "extenuation that the first volume of Professor Reilly's Pamphlets had just been published, that he had been studying these for the past week, that they bore no resemblance that he could trace to any manuscript notes of his own of any lecture whatever." An example of Reilly's lecture notes survives in its full mathematical glory – it is effectively a full textbook but printed "for Private Circulation only" – and it is easy to picture the "Classical Scholar" tearing his hair out over it.

Indeed, Reilly's nickname "Pomph" derives from these "outsize pamphlets," or "pomph-lets" as they were known, apparently from the way he pronounced the word. According to later reminiscences, Reilly did not lecture in the traditional way but in an early version of the "flipped classroom" used his classes to answer questions on the pamphlets.[56]

Reverend Joseph Wolstenholme (1829–1891)

Joining Callcott Reilly at Coopers Hill on 1 May 1871 was Reverend Joseph Wolstenholme, a mathematician who began his career as a brilliant, if unconventional, Cambridge fellow but in later life descended into a rather forlorn object of (generally affectionate) ridicule.

In 1850 Wolstenholme graduated from Cambridge University as "third wrangler," the name given to the man placing third overall in the mathematics tripos.[57] Two years later he received his MA and was elected a fellow of St. John's. Wolstenholme continued to teach in Cambridge until 1869, when he married (disastrously, according to friends) and was thereby obliged to resign his fellowship. This may have been a blessing in disguise, for "his Bohemian tastes and heterodox opinions had made a Cambridge career unadvisable."[58] During this period he published in 1867 what was to become his best-known work, a "wonderful series of original mathematical problems" covering the topics in the Cambridge curriculum, many of which "contain important results, which in other places or at other times would not infrequently have been embodied in original papers."[59] All told, he published nearly thirty original works, including what is now known as the Wolstenholme Theorem and a few books. Despite their "peculiar analytical skill and ingenuity,"[60] the general sense is that his career did not live up to its early promise.

After two years of teaching mathematics privately, Wolstenholme successfully applied for the professorship of mathematics at Coopers Hill. "Kindly, witty, illusive, he drifted through the College, practically unknown, even to his colleagues."[61] "Woolley" did not present his lectures but "in a very clear, round hand he wrote his lectures out on an enormous blackboard hoisted up and down by pulleys. When he turned his back to the exit door most of his class, having seen that 'Dobbie,' the Hall Porter had registered them as present, slipped away silently unseen by the lecturer, who finished unmoved to a sadly diminished class."[62]

From his early brilliance, Wolstenholme descended to the rather pathetic figure who "was merely a middle-aged, stooping, slippered figure in a straw hat and blazer, who, pipe-in-mouth and eyes-on-book, might be daily seen butting helplessly into the taut tennis net which seemed forever to impede his passage across the lawn."[63] Wolstenholme spent many summers with Leslie Stephen and his family, reportedly to escape from his unsuccessful marriage. From her recollections of these vacations, Stephen's daughter Virginia Woolf based her character of Augustus Carmichael on Wolstenholme.[64]

Herbert McLeod (1841–1923): A Brilliant Experimentalist

The Wolstenholmes held dinners at Coopers Hill for members of the college and guests, as noted in the extensive diary kept by Herbert McLeod,[65] professor of experimental science, who was the third faculty member appointed at the college in May 1871. McLeod's diary covers the years 1860 to 1923 in 272 small volumes of tiny handwriting, providing a detailed insight into the world of a well-connected scientist at the end of the nineteenth century.[66]

Despite his brilliance in the laboratory, McLeod first came into contact with many of the leading scientists and other citizens while collecting signatures for a declaration that scientific knowledge and biblical scripture would ultimately be shown to be consistent.[67] McLeod became a technical advisor and then friend to Lord Salisbury, secretary of state for India (1874–8), and future three-time prime minister. Salisbury maintained laboratories in his homes at Hatfield House and London. He first sought McLeod's assistance for his electrical experiments – McLeod worked with the instrument makers, Salisbury bought the apparatus, and McLeod got it working. The arrangement suited them both and friendship followed. Through the 1870s McLeod installed electric lights in Hatfield House, arranged illuminations for parties, and experimented with telephony.

Even before taking up his appointment, McLeod sought advice from his acquaintances and instrument makers on the facilities he would need at Coopers Hill. McLeod's diary describes the trials and tribulations of designing and setting up the teaching laboratory, arguing successfully with Chesney for additional funds for a purpose-built facility.[68] He finally installed two twelve-seat teaching laboratories (one each for chemistry and physics – the chemistry laboratory was nicknamed the "Stinks Lab"[69] and McLeod himself was known as

Stinks – and a research space for himself (shown as the separate small building on the 1871 Coopers Hill plan in chapter 2).

McLeod's diary gives us a glimpse into the workings of Coopers Hill as a community rather than a group of individuals. After all, most of the first group of professors lived on the Coopers Hill grounds, worked closely together in the college, and shared many related technical interests. McLeod noted in his diary his daily observations at his meteorological observatory and "was said to record the rainfall to three places of decimals, but certain lewd fellows of the baser sort clambered over the enclosing railings and on a dark night filled up the rain gauge with a mixture (it was said) of whiskey and water."[70]

With interests spanning the chemical and physical sciences, McLeod was involved in a very wide range of projects in collaboration with Coopers Hill colleagues George Sydenham Clarke and William Cawthorn Unwin. They worked on telephones and the measurement of high frequencies (tuning forks or motor speeds) and invented the McLeod Gauge for measuring very low gas pressures.[71] The ingenuity (and irreproducibility) with which everyday objects are pressed into use is fascinating. McLeod and Clarke talked about grounding their system with a clasp knife and a "little garden fork." They ran the signal along fence wires, "twenty-four feet of thin string dipped into river water," and through water – by cutting the wire and placing the ends at opposite sides of "a fountain basin filled with water."[72]

Despite all his other work, McLeod seems to have paid great attention to the students' education: "He cared for all that was best in the life of the College. His simple and ordered manner of life claimed our respect, while his unfailing courtesy and kindness claimed our affection and regard."[73]

George Sydenham Clarke (1848–1933)

Clarke, McLeod's collaborator and friend, was not appointed to a position in experimental science but to the post of instructor in mechanical drawing. In his memoirs[74] he recalls how his enthusiasm for the cello prompted the idea of using a tuning fork to measure "velocities of rotation," which led to their "long and … infinitely instructive experiments." They nicknamed the object of their quest "The Snark" after Lewis Carroll's recently released poem, because they "pursued it with forks and hope."[75]

Soon after graduating from the Royal Military Academy, Woolwich, in 1868, Clarke happened to meet George Chesney, the brother of his commanding officer, on a train. Hearing of the vacancy at Coopers Hill, Clarke applied, but placed second. However, the top candidate accepted another position and Clarke was appointed.[76]

In his 1875 *Practical Geometry and Engineering Drawing*, Clarke alluded to the challenges of his new post at Coopers Hill: "The want of a text book on the subject has been much felt by the present writer in dealing with large numbers of students. It was found impossible to keep a class of twenty-five fully employed. Time was inevitably lost in explaining away small individual difficulties and in setting problems suiting individual cases.... The present work is an attempt to embody the Descriptive Geometry course prescribed at Coopers Hill, and to meet the requirements of young Engineers generally."[77]

Clarke remained a lieutenant in the Royal Engineers throughout his period at Coopers Hill. In this capacity Chesney asked Clarke in 1873 to establish a company of volunteers attached to the Berkshire battalion as the "O" Rifle Corps Company; for more details, see chapter 4.

By 1880, Clarke was due for promotion, and feeling that he had spent too much of his time as subaltern in civilian life, he resigned from Coopers Hill. To mark the occasion, the Volunteer Corps presented him with a clock, which, in 1927, was still "a useful and cherished possession."[78] Clarke's main career was as a determined and perceptive government defence expert and colonial administrator, culminating in serving as governor of the Bombay Presidency in 1907. After his retirement, Clarke became increasingly known for his right-wing extremism and anti-Semitism.[79]

William Cawthorne Unwin (1838–1933)

Willian Unwin joined Coopers Hill in 1872 as professor of hydraulic and mechanical engineering. His interests in scientific engineering developed during his assistantship to the prominent engineer William Fairbairn, and he sought opportunities to become involved in education.[80] By the end of the 1860s he was giving a series of well-received lectures at the School of Military Engineering in such diverse subjects as theory and design of roofs and bridges; water wheels, turbines, and machinery for raising water; the steam engine; railway construction; and construction of wrought iron bridges.[81]

In his eight years or so at Coopers Hill, Unwin produced a significant quantity of original research in his own right, as well as collaborating with MacLeod and Reilly. Among these was a paper on the resistance of boiler flues to collapse that won him the ICE's Telford Medal, and a series of articles that would be the standard texts for several decades: his Chatham lectures; *The Elements of Machine Design*; his entries on hydraulics and bridges in *Encyclopaedia Britannica*; and his measurements of the tidal flow of water in the Thames related to London's sewage disposal system. For the last work, Unwin apparently purchased a steam launch, on which he used to entertain MacLeod and his other friends out of hours.[82]

Unwin features little in the life of the college, and when he did, it was with respect rather than warmth: "Again it would be presumption in me to raise my small voice in praise of one [Unwin] whose reputation stretches far beyond these shores. We were young and inexperienced when we knew him at Coopers Hill, but even then we realised he was one of the World's Giants."[83]

Unwin left Coopers Hill in 1884 to become professor at the Central Institution in South Kensington (later part of London University) and was instrumental in establishing its high reputation for technical and scientific education. He published and consulted extensively on a wide variety of engineering topics. Prominent amongst these was his service on the International Commission appointed to evaluate designs for the hydroelectric scheme at Niagara Falls.

Professorship of Applied Mathematics

In 1872, the year after Wolstenholme was appointed to teach pure mathematics, the post of professor of applied mathematics was awarded to (Alfred) George Greenhill, a Cambridge graduate with the "somewhat rare ability of applying his mathematical knowledge to practical uses, especially in dynamics and in engineering matters."[84] But Greenhill lasted just one academic session; young and brilliant, he appears to have chafed at the restrictions Coopers Hill College life imposed on him. Most of Greenhill's subsequent career was spent at the Royal Artillery College at Woolwich, where he was an expert at the science of gunnery. Increasingly eccentric, but respected nonetheless, he was knighted in 1908.

Greenhill was succeeded by Edward Nanson, yet another graduate of Cambridge, who fared no better at Coopers Hill. He too lasted just a single year before emigrating to the University of Melbourne, Australia, in June 1875.

Happily, the next incumbent of the Professorship of Applied Mathematics made up for the short tenure of the first two by remaining in the post for thirty-one years until the college closed. Gregarious, brilliant, imaginative, engaged, and idiosyncratic, George Minchin Minchin entered Trinity College Dublin in 1862 as George Minchin Smith and later changed his name, supposedly to protest his father's second marriage.[85] After winning several prizes in mathematics, he narrowly missed a fellowship under controversial circumstances and instead joined Coopers Hill.[86] Minchin was in many ways the life and soul of the Coopers Hill professoriate, and his ebullient personality shines through the years. His range of interests was tremendous, covering mathematics, hydrostatics, satirical prose, scientific poems, wireless transmission, and experimental electronics. In the words of one obituarist, "He caught the fire of life in both hands, and conveyed its benefits to all around him."[87]

Surveying

The importance of surveying in the Coopers Hill curriculum has already been mentioned. This subject was also put under the charge of a Royal Engineer, Captain (later Major-General) William Henry Edgcome. Edgcome was born in India and trained at Addiscombe. He later returned to India to work on the Great Trigonometric Survey of India and Burma.[88]

Edgcome returned to England in 1859 to take up a position as professor of military drawing and surveys at Addiscombe. However, the college closed in 1861, forcing him to go back to India, where he continued his work in education as principal of the Civil Engineering College, University of Madras. Hearing of the formation of the new college in England, Edgcome wrote to Chesney to express interest in a position. His extensive field experience in Burma must have seemed a good fit with Chesney's ideas for training practical engineers.

In autumn 1872 an instructor in surveying was recruited to support Edgcome, Major (also later Major-General) Edward Henry Courtney, again from the Royal Engineers. This pair of experienced military officers ran the surveying program for almost the entire duration of the college. Although born in India, Courtney was educated in England from about the age of eight. Instead of returning to India after completing his training at Woolwich and Chatham, Courtney spent about four years of active military service in China from 1857 to 1862, during the second China (Opium) War.[89] He was then posted to the Ordnance Survey of Great Britain working in Aberdeenshire, where his name appears, along with those of other colleagues from the Royal Engineers, at the foot of each sheet of the survey for which he was responsible.[90]

With a wealth of experience from the Ordnance Survey leading teams of surveyors, and solid military credentials, Courtney was the ideal person for drumming practical fieldwork into the engineering students. Courtney's personality probably also commended itself to Chesney for its role in raising the social profile of Coopers Hill: Courtney, "a man of society and a courtier by instinct – presided over the College festivities, balls, sports, and the like, to which he gave a certain cachet which it was important should be imprinted from the outset on the hospitalities and the functions of an Institution which was designed to take the place beside others of older standing."[91]

This description not only identifies Courtney's contributions to the college but also testifies to the college's aspirations to be in the top tier of hospitality and "manly" sports comparable with the military academies and the universities (meaning Oxford and Cambridge). For some reason Courtney was known to the students as "Q."[92] Figure 3.2 shows a photograph of Q with his surveying students, taken in 1889 after he had assumed the role of professor of surveying upon Edgcome's retirement.

Military Administrators

Chesney's choices for the academic founders of Coopers Hill therefore consisted of five academics and four Royal Engineers (including himself): a brilliant experimentalist (McLeod); two engineers with interests in education and

Figure 3.2. Coopers Hill surveying party, 1898, under supervision of "Q," Major-General Edward Courtney (© British Library Board, photo 448/1)

a belief in the necessity of combining theory and practice (Reilly, Unwin); a talented and ambitious young Royal Engineer (Clarke); excellent young professors in pure and applied mathematics (Wolstenholme, Greenhill); and two experienced military surveyors. This mixture of academics and military men must have made for an interesting college social life, with areas of common interest separated by very disparate life experiences.

To this group over the following year or two Chesney added two more military men as the college's senior administrators: John Penfold Pasco as bursar and John Ball as secretary.

Pasco joined the Royal Navy in 1851 as clerk's assistant on HMS *Vengeance* at the age of fourteen. He rose slowly through the ranks of assistant clerk, clerk, and assistant paymaster, and eventually reached the rank of paymaster in 1865, shortly before his retirement. In 1866, he was awarded the honorary title of fleet paymaster, in recognition of his seniority.[93] In his address to the Coopers Hill Society dinner in 1913, George Minchin recounted Pasco's reminiscences: "He had been a sailor and was present at the bombardment of Odessa. Many scores of times he described to me how his ship, the Encounter, sailed in a circle in front of the town and discharged a broadside on each approach; … whenever

I asked for the bombardment of Odessa, he at once brightened up, forgot his troubles and related it as if I had never heard it before."[94]

Pasco was first appointed to Coopers Hill as secretary, but Chesney quickly switched him to bursar in 1873, a position he would hold for twenty-seven years.[95] The bursar's lot in life was not made easier by the detailed accounting required by the India Office. Chesney was not exaggerating when he complained, "If a pot of jam or a glass of beer is issued to a student, the charge and the recovery have to be shown in the monthly accounts rendered for audit. This may be the best system, but the result is that the accounts are extraordinarily voluminous and troublesome to keep, and the Bursar is distinctly overworked."[96]

Pasco's administrative colleague for the first few years was Captain John Ball, another army veteran, this time with experience in India during the Rebellion and in Hong Kong, China, and South Africa. On Ball's departure in 1878, he was replaced by John George Whiffin, who, like Pasco, had been a paymaster in the navy. Unlike Pasco, however, Whiffin had led an adventurous early career aboard the HMS *Herald* circumnavigating the world, surveying the west coast of the Americas, and making three voyages into the Arctic in search of Sir John Franklin's missing expedition. Whiffin then served in the Baltic during the Crimean War, before serving in home waters aboard a variety of vessels including the famous iron warship HMS *Warrior*. Whiffin received the Arctic Medal, and the Baltic, Crimean, and Turkish medals with the clasp indicating service at Sebastopol.[97] He retired from the Royal Navy in 1873 at the age of forty-seven with the rank of paymaster-in-chief, awarded for "long and meritorious sea service."

On Whiffin's death from influenza in 1892, he was succeeded as secretary by Lieut.-Col. Walter John Boyes, another army man with seventeen years' experience in India, who had settled in Canada after his retirement but found it not to his liking.[98] Boyes remained in his post until the position of secretary was abolished shortly before the college's closure.

The combination of Chesney's personality and vision for Coopers Hill, together with his choice of military officers in key positions, gave the college the feel of a military academy. No doubt the deliberately self-contained nature of the college estate reinforced that notion.

College Servants

While we do not hear a lot about them, the college was supported by a staff of nearly thirty, ranging from the bailiff, porter, cook, and butler, to the "scouts" who looked after the students' rooms, and the boot cleaner and cook's assistants.[99] "Lecture room messenger" and later porter, John Dobson (d. 1913), in particular, became something of a college "character." "Dobson was day porter or something of the kind and bottle-washer generally, and paper-stretcher into the bargain; however what with his official duties such as reporting absentees, keeping lists of attendance at chapel, intimating punishments and being chased headlong down the turret stairs in consequence he hadn't altogether a rosy time."[100]

That said, Dobson seems to have found at least one way of taking advantage of his position:

> He was a great character. He also kept the attendance register at lectures, where he picked up many technical and scientific terms. Once he was heard, when trying to balance an opal shade on a chandelier, to say "I must find the moment of inertia of this thing."
>
> For a modest but unsolicited contribution in beer – then unrationed, cheap, of good quality and plentiful in supply – he was often willing to overlook temporary absences at lectures, or at night on late arrivals, by those who could not attain entrance to the College after 9 p.m. at the Porter's Lodge.[101]

Following the theme of beer, another former student recalled, "At the entrance I see Dobson with his coat tails ready parted, thirsting for his bottle of beer."[102]

Challenges and Change of Leadership

On the academic side of the college, things were generally going well – good professors, enthusiastic students, a modern curriculum, and a demand for engineers in India. On the financial side, however, the picture was not so good.

During the 3 March 1871 House of Commons debate on the new college, Grant Duff promised the House that there would be no charge to the Government of India for Coopers Hill. However, in the usual way of the world, the real expenditures proved to be higher and the revenues lower than predicted. By 1876–7,

- Capital costs – including the purchase price, new accommodation wing, chapel, boathouse, furnishings, etc. – amounted to £119,000, over 30 per cent higher than Grant Duff's original figure.[103]
- Receipts from student fees were almost £19,000, falling short of the projected £22,500 because the enrolment had not reached the anticipated 150 students.
- Expenditures were 20 per cent higher than revenues, at nearly £24,000 compared to the estimated £18,350.[104] This was broken down roughly as professors' salaries £8,000, teaching costs £2,000, staff salaries £2,250, food £6,500, maintenance £1,400, and Practical Course fees £3,500.[105]

Each year, therefore, the college was costing the Indian taxpayer £5,000 plus £3,500 interest on the capital. This had been going on since the college opened, so by the end of 1877, a cumulative operating deficit of £55,000 had been amassed (in addition to the £119,000 capital). Instead of the small surplus promised by Grant Duff in February 1871, there was a real cost per student. By July 1877, 180 men had been appointed in India so the cost per student (not including buildings) was over £300.

Parliament and the Government of India notwithstanding, Chesney himself had never expected the college to be entirely self-supporting. Indeed, he had initially favoured lower fees of £110–120 and believed that the resultant cost to the Government of India of running the college was amply justified. Of his financial estimates, he commented,

> It makes the net charge to the State £2,750 a year, exclusive of rent or cost of building, and of repairs. Even if it were double or triple that amount, it would be but a trifling per centage on the departmental operations. The salaries of the Public Works Department cost one million sterling a year, and it controls an expenditure of six millions. It seems obvious that if only a very moderate degree of increased efficiency be imparted to the Department by the training thus to be given to its younger members, the outlay will be well bestowed.[106]

Here Chesney was alluding to the money that would be saved in India by better engineering. Argyll was of the same opinion, saying in the House of Lords, "The Indian Government had lost annually enormous sums of money by the carelessness and incompetence of many of the civil engineers in India … the Indian Government were annually losing hundreds of thousands of pounds in bad engineering."[107]

Unfortunately from the college's perspective, Gladstone's government had been defeated in the 1874 election and Argyll had been replaced as secretary of state for India by McLeod's friend Lord Salisbury. On 11 October 1876, the India Office wrote to Chesney that "Lord Salisbury is desirous that the College should be made self-supporting and that the receipts should cover the entire charges, inclusive of interest on the buildings & c." and requested a report on how it might be achieved.[108]

If the government expected the college's revenues to cover the capital costs of the college's buildings, Chesney frankly stated, "I am quite unable to propose any plan for satisfying it. Nor, I believe, can any case be cited where such a thing has been accomplished."[109] Chesney also recognized that most of the costs were independent of the precise number of students. Nevertheless, after much thought, Chesney proposed three measures for improving the operating performance of the college. Two were expressed in a letter to the secretary of state on 16 March 1877.[110]

First, he proposed an increase in the annual tuition fee of £30, £22 of which would go towards sustainability and the rest in support of scholarships; Chesney felt that this relatively modest increase would not significantly reduce the chance of "securing a supply of eligible candidates."

Second, Chesney noted that the Practical Course was "a very heavy drag on the finances of the College." He therefore argued for change in the course of study such that students spent a full three years in college, followed by a "fourth year" of supervised practical work. For those appointed to the Indian service at the end of the three years, the salary and fees for the practical year would be covered by the PWD. This not only removed £3,500 from the college's expenses (by transferring them to the PWD), but also strengthened the students' practical knowledge. Chesney hoped that by providing supervision in the workplace the students would gain a better experience, since students in the previous Practical Course had suffered from the neglect sometimes found in the pupilage system.

Why were students not sent to India for their practical experience? Richard Strachey, supported by other members of the Council of India, believed they should: "The great works there in progress such as Railways, Irrigation Works, Docks & Harbours, will supply in my opinion far more useful schools for young Indian Engineers than any works going on in England."[111] From the establishment of the college, however, Chesney had been opposed to this idea, saying, "The particular sort of practice which may be gained in a first class workshop, or on some great English engineering work, is seldom to be had in India, where the young engineer is liable to be at once detached to some out of the way place where he is deprived of the lessons derived from other people's experience."[112] Chesney's proposal to continue the Practical Course in England was ultimately approved.

In 1879, Chesney additionally advocated that a third measure should be taken: "More candidates should be admitted at the preliminary entrance competition than there are guaranteed appointments, and that these appointments should be competed for [at the college]."[113] While this removed one of the big attractions of the college, it meant that the competition for PWD positions was now effectively at the end of the program. In addition to maintaining student numbers, this eliminated Chesney's concerns about student diligence during the course and improved the quality of engineers going to India.

In these second and third proposals, Chesney went some way towards meeting the recommendations that would shortly be made by Sir Andrew Clarke (Public Works member of the Government of India) for reforms at the PWD. Over the previous decade the PWD had swung from a deficit of engineers to a surplus. This had come about partly because of a reduction of public works projects because war and a series of terrible famines throughout the 1870s had drawn funding away. The situation was exacerbated by senior engineers choosing not to retire, so the prospects of expanding the number of students seemed slim. Indeed, the number of first-year places was reduced to forty-five in 1877[114] and then to thirty-five;[115] the number of engineers joining the PWD remained at over forty per year between 1874 and 1879, after which it declined further.

As reported by the *Oracle*, Clarke's proposals included encouraging senior officers to retire by means of bonuses and improved pensions and making "Coopers Hill a national school of Engineering, supported by England and aided by India and the Colonies."[116]

The initially optional fourth year was well received by the editorial staff of the *Oracle,* who wrote in October 1877 of the professional, social, and sporting advantages:

> A year's reprieve, before departing to a country, unknown to most of us, and from whose bourn we may not return for many years, seems in itself a sufficient induce-ment to stay, but when this is also sugared over with an extra temptation in the way of pay, it forms a bonne bouche which few have been able to resist.
>
> Apart, however, from the professional advantages to be gained by a year of prac-tical work in England, we must not forget to notice the great gain in the arena of football and cricket.[117]

Chesney was responsible for two more changes before the end of the dec-ade. The first was to offer a course for training Indian telegraphers through Coopers Hill, initially of one year's duration but quickly increased to two. Tele-graph officers for India had previously been taught at Southampton by William Preece. Between 1867 and 1871 he taught seventy-two men, who were collec-tively known as "Preece's Lambs."[118] This strange moniker had been bestowed on the students by "a somewhat exuberant lady of Southampton" of their ac-quaintance, and the name had stuck.[119] The transferral of telegraph training to Coopers Hill had little immediate financial impact, since it passed just five to seven students per year in 1877–9, and typically only two or three per year thereafter.[120] However, the addition of telegraph students further increased the relevance of the college in the short term, and it would become the foundation for expanding electrical engineering instruction.

Chesney's remaining action was to resign from the presidency of Coopers Hill in 1880 to take up the "important position" of military secretary to the Government of India. By this time he had already exceeded the seven-year con-tract originally approved by the Council of India (starting 24 April 1870).[121] He had been seeking new opportunities but had been turned down for several high-profile appointments.[122] To make matters worse, the Government of India had sent a telegram to the India Office in June 1879 advocating the closure of Coopers Hill, despite the college's academic and social success and Chesney's own efforts towards its self-sufficiency.[123] Chesney was ambitious for his career and for appropriate recognition in the honours lists. Shrewd politician that he was, he probably recognized that it was an opportune time to move on from Coopers Hill.

George Chesney served as secretary to the Military Department of the Government of India, concerned with finance and logistics of the armed forces, until 1886, when he was appointed military member of the Governor General's Council responsible for Indian defence policy. He was knighted for his services in 1890. On account of his wife's ill health, the Chesney's returned to England in 1891, where he was elected as member of Parliament for the City of Oxford on an agenda of army reform. He died suddenly in London on 31 March 1895.

When Chesney's successor was appointed, the president's salary was immediately lowered back down to £1000 – it had been increased to £1,500 in 1873[124] – reducing what Chesney himself recognized as "a heavy item in the estimate."

Chesney's departure signalled the close of a crucial era in Coopers Hill's history. Since 1870, Chesney's hand had been in everything, ranging from the buildings to the curriculum, from the professors to every organizational detail of the college, even the laundry facilities. The new president and the changing times would inevitably alter the character of the college, but its successes during the first decade, and its very existence, were attributable in great part to his vision, skill, and determination. Of George Chesney, the Duke of Argyll wrote, "Major Chesney combined every possible qualification for a new and a most difficult task. Very able, highly cultivated and accomplished, a thorough gentleman, and with a personal manner of great modesty, gentleness, and charm, he was just the man to start a new college for educating young men in a great profession, under the best and highest influences for the future."[125]

President Alex Taylor

Implementation of the most controversial of the reforms proposed by George Chesney in late 1870s – holding the competition for Indian posts at graduation – fell to another Royal Engineer, Alexander Taylor, who was appointed president of Coopers Hill in 1880.

Taylor was not perhaps the obvious choice to replace Chesney. At fifty-six, some thought him too old and lacking in physical energy, and he had no previous experience in education. He was famous for his actions in the two Anglo-Sikh Wars and at the capture of Delhi, which was a turning point in the Indian Rebellion. Although an engineer responsible for bridges and transportation, Taylor fought in the fiercest battles between the East India Company and the Sikh Empire that ultimately led to the annexation by the British of the Punjab in 1849.

In June 1857, Taylor was ordered to join the siege of Delhi where, because his superior officer was sick, he performed the bulk of the reconnaissance behind the enemy lines, and evolved the "bold plan of attack." Once the British entered Delhi they encountered fierce resistance from the city's tall buildings. Taylor "again showed extraordinary fertility of resource" by proposing that the engineers should "break internal passages from house to house" rather than exposing the British troops to gunfire in the streets.[126] Thereafter, the city fell quickly.

After the fighting, Taylor held a sequence of high-level positions in the Punjab (as chief engineer) and the Government of India (responsible for Indian defence). He was knighted for his services in 1877, shortly before his retirement.

The year after "Alex" Taylor's death, his daughter Alicia Cameron Taylor published her two-volume hagiography *General Sir Alex Taylor G.C.B. R.E.: His Times, His Friends, and His Work.* It contained a chapter on Coopers Hill based on information from Chesney's diaries, Arthur Hicks (secretary of the Coopers Hill Society), and professors. She described what happened when Chesney went to ask his former comrade-in-arms whether he would be interested in becoming the new president of Coopers Hill: "On approaching an eminence in the neighbourhood of Surbiton where Sir Alex was staying, [Chesney] saw a stalwart figure and two smaller ones speeding down the slope, each perched on the summit of one of those alarmingly tall and slender wheels which preceded the modern cycle. Suddenly the largest of these machines shot forward, and ... the rider steered for a crossway hedge, breasted it, parted company with his machine, and landed gracefully on the other side. This was Taylor, who was being initiated by his boys into the art of bicycle-riding."[127] Taylor initially turned Chesney down, saying, "I am no schoolmaster; I have no gift that way." But Chesney persisted, introduced Taylor to the college and its staff, and eventually persuaded him to take the post, commencing November 1880.

One of the students at the college in 1880, Harold B. Taylor (CH 1879–82), affirmed Taylor's physical energy, describing how the new president hung off the back of a brougham driven by the students as they raced the departing Chesney's official conveyance down to Egham railway station via the "perilous" shortcut.[128] Another student later recollected, "When he was not far short of seventy years of age he was seen to climb up to a third-floor window of what is known as the 'new block' at Coopers Hill with the object of personally determining whether it were possible to obtain ingress to the College by a route other than that of the Porter's Lodge."[129]

The New Scheme

George Chesney's proposed £30 tuition increase (from £150 to £180 per year) came into effect for students starting in October 1883[130] and the fourth year of practical experience was already running. The college now administered the entrance process. A candidate whose initial application was acceptable took an examination in English and mathematics and had to demonstrate "a fair general education, by certificate from their school or college, or by undergoing an examination in some classical or modern language, and in history or geography."[131]

Effectively the entrance examination now became pass-fail rather than competitive. In 1883, for example, fifty-six people applied and forty-nine were accepted,[132] a far cry from 1871 when more than two hundred applicants sought the fifty places. The New Scheme was extensively advertised to prospective students; in each of 1881–2 and 1882–3, over £1000 was spent on advertising, about five times as much as normal.[133]

The new PWD appointment process came into effect in the 1880–1 session. At the same time, the provisions of Goldsmid's resolution were removed from the calendar, even though Coopers Hill remained the sole point of entry into the PWD. By disconnecting entry into Coopers Hill from positions in the PWD, the New Scheme, as it was called, dramatically altered the college's original raison d'être. Although Chesney had argued that the Government of India still gained longer-term benefits from training its own engineers, it was an uphill struggle to justify charging the college expenses to the Indian revenues when so few joined the PWD (with just thirteen places for students who entered in 1883).

These changes had not gone unnoticed in Parliament, and the college came in for vigorous criticism in a House of Commons debate on 21 April 1882 where Edward Gibson, MP for Dublin University (and therefore not disinterested in the matter, as supporters of Coopers Hill were quick to point out) called on the government to appoint a select committee to "inquire into the working and expense of the Cooper's Hill College."[134] Speakers in favour of Gibson's motion, including Mr. Plunket, brought up the now-familiar objections – the cost, the Government of India's original objections, and other universities could do the job equally well (especially Trinity College Dublin, observed Mr. Gibson). In a debate of long speeches, Lewis Pugh summarized the basic issue succinctly: Coopers Hill "was not a College for India only, but for home requirements. It was only Indian in this respect, that India had to pay the whole cost of it."[135]

Despite the vigorous, and one must say well-founded, arguments for the enquiry, the secretary of state for India, the Marquis of Hartington, refused to establish the requested select committee. He argued that a review of the college had already been done, that changes had been made, and that it would be premature to judge the effectiveness of the New Scheme at that time. It was a fair position, but hardly a ringing endorsement of the college.

The press jumped on the bandwagon. In a scathing editorial following the parliamentary debate, the *Engineer* criticized both the cost and the concept: "All such establishments as that of Cooper's Hill stand or fall ultimately on their own merits, and the evidence is not wanting that Cooper's Hill College is practically doomed.... Not one-third of the number it can accommodate can obtain berths in India, and an attempt is being made to keep the College books full by offering a very good but very expensive education to many men who have no intention of going to India."[136] There was an element of a "young engineers these days can't do anything" sentiment in the article. In fact, the *Engineer* had also made a statement to this effect a year earlier and in response to the ensuing letters of protest had been forced to explain itself more clearly:

> Men trained in colleges know nothing about administrative details. The best man ever turned out by Cooper's Hill College could not take charge of a large contract, manage his men, buy his materials in the cheapest market, arrange for the transport of them at the smallest cost, or attend to any one of a hundred different things which claim the constant attention of the competent and conscientious engineer. Our correspondents appear to think we disparage Cooper's Hill men because we have implied that they are ignorant of much that the fully trained engineer has at his fingers' ends; but we do not blame the students for this. We do not blame the College. What we assert is that the system is not a good system, and that it was specially unsuitable to India.[137]

The question of what a twenty-two-year-old engineering graduate can be expected to know is still debated. There is simply not enough time to cover everything. Are we training engineers or educating them? What is the right balance between the "fundamentals," the latest technologies, professional skills, workplace experience, workshop training, understanding business, and a host of other important skills? Every student, employer, administrator, and professor will give a different answer.

Negative publicity such as this must have had its effect on the staff and students of the college, as well as on President Taylor personally. After all, it was only three years since the Government of India had advocated closing Coopers Hill. The secretary of state had already taken measures to forestall further such criticism; a board of visitors[138] had been appointed in 1880 to oversee the work of the college and to maintain the relevance of its education to engineering in general, and to India specifically. The board consisted of a chairman, four civil engineers, and four members with Indian experience. Its mandate was "to provide a highly qualified authority to supervise, in communication with the President of the College, the course of study to be pursued, and to advise as to the internal management of the College."[139]

The board made formal reports in 1882 and 1886, but there was then a big gap until reports in 1892, 1895, and 1897. It appears, however, that the Board of Visitors did pay annual formal visits to Coopers Hill and presumably gave verbal feedback to the president on their findings. One of their early recommendations was to remove the Indian history and language content from the curriculum because it was no longer relevant to the majority of students.

The Trees for the Wood

The decade of the 1880s saw one other major change in the academic life of Coopers Hill – the addition in 1885 of the Forestry program, together with its professors and facilities. Until this time, there had been no establishment in Britain providing training in scientific forestry, a subject in which students were trained "to observe the action of nature in a forest, and to follow it, or to utilise it for our advantage, when we are able to do so."[140]

The priority of Indian forestry was primarily commercial and secondarily environmental. In the absence of any controls on the exploitation of its timber, India's forests had suffered extensively in the first sixty years of the nineteenth century from the combined onslaughts of commercial exploitation, the expansion of agriculture, and railway expansion (using wood for something like 25 million railway sleepers, as well as fuelling the steam engines).[141] The administration was concerned by both the environmental damage from deforestation and the rising price of wood (which rewarded further exploitation):[142] "The natural forests in most parts of the country had been ruthlessly wasted by felling and burning, and no system had been adopted to regulate the cutting, or to provide for the wants of future generations by preserving the existing forests or forming new plantations. Magnificent trees were sacrificed for insignificant purposes, and planks were not sawn, but hewn with an axe, one tree furnishing a single plank."[143]

These are the words of Hugh Cleghorn, who together with Dietrich Brandis effectively established the Indian Forest Department in 1864. France and Germany were leaders in forestry at the time, Britain having not seen the necessity for extensive forest management on account of the convenience of coal and inexpensive imported lumber. Cleghorn, a former doctor who had developed an interest in botany, had pioneered forest conservation in the Madras residency. Brandis was a trained scientific forester advocating the "German method" of conservation, centred on selective and rotating tree felling.

British foresters for the Indian Forest Department were trained in the French college at Nancy, and Brandis had established a forest research institute at Dehra Dun in 1878, so by the early 1880s calls for a British school of forestry were mounting. In the summer of 1882, the Society of Arts presented the secretary of state for India with a memorial requesting the establishment of a forestry school at Coopers Hill.[144] Eventually, the Forestry course was granted approval to go ahead, and the first Forest students entered Coopers Hill in 1885 for their two-year program. The delay was caused by a discussion about whether Coopers Hill would train foresters just for India or for British needs more generally. Ultimately, the decision was made to proceed in the way the college originally started: a stiff entrance examination held in London by the civil service commissioners, and with guaranteed positions for the men who qualified for the diploma.

The foresters took engineering, mathematics, surveying, geometrical drawing, and accounts with the engineering students, thereby helping to defray the costs of offering those subjects. Specialized courses were introduced for the Forestry students, taught by two new professors, and purpose-built facilities were constructed in 1888 for a cost of £6,000.[145] Consisting of classrooms, a museum, and a laboratory, "it may safely be stated that there is no other centre in the Empire where so thorough and excellently designed a curriculum for a forester or planter can be obtained."[146]

Forestry was under the guidance of Professor William Schlich,[147] formerly inspector general of Indian forests (having taken over from Brandis in 1883). He had previously worked for the Hesse state forestry in Germany and received his doctorate for a study of the financial aspects of forestry. When he was made redundant by changes resulting from the Austro-Prussian war of 1866, he went to Burma and India. He was appointed to Coopers Hill in the autumn of 1885[148] to teach the cultivation and nurturing of forests in the students' first year, and the commercial utilization of forests and forest law in the second year.

In January 1886 thirty-two-year-old Harry Marshall Ward joined the Forestry school at Coopers Hill as professor of botany. Ward took the natural sciences tripos at Cambridge University, graduating with a first-class degree in 1879. He then spent two years in Ceylon (Sri Lanka) attempting to find a cure for the coffee rust fungus that was ravaging the crops of Britain's main source of coffee beans. Ward was successful to the extent that he revealed the life cycle of the fungus and correctly determined how and when to prevent the infection. However, the fungicides available at the time were inadequate to kill the fungus and the plantations were destroyed. By 1890 coffee production had been replaced by the tea plantations, with which we now associate that island.[149] Ward accepted the post at Coopers Hill, although "the utilitarian atmosphere in which he found himself was not very congenial to him."[150]

Although the Forestry program at Coopers Hill was intended to educate just ten students per year, its establishment made a significant impact in the field, partly because it was the first such school in the country, but principally because of the high profile of William Schlich. As time progressed the school's reputation was maintained by Ward's increasing eminence, and the quality of the education. Four students completed the diploma in 1887, with typically ten or eleven per year for the first decade, and half that number thereafter.[151] The Forestry curriculum itself will be considered in chapter 6.

Financial Questions

1888 Report

Even before the impact of the Forest School on the long-term sustainability of Coopers Hill could be assessed, the extra expenses incurred by the additional staff and facilities once again drew parliamentary attention to the immediate financial situation. The context for this was the long-standing goal of the secretary of state to reduce the Government of India's "home charges" – the costs it incurred in England, including the Royal Indian Engineering College.[152] A Special Committee on Home Charges was convened, and it produced a series of fifteen reports between July 1888 and April 1890. The fifth of these, on 26 October 1888, concerned Coopers Hill.[153] The report considered the decreasing numbers of students appointed in India, the costs of the teaching staff, the revenues from student fees and technical testing services it offered, and the costs of establishing the Forest School. In particular, it highlighted the high cost of the practical training course, which it believed could be offered more effectively in India. The general tenor of the report was that, while the individual salaries of the professors were "not excessive," the overall cost of the college was too high for the number of engineers. The committee therefore thought it might be time to "relieve the Secretary of State from the responsibility involved in maintaining the College at Cooper's Hill as a Government institution, either by transferring the establishment to some public body or private corporation, subject to a return on the capital outlay of 128,000l. in the form of rent, or, failing such an arrangement, by disposing of the land and buildings upon the best terms that can be obtained."[154] Nevertheless, as George Chesney observed in 1871, the costs of running Coopers Hill were relatively small in the grand scheme of things. The operating deficit in 1887–8 was only £132, exclusive of the Practical Course (which cost £4,600) and the interest on the capital cost of buildings (which at 3 per cent of £129,000 was around £3,900). In contrast, the total annual India home charges were almost £15 million. Perhaps for this reason, no immediate direct action seems to have been taken as a result of the Special Committee's enquiry.

In other ways, however, the changes made under the New Scheme, and the addition of the Forestry course, seem to have had the desired positive effect. By the end of the decade, in a complete reversal of its earlier opinion, the Government of India wrote to the secretary of state for India that "we have no hesitation in saying that, having been created, [the college] should be maintained as long as possible, if only for the purpose of avoiding radical and frequent change."[155] This grudging support was softened by the statement that they "fully appreciated the College as a means of training recruits and considered it far superior to the methods of recruitment formerly adopted and probably so to any other system that could now be devised." The fortunes of Coopers Hill were on an upswing, for the moment at least.

1895 Report

Towards the end of Alexander Taylor's tenure as president, the issue of the college's finances was again taken up by the government. After seven years with operating surpluses (albeit small), it became apparent that the college would report a deficit of more than £3,000 in the 1893–4 fiscal year (it would also report deficits of over £1,500 in each of the two subsequent years). In response, the undersecretary of state established a six-person Committee of Enquiry in February 1894 to look into the college's situation and recommend economy measures.[156] The committee's first substantive observation in its 1895 report was that the costs of the forestry students' continental tour had not been included in the expense figures, and so there had never been any true surpluses.

The committee attributed this financial situation to a decreasing number of students in residence, combined with an operational cost that had almost doubled since 1871. This had been driven primarily by an increased complement of teaching staff, which had in turn been necessitated by improvements in the curriculum, the opening of the college to non-PWD candidates under the New Scheme, and the establishment of the Forest School (for a visual contrast, compare the staff photographs in figures 3.1 and 8.1). They considered options for increasing the student numbers, including training members of the Indian Police, but discounted them because it would have been "an attempt to prop up a Government institution which is essentially an Engineering College, by making it the door of access to services which have no connection and ought not to be tied up with the Public Works Department."[157]

Reiterating George Chesney's 1877 conclusion, the Committee of Enquiry was unable to suggest any specific measures to reduce costs other than "minor economies": "To sum up, we cannot suggest any measures which will result in restoring equilibrium to the College's finances in the immediate future. After a course of years an approach to equilibrium between current receipts and expenditure may be attained. But no return on the heavy capital expended on land and buildings is to be expected."[158] The report concluded that in view of the small number of annual PWD appointments there was "grave doubt" whether Coopers Hill was still required.

Sir Alex's Retirement

Alexander Taylor retired as president in autumn 1896, a few months after his seventieth birthday. Sir Alex and Lady Taylor built a house in Englefield Green to be near their daughter Mildred Bethune Woods, who was married to Richard Woods (see chapter 7), then an instructor at Coopers Hill. Throughout his time at Coopers Hill, Taylor was an avid yachtsman and he continued to sail his boat on the Thames until forbidden to do so by his doctor.[159]

The presidencies of George Chesney and Alex Taylor saw Coopers Hill through its first twenty-five years. Both men were deeply committed to the college and its ideals. Chesney was the politically savvy man of details who ran the "tight ship" necessary in the first decade. The students admired and respected

Chesney, but rarely spoke of him with great affection. On his death in March 1895 of angina, the students and staff of the college erected a handsome memorial in the churchyard of St. Jude's, Englefield Green.

Taylor, at the end of his career but still active, was more relaxed and approachable. His obituary in the *Coopers Hill Magazine* summed up the general feeling of the students: "Sir Alexander was … a firm friend and always did his best to help those who were in any sort of trouble. He was a strict disciplinarian but generally had a strong tendency to take a merciful view of a case."[160] Taylor's adaptability helped Coopers Hill navigate the changes and compromises necessary in the 1880s in a way that Chesney might have found very difficult. So, with similar beliefs but different styles, both presidents were essential to the shaping of the college's character.[161] These were the prime years for the college, and under their leadership the majority of Coopers Hill engineers graduated and set sail for India.

Accordingly, the next chapter examines the lives of students at Coopers Hill and the way in which their environment shaped them into the gentlemen engineers Chesney and Taylor believed were needed in India.

4

Student Life

College Life

Life at Coopers Hill does not feature in the literary record, and it appeared rarely in the newspapers of the day. Unlike the ancient Oxbridge colleges, the nearby Royal Holloway College, or other venerable places of learning, no modern students walk its halls or observe its time-honoured traditions. No one alive experienced student life at Coopers Hill first-hand, and it is unlikely even that there are those alive who as children heard stories of their fathers' student days.[1]

But Coopers Hill is where, over the course of thirty-five years, more than 1600 young men went to study, develop their esprit de corps, and have some fun. "It might be safely prophesied with regard to most of us that no three years of our future life would be more replete with new ideas, or fraught with deeper and more lasting associations and friendships, than those we are now spending at Coopers Hill."[2] Their academic, social, and sporting lives at the college defined what it meant to be a "Coopers Hill man" and shaped their future actions in India, Africa, South America, and around the world. These experiences were designed by George Chesney to produce not only technically proficient engineers but a corps of men with shared values of service and integrity. This chapter therefore explores student life at Coopers Hill with the goal of understanding how and to what extent this vision was realized in practice.

Sources

The heart of student life still beats strongly from the pages of the *Oracle* from 1874 to 1880, and the *Coopers Hill Magazine* from June 1897 to 1961.[3] As readers encounter stories of sports fixtures, news of former students, and other events of the college, they are carried along by the flow of students' lives, and their names become familiar. A sense of the college's character emerges.

But only so far. We are not living their lives and we cannot immerse ourselves in their thoughts, culture, or expectations. And unlike them, we know the future and the fates of those young men. The college will close and the British Empire itself will end. Moreover, these magazines could not be warts-and-all accounts of college life. While the *Oracle* attempted to be the "medium through which members of the College may ventilate their grievances,"[4] the same could never really be said for the *Coopers Hill Magazine* because it was also sent to a large number of alumni and others, with a circulation of more than five hundred copies in 1900.[5] It had to be a respectable voice for the college – too much so, in the opinion of one early correspondent, who wrote that the "first number of the magazine is too respectable and serious. The Editor should compose the 'Gossip' (if nothing else) after a big drink."[6] It seems that the content was to some extent influenced by the president; on one occasion, an issue of the *Oracle* was published without its lead editorial because Chesney objected to a single sentence.[7] On another, the editor of the *Coopers Hill Magazine*, against his better judgment, was prevailed upon by then president John Pennycuick to publish a letter concerning the latter's resignation.[8] Less public goings on at the college are hinted at between the lines in the "Correspondence" and "Questions and Rumours" sections.[9]

Coopers Hill was not alone in having a newspaper. Between 1839 and 1858, students at the East India Company College produced the *Haileybury Observer*, which aimed to be a literary magazine rather than a college newspaper.[10] Later, the Royal Military Academy's *R.M.A. Magazine* was published from 1900 onwards.[11] In both cases, as at Coopers Hill, these publications had short-lived predecessors.

However, the *Coopers Hill Magazine* was unique in continuing to report on college affairs for nearly sixty years after the closure. Much of this longevity was due to Arthur Hicks, who was an editor of the *Oracle* as a student and of the magazine as a lecturer – a role he would maintain almost until his death in January 1928 (see chapter 9). In those later years, the aging alumni started sending their reminiscences of the college to the *Coopers Hill Magazine*, no doubt worn smooth with frequent retellings and reflecting the shared mythologies of each student generation.

Fortunately, at least one contemporary student diary survives to record everyday life at Coopers Hill.

James Bonnell Ernest Hudson

Ernest Hudson[12] (16 May 1866[13] – 10 January 1952) kept his diary, work notepads, and other documents between 1887 and 1908, taking him through his last two years at Coopers Hill, his departure for India, and his working life in the Indian Telegraph Service. Like many of his contemporaries he enjoyed taking photographs and left several elegantly labelled photograph albums. These documents will be used in chapter 5 as a source for the experiences of a telegraph officer in India and chapter 7 as background for the chapter on engineering.

Ernest was the second of seven children of Thomas Keith Hudson, an officer in the Royal Navy,[14] and Mary Jane Bonnell (whence his middle name), who was from a well-connected family with an estate at Pelling, very close to Coopers Hill. At the time he attended Coopers Hill, the Hudsons lived on the Isle of Wight, where they were a well-known local family. Growing up on the Isle of Wight with a naval father, Ernest was inevitably a boating enthusiast. His spare time at home was spent on the ocean, or shooting on the farms around their house, The Dell, and for much of his free time at Coopers Hill he was rowing, punting, or sailing on the Thames.

It is not clear what motivated Ernest Hudson to choose an education that would lead him to India. It may have been at the suggestion of his uncle, Lieutenant-General Sir John Hudson, who was a prominent soldier in India.[15] Or perhaps it was simply that he was familiar with Coopers Hill because of the proximity of the Pelling estate.

Hudson's 1887 Diary

Hudson's diary for 1887 covers the second half of his second year and the first part of his third year at Coopers Hill. That of the following year records his last two terms and the voyage to India. The 1887 diary is in fact written in an 1898 diary from information in "an old journal and rough entries on scraps of paper + c." It should therefore be kept in mind that some youthful escapades, embarrassing now to the professional man, may have been edited out of the transcribed version. Indeed, some "tidying up" may have taken place after 1898. because there are stylistic differences between the transcribed entries of 1887 and the more student-sounding "live" entries of the following year.

Ernest was a mediocre student (see below) and comes across in his diaries as conscientious and careful rather than gifted. Social relationships were important to him and he cultivated and recorded a wide range of acquaintances. Hudson's diary entries were fairly brief and focused mainly on factual details of his leisure time, friends, and events in college. He did not talk much about the work at Coopers Hill, beyond occasional references to deadlines for assignments and exams and his standing in the year. He certainly would have spent a lot of time on the work and perhaps even enjoyed it, but that did not make it interesting for him to write about. Nevertheless, the backdrop to these events is the tough and rigorously enforced schedule of work discussed in chapter 3.

Hudson returned to Coopers Hill on 22 January for the Easter (second) term of his second year. During the first week back, he had exams in Descriptive Engineering and Geology. Storms had flooded the meades (the flat, grassy area

Figure 4.1. Some of Hudson's year at Coopers Hill, taken at Windsor, 1885: (*back row*) Hudson, Cather, Doughty, Hill, Ricardo, (*middle row*) Sutherland, Mahon, Cumming, Smythe, Romilly, Cape, (*front*) Couper (Hudson Collection, CSAS)

between Coopers Hill and the Thames) and there was "4″ water inside boat-house." He and his friends[16] canoed along the river on several days that week and, against regulations, "shot with an air gun in the woods." Hudson later found out he had placed first in the Geology exam, "licking both 1st and 2nd years."

Figure 4.1 shows a group of Hudson's friends at Windsor in the autumn term 1885, soon after Hudson started at Coopers Hill. At least some of the party are dressed for hiking, with walking sticks and stout boots, presumably having walked through the Windsor Great Park. The records of the Coopers Hill Society include numerous albums donated by its members containing similar photographs of sports teams, athletic events, and the bal masqué.

On one Sunday, Hudson recorded that he "went to tea at the Dukes." This refers to President Alex Taylor, and Hudson implied that the Duke meant Taylor personally, a view that Taylor's daughter reinforces in her biography of Taylor.[17] But according to John Cameron's *Short History*, "Duke" was the name given by the students to whoever was president at the time.[18] Cameron attended Coopers Hill from 1904 to 1906, so it seems likely that the tradition started with Taylor and continued to be applied to subsequent presidents.

"O" Rifle Corps

On 11 February Hudson wrote that he attended the first company drill of the term, referring to "O" Company of the Company of the 1st Volunteer Battalion, Royal Berkshire Regiment. The establishment in 1859 of the military volunteer movement was prompted partly by the combined military shocks from the Crimean War and the Indian Rebellion.[19] It was natural for Coopers Hill, an institution with military influences, to form a volunteer company. After all, George Chesney was an army man himself, and the narrator of *The Battle of Dorking*, written at the same time as he was designing Coopers Hill, was a volunteer. "O" Company was established in 1873 by George Sydenham Clarke at Chesney's request.[20] After Clarke's departure, the company was led by the popular Thomas "Tommy" Eagles, expert rifle shot and instructor in technical drawing, until his death in 1892.[21] At the time, the British Volunteer Force had a strength of over 170,000 men, a number that would increase to more than 250,000 by the college's closure in 1906.[22]

Military exercises at Coopers Hill became mandatory in 1877: "Every student will be required to go through a course of exercise in the gymnasium, and of Military exercises, including the use of the Rifle."[23] Marks were available for drill and gymnastics, and a prize was initiated for rifle shooting.

The Volunteer Company took part in drill exercises at the college and underwent periodic inspections by the adjutant of the Berkshire regiment. More practically, they exercised their hands-on skills by building earthworks, trestle bridges, and pontoons across the Thames. They also attended ceremonial events, drill competitions, and an annual camp where they participated in war games with other units. The sight of the students and their officers arrayed in military uniform was impressive.[24]

The college's enthusiasm for the Volunteer Corps waxed and waned over the years. One recurrent difficulty, as the *Oracle*'s leading article lamented in November 1878, was that each year a significant portion of the corps left, including almost all the non-commissioned officers. It went on: "Last year it was at first feared that the efficiency of the corps would hardly come up to the hitherto high standard, particularly as the adjutant was not altogether satisfied when he came down to inspect it, but we may safely say that the review at Reading, and the encomiums received, fully dispelled this idea." However, the *Oracle*, living up to its self-imposed role as guardian of college standards,[25] could not leave the matter without issuing a warning: "Still it must be admitted that there is great room for improvement, especially at the attendance at company drill...."

The greatest evil is the desultory drill which has been too often witnessed upon the playing field, where hardly sufficient men have been present to render it possible to go through many of the most important exercises." The same article noted with approval the new prize for shooting and the awarding of marks for drill: "Time will shew whether this system will work well or not."[26] Apparently not, because a year later, in October 1879 the *Oracle* wrote that the new system had "tended very slightly to uphold the reputation of the Volunteer Corps."[27]

There were, however, high points for members of the Volunteer Corps. For example, on 13 March 1879 the company marched from the college to Windsor Castle to form part of the honour guard for the marriage of Prince Arthur, Duke of Connaught, seventh child of Queen Victoria, to Princess Louise Margaret of Prussia. A large part of their time was spent keeping onlookers at bay, but the spectacle "was of the grandest description, and such as to leave a lasting impression on every one who beheld it." After lunch, the company marched back to Runnymede, where it was drawn up in a double line to await the royal carriage: "As the *cortège* passed, arms were presented, the company being duly acknowledged by the Duke and Duchess."[28]

By 1899 the familiar refrain about the lack of attendees was still evident:

> There was a church parade of the Volunteers last Sunday [16 July], at which about half of the members of the Company were present.... Of course it is entirely a voluntary parade, but perhaps the voluntary principle would receive a considerable impetus if tutors were not quite so lavish of their signatures[29] on the Saturday preceding church parade.... The worst possible form was shown by members of the Company who attended service in mufti,[30] and it is to be hoped that on future occasions of the sort we shall be saved the pain of witnessing such a disgraceful exhibition of slackness.[31]

A week before, the company had been put through its paces by Col. Dickinson, commander of the 49th Regimental District in the presence of Col. Walter, commanding the battalion, Capt. Feetham, the adjutant, the president (having cancelled an important engagement), and "a representative selection of College youth and beauty." After the drill, Col. Dickinson "noticed the improvement in the Company since last year, and a considerable improvement on the year before, when he first inspected the Company, but that the Company was still very far from being as smart as it ought to be."[32]

On the whole, the regimental camp was not especially popular with students because of "the time ... at which the camp is held, *viz:* at the end of a long and busy term culminating in three weeks examination – calculated to unhinge temporarily the soundest brain."[33] Parading in the heat was also uncomfortable, and Ernest Hudson wrote that Strachey[34] and Sears fainted from the heat during the Volunteer Company drill. Nevertheless, Hudson attended the Volunteer

Figure 4.2. Volunteer camp, Hampstead Park, Berkshire, 1888: (*back*) Hudson Sergeant Lanning, Walsh, (*front*) Turner, Ellis, Humphreys (Hudson Collection, CSAS)

Company in 1886, a review at Aldershot in July 1887, and the 1888 camp at Hampstead Park (figure 4.2).

At about this time, an anonymous poem "With the Cooper's Hill Volunteer Company at Aldershot" was published "by one of the Awkward Squad" about their experience at Aldershot, telling us something of the mood and activities at the camp.[35] Selected verses give the idea:

> Prepared to march, all stiff as starch.
> "March!" was our captain's cry;
> In such a sun, thought everyone.
> He ought to call "*July!*"

> Yes, martial zeal we always feel,
> And martial swagger try;
> No sloven slouch or careless crouch
> Beneath an *Eagle's* eye.[36]

> When food was served, our neatness swerved,
> A change we underwent;

There was in fact, to be exact,
A *mess* in every tent.

In 1901 "O" Company was converted, appropriately enough, into an engineer company, becoming the "I" Company of the 1st Middlesex Royal Engineer Volunteers, comprising about 120 officers and men.[37] I Company represented the college at the coronation of King Edward VII in August 1902, "taking up a position in the Mall, next to the Horse Guards' Parade," after some of them spent a night in a railway waiting room.[38]

The last inspection of the Coopers Hill Volunteer Company was performed on 15 March 1906 by Col. Lake R.E., chief engineer of the London District, who "expressed his satisfaction with their turn out, marching past, and drill" and expressed his regret that this "was the last Inspection parade of the 'I' Company."[39]

Indian Volunteer Corps

After leaving Coopers Hill, it was a natural progression, especially in later years, for graduates to join the volunteer forces in India. The first of these corps was established after the 1857 uprisings, but most were raised in the 1870s and 1880s.[40] Their purpose was initially to help protect the outlying stations in recognition of the vastly outnumbered British population. In 1883 the enrolled strength of the Indian Volunteer Force was 12,213, of which 9,421 were considered "efficient," or capable of active service. By their nature these auxiliary units were small and numerous.[41] In the early twentieth century this number had grown to more than seventy-five volunteer companies[42] comprising 32,000 men, around half of the population eligible to serve.[43] The duties of the volunteers had evolved principally to "hold fortified posts and railway stations, and to guard the lines of communication in India." They comprised corps of Light Horse, Artillery, Engineers, Mounted Rifles, and Rifles. Most units were regionally based, although several were raised by the railways (e.g., the Bengal-Nagpur Railway Volunteer Rifle Corps).

Little was written about the involvement of Coopers Hill men in the Indian Volunteer Movement before the turn of the century.[44] Forester James Best commented that few forest officers attended the annual volunteer camp, which "was not very serious soldiering, it was more of a yearly reunion and ten days under canvas for those whose lot lay in offices." In contrast, at the opening of the Bengal-Nagpur Railway in March 1891, Sir Trevredyn Wynne proudly described the 600-strong B-NR Volunteer Corps, established "by the way of a little relaxation."[45] However, engineer Stephen Martin Leake (CH 1880–93; see chapter 7) recalled that there was little volunteer activity in his district of the railway.[46] This was supported by Ernest Hudson's experience while constructing a telegraph line in the same area (see chapter 5). In March 1890, he described being enlisted in the B-NR Volunteers just so he could shoot in a rifle competition taking place that day.[47]

But by the late 1900s, the volunteer movement had gained momentum, and a direct connection was made by Lord Wenlock at the 1898 Coopers Hill Prize Day between the "efficiency and discipline" learned in the college's corps and the training of reserve forces being raised throughout India.[48] John Ottley concurred, saying that in India "every man ought to be a professional soldier or a volunteer."[49]

Starting in late 1908 the *Coopers Hill Magazine* added a "Volunteer News" section to its summary of the Indian gazettes to list the Coopers Hill appointments, promotions, and transfers. One update in 1911 was a brief account of the camp exercises of the Southern Provinces Mounted Rifles that included a photograph of sixteen men from Coopers Hill who were members (including Anthony Wimbush [see chapter 6] and Hudson's friend Frank Cowley-Brown). The magazine stopped publishing this section after September 1916, just before conscription into the Indian Defence Force began the following March.

Summer 1887

College Regatta

Hudson's March and April 1887 were taken up with rowing, the windy and snowy weather, and end-of-term exams. The only mentions of work are the starting of his prismatic compass survey and staying up until three in the morning to finish his "G.D. plates," presumably geometrical drawing, which together with his "Survey plates," were finished a few days later. Hudson placed tenth in the exam list.

The rowing practice led up to a series of college races in mid-June and the college regatta on Friday, the seventeenth. Races were typically run for a mile upstream from "Magna Charta Island," past the foot of Priests Hill and the Bells pub at Ouseley.[50] The regatta also allowed for some fun events, including in 1899[51] the "mop fight in canoes" won by Holman-Hunt and Todd, although Bidder's "tramp costume caused a good deal of amusement." There was also a tug-of-war in punts, won by Holman-Hunt's team, and a dongola race,[52] won (again) by Holman-Hunt's team.[53]

At the time, rowing regattas were a popular pastime, teams from towns and other organizations, as well as colleges and schools. Coopers Hill regularly attended nearby regattas at Staines, Walton, Egham, Moulsey, and Kingston, as well as the more famous Henley. They also raced against other institutions such as Beaumont College and Eton school.

Choice of Career

Hudson's entries for the rest of the summer term primarily concern rowing, sailing, and punting with his friends. He went home on 27 July, with the comment that he placed fifteenth on the exam list.

Hudson made an interesting remark in passing on the day of the college dance. He seems not to have attended, as he states he went punting up the "old river" and did not return until midnight, but he wrote "Library lit up with electric light," the first mention we have of electric lights at Cooper's Hill. By contrast, Englefield Green was not lit by *gas* lamps until 1901, and the "Lawrence" and "Mayo" corridors[54] of the college were fitted with electric lighting by the students themselves in 1903 and 1904, respectively. The lighting of the library may have been the work of Prof. McLeod, who had installed similar systems for Lord Salisbury as early as the mid-1870s, and still maintained an active interest in electric lighting at this time.[55]

Over the summer vacation, Hudson noted that he "heard from Q that I had got a Telegraph appointment." The fact that Q, Captain Courtney, informed Hudson about the position suggests he was Hudson's college tutor. Hudson heard more details about the appointment from Whiffin, the college secretary, about five weeks later, and by the time he returned to college on 27 September he had decided to accept the position. The second appointment went to William Henderson.

Hudson and Henderson were both in their third years at Coopers Hill, but that is odd because the telegraph course was only of two years' duration. Students normally indicated their interest in a telegraph appointment at the end of first year by writing to the college secretary in response to a posted notice.[56] However, the exam results listed in the college calendar show Hudson's and Henderson's names in the first-year list at the ends of two consecutive years, 1886 and 1887. This suggested that they repeated their first year.[57] From the 1901 testimony of Prof. Stocker, who was in charge of Telegraph instruction, it seemed generally known that the more able students preferred the PWD.[58] So it appears that Hudson tried to improve his placement by repeating the first year. Indeed, his rank went from joint thirty-first in 1886 to joint fifteenth in 1887. Nevertheless, it is likely that Hudson ultimately opted for the Telegraph appointment because, even with his improved standing, he would probably not have been selected for the PWD.[59]

Crime and Punishment

In Hudson's diary entry for Monday, 14 March 1887 we get a glimpse of the hidden goings-on in the college: "Oliver rusticated for a fortnight and Hart and

M. Hill close arrest for 3 Saturdays for rux on Saturday night."[60] The term "rux" was apparently derived either from "ruction" or "ruckus," to denote reprimanding or scolding someone and was in popular use at England's private schools in the 1880s.[61] For whatever took place, Oliver was suspended from the college – rusticated – for two weeks. Students under "close arrest" were probably confined to their rooms, in contrast to the less serious "gating," which meant being forbidden to leave the college grounds.

As early as October 1871, McLeod recorded in his diary instances of late-night parties and other hijinks, such as when the local policeman stopped a group of McLeod's students for singing and making a row in Egham.[62]

College Rules

Student life in college was governed by a set of printed rules that covered topics including the amenities in each student room, where smoking was permitted, and the arrangements for meals. "Academical dress" was required at all lectures (except surveying, workshops, and geometric drawing), chapel, dinner, and when attending the president or professors. It was also required in the front hall and adjoining areas before 3 p.m. Moreover, "Black coats must be worn at dinner, and in chapel on Sunday. [Guests in college are expected to conform to this rule.] Boating or cricket flannels are not to be worn with gowns. Slippers must not be worn in any of the public rooms and passages after breakfast. When in front of the College, coats are to be worn with boating and cricket dress, and knickerbockers are not to be worn without stockings. Students must appear properly dressed on all occasions in the public rooms and passages and in the College grounds."[63]

Students were expected to attend chapel on Sunday mornings, as well as on at least four mornings per week. Students wishing to miss chapel, be out of the college grounds after 9 p.m., away at the weekend, or absent from lectures were required to apply to their tutor for special leave and to report themselves to the Porter's Lodge on their return. College rules also covered the library, the election of officers of college clubs, and the loan of items such as drawing boards.

The rules were particularly clear about the need for "perfect quiet ... throughout the buildings" during the hours set aside for private study and after the gas was turned out at 11 p.m., and the "strict observance of good manners and gentlemanly conduct."

The abolition in 1873 of free beer must have gone a long way towards achieving this aim: "Beer has been supplied at luncheon and dinner *ad libitum* without charge. The arrangement, however, does not work well; it leads to excessive consumption and waste, and also works very badly for the servants. More beer is drawn than can be consumed at table, in order that the surplus may be drunk afterwards. It is almost impossible to prevent this, in practice, and the result is that several servants have had to be dismissed for drunkenness." Chesney and the college tutors decided it was time to charge for beer, which would not only "lead to great economy, but will also conduce to the greater efficiency of studies and to the better conduct of the Establishment."[64]

The college rules changed little between 1885 and 1900, at which time they were made more detailed and renamed "Standing Orders."

Further Ruxes

After starting the autumn 1887–8 term in the usual way with rowing and sailing, Hudson comments cryptically on the first of a series of disruptions in the college. After a smoking concert, there was a "big rux in the evening and Lodge raised," meaning that Professor Alfred Lodge[65] was roused from his bed.[66] Ten days later Lodge was angered for being awoken again by a group of rowdy students after another smoking concert, which was, Hudson wrote, a "rather a noisy affair but great fun." The next day, "All men who were at Egham smoking concert had their leave stopped for making noise coming up through Egham and in College when they got back."[67] It seems that Lodge, along with Mr. Stewart, had reported the students, who retaliated with fireworks into the early hours and during dinner in the hall. A lot of men were "hauled up" in front of the Duke, but he seems to have been unable to identify the perpetrators.

In an odd contrast, Hudson started sailing with the Duke the following spring, for Taylor was renowned for his enthusiasm for yachting. Soon after returning to Coopers Hill, he and Court "helped the Duke rig up his boat ... + then went for a sail with him. Dined with him in the evening."[68]

In the meantime, feelings against Lodge seem to have festered because he was again the victim of ruxing. A few entries contain wonderful juxtapositions of academic life, Hudson's boat club responsibilities, and the adolescent out-of-hours behaviour:

Friday April 6, 1888
    The exams commenced, with accounts + signalling.
    Big row on in the evening and Lodge's room was ruxed the handle being taken of [sic] the door + door smashed in. Had a [boat club] committee meeting over at Pomphs + arranged to buy two new skiffs[69]

Saturday April 7, 1888
    The duke sailed his new boat for the first time + would have won if he had not run into the Vixen + smashed his boat-sprit.... We [the house committee[70]] went to the duke in the evening about a notice he had put up that all college were under close arrest for ruxing Lodge.

Sunday April 8, 1888
All the college under close arrest from 8 in the evening to 7 the next morning.

Monday April 9, 1888
The Duke had the house committee up again + agreed to take down the notice to the effect that the college was under close arrest if an apology were sent to Lodge.
Clayton called a general meeting in the evening + we wrote an apology + sent it to him through the Duke.

This seems to have been the end of the matter, either because the students found more interesting things to do during the summer months or because Lodge moved out of college.[71]

Punishment for what might be called moderately serious breaches of the college regulations, such as disturbances in the college and insubordination was frequently rustication. This could also be part of a sequence of progressively escalating punishment for offences like drunkenness – warning, rustication, and ultimately expulsion. For less serious infractions such as smoking in public rooms or corridors, or not wearing academic dress as required, there was a system of fines of a few shillings.[72] Money collected from fines was distributed among the college clubs.[73]

When Col. John Walter Ottley was appointed as college president in 1899 he made a point of stressing the importance of discipline, which was "absolutely essential on the cricket and football fields if success was to be attained, and so it was in all the business of life." He went on to say that he disapproved of the system of fines for the breach of the rules, for two reasons: "In the first place, as the fine was paid by the parent, the punishment, such as it was, appeared to fall on the wrong person. In the second place the system seemed to be almost an invitation to the breach of rules. It was tantamount to saying: 'you are forbidden to do such and such a thing, but you may do it on payment of a small fine ranging from say one to five shillings.'"[74]

In the keeping of curfew, letters of permission from tutors, and fines, regulations and discipline at Coopers Hill were in keeping with the times; similar rules were in force at Oxford and Cambridge. Writing about Oxford of the 1830s, James Pycroft showed why Ottley's objection to fines missed the point: "The Dons visit the sins of the sons on the pockets of the fathers, for not teaching their sons better, and then the fathers are pretty sure to make the home for the time rather hot for the sons."[75]

In comparison, the ordinances of the University of Cambridge in 1886 required all lodging houses offering accommodation to students to be licensed. To get a licence, the owner had to promise not to rent rooms to "any Student without written permission from his Tutor," to keep the house locked between

10 p.m. and 6 a.m., and report to the university the names of students who arrived or left between those hours.

The Signpost

One of the most enduring shared mythologies at Coopers Hill concerned the iconic "signpost." As late as 1960 Cameron felt it was worth mentioning "the famous sign post" in his *Short History*.[76] But he explained no further, presumably thinking that members of the Coopers Hill Society to whom the history was circulated would know exactly what he meant.

It is not until George Minchin's helpful speech in 1913 that we get some idea of what was going on:

> The tale relates to the successive cuttings down of the sign-post on Englefield Green – in itself an indefensible episode. The post had been cut down several times, and as often repaired by the road surveyor. But no culprit was ever caught, and the Coopers Hill men suggested that the guilt lay on the lady students of Holloway College....
>
> The thing remained a mystery, and at last it was taken up by that astute body, the Egham police, who set an ambush one night in the shrubbery of the corner house on the Green, a few yards from the sign-post. Somehow the Coopers Hill men got wind of the ambush; and some of them sallied out after midnight and approached the sign-post. Now the hearts of the police beat high, but the students passed the post without paying the slightest attention to it, and quietly entered the grounds of the corner house whose inmates they informed of the presence of burglars in the shrubbery. Out came the people of the house armed; and, under the guidance of the students, they rushed the shrubbery, where they found the police. The loud protestations and explanations of the police were drowned by the virtuous indignation of the Coopers Hill men, who insisted on the instant arrest of the police![77]

But the students did not always get the better of the police: Hudson received a letter from Horace Turner (CH 1886–9) in early 1889 with news from the college that "a lot of fellows had gone down to the B.M. [probably the Barley Mow pub], pulled up the sign post on the way back and been caught by a bobbie &c."[78] In 1897 an ornate drinking fountain was installed next to the signpost but this was reportedly also damaged by one of the students.[79]

The invitation to the 1901 Bal Masqué, which traditionally comprised sketches and caricatures of college life,[80] included the small sketch of students attacking and burying the signpost (figure 4.3), accompanied by a ditty.

An Ordnance Survey map surveyed between 1865 and 1870 shows a "guide post" at the junction of what is now Bishopsgate and Saint Jude's Roads.[81]

Figure 4.3. Coopers Hill students attacking the Englefield Green signpost. "We covered them in at dead of night; The sods with our coal shovels turning. No signs of Signboards left we in sight; Well satisfied Collwards returning." Sketch and words from invitation to the 1901 Bal Masqué (Canning Collection, CSAS)

A number of paths cut across the common land from the post, and one fingerpost presumably pointed along the track to the college lodge. Perhaps the first generation of students to remove the signpost took it down to add to the mystique of the college by keeping its location secret and confusing visitors?

Nowadays at that road junction a multiplicity of signs show street names, speed limits, and directions to places of interest. There is also a water trough labelled the Metropolitan Drinking Fountain & Cattle Trough Association. At the location of the famous sign is an attractive black Narnia-type lamp post. Attractive, but not a signpost to Coopers Hill – the students were apparently victorious after all.

Medical Complications

Hudson's Medical Examination

Hudson marked Monday, 16 July 1888, with the victorious words "Last day of my exams." Speech Day was on the 25th and more than 150 guests visited the college. Hudson notes, "I had got my [Telegraph] appointment + licked Henderson." Unfortunately, it "poured hard nearly all day," or as the *Times* announced,

"A large Company visited the College on Wednesday afternoon, in most inclement weather, for the purpose of witnessing the annual distribution of prizes."[82]

On the day following the prize giving, Hudson and a few others went to the India Office in London for their medical examinations, before they went to the Volunteer Company's camp at Hampstead Park (as already described). On his way home after the camp, he stopped in at the India Office to hear that he had been "disqualified for acute varicosele." The thoroughness, not to mention intrusiveness, of the India Office's examination is shown by their observation of this enlargement of the veins in the scrotum.

Hudson's father immediately used his network of family friends and connections to convince the India Office not to give Hudson's position away, as they had evidently planned, to PWD man John Ridout. Instead they agreed to await the outcome of an operation to fix the problem. Mr. McKellar from St. Thomas' Hospital performed the operation at 11 a.m. on Tuesday 21, and it took until 2 p.m. for the chloroform to wear off. It was three weeks before Hudson could leave his room, but thereafter his recovery was swift. McKellar's bill for the operation and convalescent care was £40.

It was not until 25 October, after Henderson[83] had already departed for India, that Hudson finally received word from the India Office telling him to reappear for his medical examination. There, to his surprise, he met many of the Coopers Hill Forestry men who had just returned from their practical work in Germany, as well as "several old Indian Colonels +c." who were there for their routine examinations. Sir J. Fayrer, president of the Medical Board, performed Hudson's examination himself and led him "to suppose that he was now fit." The much-anticipated confirmation letter came from the India Office on 8 November 1888.

Hudson embarked for India three weeks later. In the intervening time he bought supplies, a trunk and Gladstone bag, and clothes, including a top hat, shirts, ties, a cash box, and handkerchiefs. On the 21st he went to the P&O office to get shipping labels, the departure time of the train from Liverpool Street Station, and the sailing time of the SS *Nepaul*.[84] Hudson also visited Coopers Hill to bid farewell to his comrades, as other friends had done the year before, going down to the river and seeing the Duke and Pomph.

Arsenic and Sports

Continuing the medical theme, Coopers Hill's long-serving medical officer came across some interesting cases, in addition to the routine ailments and injuries of 150 active young men. He was Henry Edward Giffard, a Scotsman who trained in London and was admitted into the Royal College of Surgeons of England in April 1877.[85] Giffard, who would also have maintained his private practice, was paid according to the number of students resident in the College, at 10s. 6d. per student per term (amounting to a sum of £181 in 1894–5).[86]

In 1888, as Hudson was writing his diary, it was being reported in the press and at the Prize Day that Giffard had detected signs of arsenic poisoning in some of the students at Coopers Hill.[87] "In every case the source of the poisoning was traced either to a cretonne or to an imitation Indian muslin, used as a decoration by the student." Arsenic was used in paints and fabric dyes, especially "Scheele's Green," which was made with copper and arsenic. Many of these colours flaked off or generated dust that could be inhaled. In the 1850s fashionable but poisonous green wallpaper decorated the nation's homes, releasing its poison into the inside air.[88]

It is not clear how many students were affected, but Dr. F.E. Matthews, the demonstrator in chemistry at Coopers Hill, tested forty-four samples of cretonne, a heavy, double-sided printed fabric often used in upholstery. It seems that McLeod was also involved in the testing, as his diary mentions tests in the rooms of Ernest "Easy" Bell and Edward Oliver.[89]

Matthews's tests revealed that none of his samples was "absolutely free" of arsenic, and twenty of the forty-four samples contained poisonous levels. Matthews commented, "It is quite a common occurrence to have enough of these substances in a room which would contain enough arsenic to give 100 people a fatal dose."[90] Five samples of imitation Indian muslin also contained poisonous levels. The *British Medical Journal* closed its article on the Coopers Hill arsenic tests with the following: "The National Health Society carried out a very full investigation of the matter, and prepared a conclusive report, drafting also a Bill for Parliament, designed to avoid these dangers; but neither legislators nor officials were prepared to assent to such a measure. Perhaps in time and by frequently reading disclosures of the existing perils to health, public opinion may be ripened."[91]

Through Giffard we also know of a horrific accident suffered by a Coopers Hill student during a rugby football game just a few weeks after joining the college. Giffard reported the case to Julius Althaus, a German-English neurologist and author of many scientific and popular books,[92] who included it in his 1885 *On Sclerosis of the Spinal Cord*. "The lad, aged nineteen, was thrown heavily on his back, with the result of becoming immediately completely paralysed and anaesthetic from the waist downwards, but not insensible.... He was unable to cough, to clear his throat, to sneeze, or blow his nose; but could move his head from one side to the other, and had no difficulty in speaking, masticating, and swallowing, and could put out his tongue."[93] This unfortunate young man was probably E.A. Lewis, who died some four months after his injury and was commemorated with a window and plaque in the chapel.[94]

Student Deaths

Three other students are known to have died at the college. The death of Ernest Pickwoad was announced in the *Oracle* in July 1877, with an obituary in the following month.[95] Pickwoad, a mainstay of the football team, was taken ill with an unspecified sickness (later reported to be typhoid)[96] soon after returning for the Easter term. He was buried in the nearby community of Clewer, "and his remains were followed to the grave by most of the members of the College."[97] A subscription was raised for a tombstone and for a memorial window in the chapel, the second one to be installed. When the college was closed, the memorial was sent to Pickwoad's brother in Ireland.

About two years after Pickwoad's death the *Oracle* reported that twenty-one-year-old W. Swinfen Huskisson had died after a painful illness at the college.[98] At the other end of the college's history, Francis Hare, died in 1906 of complications arising from meningitis, after an illness of just three days.[99]

These stories are included as a reminder that Victorian England was a dangerous place. Injuries and diseases that are straightforward to modern medicine could be debilitating or fatal. But it was not just that medicine was less advanced; the lack of suitable transportation meant that an incapacitated victim had to remain on the spot and be treated with the medical assistance that was – or was not – locally available. For those of us living in modern cities, the significance of this difference can be hard to grasp. Leaving aside conditions resulting from Indian service (disease, sunstroke, big cats, etc.), Coopers Hill students and staff died from disease, appendicitis, transportation accidents (horse, car, motorcycle), being crushed by a railway wagon, gunshot wounds, "tin poisoning," accidents while bathing, falls, robbery, and murder.

Poetry

The college magazines often printed students' poems (and a few from professors too). Some clearly represent "in" jokes, but others stand the test of time quite well, particularly those describing late-night studying. The following example from an early edition of *Coopers Hill Magazine* supposedly described a second-year student working into the night before his second exam in "Pomph."

Not a sound was heard, not a single note,
His expression seemed anxious and worried,
His thoughts, p'raps had wandered to subjects remote,
For his head in his hands he had buried;
He sat there alone in the dead of the night
The pages of "Pomph" slowly turning,

By the College lamp's flickering morbid light,
And an un-snuffed candle burning.

He knew that his friends who'd retired to rest
Would certainly have jeered had they found him,
And made him for ever a subject of jest,
As he sat with a wet towel wrapt round him;
Not a sigh escaped his lips, not a word he said,
Not a visible sign of sorrow,
But he anxiously glanced at his watch
And bitterly thought of the morrow.

He thought of the comfort and warmth of his bed,
And the sleep that alas was denied him,
Then he turned with a sigh to the "Pomph" that he read
To the Minchin and Lodge just beside him.
He thought how his friends had grown distant and cool,
For his rudeness had forced them to "chuck" him,
And he knew very well they would think him a fool
If the Examiner decided to pluck him.

 Not half of his self-imposed task was done,
Not a part of what he was desiring,
When the Holloway Clock in the Distance boomed one
And he'd not a thought of retiring.
Slowly and sadly he worked, and we see
On success he has certainly reckoned,
So we'll leave him alone to his "Smugging" for he
Has resolved to "Q" in his "Second."[100]

Balls and Bal Masqué

Hudson's diary mentions the college dance and one or two smoking concerts but is generally silent on the formal social life of Coopers Hill. For descriptions of the events we therefore have to rely on the *Oracle* and *Coopers Hill Magazine*, accepting the gaps in those records.

In the 1877–8 academic year the *Oracle* published accounts of two events that were mainstays of the college calendar, the Ball and the Bal Masqué. Both events had been running for several years. Future articles on the Ball would be occasional but there would be a piece on the Bal Masqué almost every year.

The 1877 college Ball took place on 20 December and was a grand affair attracting between two and three hundred guests, many of whom travelled from London on a special train. Dancing started around nine o'clock with twenty-two waltzes on the dance card, played by the Sandhurst Band. The library, theatre, and Salisbury Room were given over to the dancing. Supper was served at midnight in the Victoria Room, before dancing resumed and continued until 3:30 a.m., when many of the guests had to leave to get the train back to town. "The decorations were more complete than usual, a feature being a fountain that played the whole evening in the middle of a fernery which had been put up in Salisbury. The Library and Theatre were gaily decorated with flags, the College colours being especially prominent, and the time honoured iron girder bridge[101] was displayed in the Theatre to the admiring dancers."[102]

Twenty years later, the Ball was still going strong, held in both summer and Christmas versions. In summer 1898 the dancing was in the library, the Salisbury Room, and the Reading Room, while supper was in the dining hall. A "tent was pitched on the lawn for sitting out and looked very pretty lit up with Chinese lanterns." This was Courtney's last Ball, and it was "very sad to think that we shall no longer have him to 'run' the College Balls as he has done for so many years."[103] The *Coopers Hill Magazine* carried only brief reports in December 1903 and 1904, although the events, still labelled "Annual," continued to attract two hundred or so guests.

In contrast, the Bal Masqué was still reported extensively in the magazine and was a highlight of the student calendar. Only college students attended it and, as the name suggests, the attendees wore fancy dress. A large part of the fun came from the humorous re-enactments of recent college events. Each year's invitation was a collage of cartoons usually depicting humorous events or caricatures of college personalities. Some are recognizable, such as the prawn-headed figure of the surveying instructor Captain Stewart (nicknamed the Prawn) and Harry Seely (lecturer in geology and mineralogy) shown as a fossil. Most of the references are now obscure, so the modern reader has the distinct feeling of missing "in" jokes. Each year's Bal Masqué had its own character, or characters, but followed much the same format.

The first description of a masked ball dates from November 1877:

The first waltz was struck up, and every body flocked in. In the room might be seen Mr. Pickwick trying his hand at a violin, Blue Beard, and a veritable Paddy; Punch (who, for some unaccountable reason, forgot to bring his wife) was there, as also were Don Caesar, Falstaff, Hamlet, Mephistopheles, Old Mother Hubbard, a Chinaman, a most obliging Waiter, and a Dragoon, not to mention three Sailors, a Pink Domino, "Robert" A.1. who was in great demand in quelling disturbances, a Harlequin in his silvery attire, and his companion the Clown, besides thousands of Indescribables....

Figure 4.4. 1901 Bal Masqué participants, all male (Canning Collection, CSAS)

After a few more waltzes, the College clock struck eleven, and everyone adjourned to his room; some to talk over the amusements of the evening, while some to sleep, perhaps with a horrible nightmare of the Fancy Dress Ball of 1877.[104]

The numerous photographs donated over the years to the Coopers Hill Society and preserved in the British Library, such as that in figure 4.4, attest to the centrality of the Bal Masqué in college life. Notable costumes included the three blind mice, a large moth, and playing cards. Soldiers of all historical periods were common, as well as stereotypes of Indians and other nationalities now considered unacceptable. In any given year about half the attendees were in drag: "Space fails us to describe the dresses of the bevy of beautiful damsels also present, many of whom, such as Miss Ramsay and Miss Appleby 'came out' on this occasion while one or two 'came down,' but special mention must be made of Miss Walker who looked charming in her simple décolleté home-made dress of white muslin."[105]

Sports

Team sports, particularly Rugby Union, was a central element of life at Coopers Hill because it instilled self-discipline, loyalty to one's comrades, "grit" and determination, and pride in the traditions of one's team. The attitude was summed up in the Duke of Wellington's supposed statement that "the Battle of Waterloo was won on the playing fields of Eton."[106] There was such a fierce sense of

Figure 4.5. Coopers Hill Football Team 1886–7: (*back row*) Simpson, O'Donoghue, Oliver, Trapman, Smith, Heaton, Buscarlet, (*front row*) Rogers, Langlands, Holms, Keeling, Sutherland, Hart, Tottenham (Hudson Collection, CSAS)

identity with the team that when, for example, in 1877 the current members of the college changed the rugger jerseys from purple and orange horizontal stripes to the monogrammed white, as shown in figure 4.5, there was an outcry from old members in India: "Under the old colours was laid the foundation of that reputation for Football which the College has always sustained."[107]

Many obituaries and testimonies to Coopers Hill engineers contain variations on the words "he was a true sportsman." While the specifics could vary from rugby to hunting to polo, the notion was that a good sportsman was a true gentleman – reliable, courageous, honourable, dedicated. This is reflected in an article on the history of sports at Coopers Hill, which concluded, "All who know what Coopers Hill is must admit that it has always turned out sportsmen in the truest sense. The authorities of the college have always encouraged sports, and more specifically those in which combination is essential, as these have been found most beneficial in producing men of character, able to take up and maintain any position they might afterwards be called upon to fill, in India or other parts of the world."[108]

Even a non-player was expected to lend his patriotic support to the players. In 1879 at the end of a decade of outstanding rugby success, a letter to the editor of the *Oracle* on "the grave evil" of loafing prompted the leading article to comment, "Two years, or even one year ago, it would have been considered an absolute disgrace to a man if he took so little interest in the doings of our football team as to absent himself, without sufficient cause, from the college ground when a match was being played."[109]

It was almost as bad to appear at the match dressed inappropriately: "While the last two football matches were being played, a tendency was observed, on the part of certain members of the College, to appear on the ground as spectators of the game, attired in such eccentric costumes as ulsters and smoking caps of various gaudy hues.[110] ... [H]itherto it has been the invariable custom for men who come out to watch the matches to dress themselves as they would if they were going to walk outside the College grounds."[111] Standards must be maintained!

The first decade of rugby at the college was certainly noteworthy: "At this time the college certainly had the strongest side in the South of England, and was probably the strongest club in the country."[112] Eleven of its players were members of international rugby teams for England, Scotland, or Ireland in the 1870s (five at once in the 1877–8 college side), with a further three in later years. One of these was Stephen Finney (representing England 1872–3), who appeared as a successful railway engineer in chapter 3.

Over the history of Coopers Hill, the college's record stands at: played 491, won 268, lost 155, drawn 68. "Such a record is one that few clubs are able to show, and is a just source of pride to all who have had the privilege of being, at some time, members of the college."[113] Yet for all this blood, sweat, and tears, what is there to be said now about these half a thousand games? We can honour the memories of those young sportsmen, their subsequent trials and tribulations, and learn the meaning the game had for them.

Joining rugby as the most prestigious sports were rowing and cricket. In these sports, however, the college was never as successful as at rugby. In rowing, this was probably because many of the students were new to the sport when they arrived at Coopers Hill. For cricket, the constraints term dates impose on their season and the distractions of the annual exams were factors. Nevertheless, the college magazine of the time printed long descriptions of the matches and games, often with frank analyses of each team member's strengths and weaknesses.

Lawn tennis was introduced fairly soon after the college was established, "as an interlude for those men who did not care to row, or play cricket, during every afternoon of the week." For the traditionalists it "enrolled among its members a class of men who have neither the inclination nor the energy to support those clubs which really represent the College." To emphasize the point that the "manly" sports served a purpose, "Lawn Tennis men must bear in mind that the game, being a purely selfish one, and only to be looked upon as a means of exercise, must always be considered as secondary to the leading clubs of the College."[114]

Tennis was already a social game; professors as well as students played (there was a separate court for the professors too), and ladies were also invited to join in. Some of the professors were quite accomplished. Professor Minchin boasted in a letter to his friend Oliver Lodge, "I am now simply invincible at lawn tennis."[115] Lodge in his turn commented in his obituary for Minchin, "His extraordinary exhibition of energy and skill in the lawn tennis of those days, which consisted chiefly in long spells of lobbing from the back line, always attracted a gallery."[116] By the turn of the century the Coopers Hill Tennis VI was flourishing and competing regularly (but with mixed success) against tennis clubs from the south of England.

The author above who dismissed tennis so completely would have soon discovered that tennis was very much more popular in India than rugby football. In the early years, the college had raised a subscription to purchase a trophy to be presented to the winner of a rugby competition between the presidencies of Calcutta, Madras, and Bombay. "This competition was never a success in India, as the climate is hardly in favour of rugby football, and after a very short time the inter-Presidency matches were discontinued." Instead, the Calcutta Cup, as it is called, is now awarded to the victor of the annual England versus Scotland six-nations rugby match.[117] On the other hand, the British community throughout India, including Ernest Hudson and his wife, played tennis because it was much more amenable to the climate, could be played with fewer people, and was a pleasant social event.

The college also ran athletics days, during which the students competed in the usual trials of speed and agility. Less usually, however, there was also a donkey race. On 23 May 1888 Hudson mentions that he took part in this event, in which riders wore fancy dress (see figure 4.6). In June 1899 the runners were given spoof names in the style of racehorse pedigree descriptions. So we have, for example, "Back Hair, by Coming Backwards out of Hedge," "Hurled, by Fair Lady out of Datchet," and the optimistic "First Lecture, by Miracle out of Bed."[118]

Esprit de Corps Revisited

For George Chesney, a – perhaps even *the* – major purpose of Coopers Hill was to build esprit de corps among the engineers serving in India. Having seen something of life at the college, was it successful, and if so, how?

First, we need to recognize that the "atmosphere" of the college changed over its three decades. In the first ten years under Chesney's firm hand, it was purposeful; students were pioneers together, and they knew they were going to India and what their role there would be. The second decade could be categorized

Figure 4.6. Riders at the start of Donkey Race, 1888 (Hudson Collection, CSAS)

as "uncertain" – the lack of guaranteed PWD positions, the addition of new courses of study, and a partial dilution of the military ethos, led to increased freedoms but also to the reduction of the common purpose. Coopers Hill rediscovered its internal purpose in the final fifteen years of its existence with an increased sense of professionalism, and an identity within the global and increasingly diverse engineering profession.

This reality led to one former student, "O.C.H.," in 1906 to state boldly in a letter to the *Coopers Hill Magazine*, "It is a hard thing to have to say, but nevertheless it is true that the much vaunted 'esprit de corps' and 'camaraderie' which is supposed to exist the world over among C.H. men, exists only among men who have actually known each other at the College. I believe that it exists among C.H. men as a whole, only on paper in the C.H. Magazine. There is not much in common between the C.H. man of the present day and 30 years ago."[119]

In an earlier age, the *Oracle* certainly saw itself in this role: "The Oracle … appeals to two classes of readers – the past and present members of Coopers Hill. Not the least pleasant of our three years course here is the love of 'alma mater,' and we have had many and recent proofs of how strong that love is. The Oracle tends to cement that union by supplying to those absent a detailed account of those present, and by marking the various occurrences in the place which is common to both."[120]

O.C.H.'s view drew swift reaction, and the lead article of the next edition notes that they received several responses and that "the writers one and all refute the statement, while some have written in quite an angry spirit." One correspondent calling himself "Cooperzillian" wrote, "It is possible that an O.C.H. has been unfortunate during his short career, but the thought forces itself upon me that perhaps some of the fault lies with him."[121]

There was a measure of truth in this because the same issue of the magazine indicates that 550 old boys had already signed up for the new Coopers Hill Society, intending to keep the college's spirit alive after the impending closure (by 1908, membership was more than 800). That the society should continue for nearly sixty years after the college's closure is a strong testament to the sense of identity people felt with the college. Clearly there was an esprit de corps.

The sense of identity with the college is the essence. Men who knew each other in the college forged a special bond, but the "corporate sense of honour" Chesney was seeking referred to the honour and reputation of the institution as a whole, not just the individuals. Old C.H. Men would have also been linked by shared experiences – professors and staff, college grounds,[122] long hours of tough work, sports teams, the Volunteer Company, a common mythology of ruxes, signposts, and other misdeeds, and a reinforcement of cultural and social identity. Hudson's diary in later years certainly supports this notion.

C.W. Hodson,[123] speaking at the 1907 half-yearly Coopers Hill dinner, summarized his view of the benefits of the college's esprit de corps:

> Though Coopers Hill had been abolished as a College, it still exists as a fraternity and a pass-word amongst its members....
>
> The strong sense of good fellowship and esprit-de-corps which has characterised Coopers Hill men has been most beneficial to the Service. One of its consequences have been that juniors going out to a new country, or being transferred from one part of the country to another were able to feel that they would find friends and comrades to welcome them; and seniors could feel that by taking an interest in juniors posted to them and training them in the way in which they should go, they would not only be making personal friends for themselves but also doing a service to our old College and helping to add to its high record of efficiency and zeal for the service.[124]

By the time Hodson made these remarks, new PWD engineers would no longer be from Coopers Hill. Professor Unwin, in his address to the same audience, described the great advances in engineering education that had been made at universities since the establishment of Coopers Hill. Noting that he opposed "to some extent the convictions and perhaps prejudices of old Coopers Hill men," he pointed to a different sort of camaraderie that would emerge under the new system of PWD recruitment: "I think they will be found very amenable to the discipline of a public service, and men with whom it will be pleasant to work. I think they will readily learn the traditions of the department they enter and acquire an *esprit-de-corps* of the service which will be some equivalent of the *esprit-de-corps* of a common *alma mater*."[125] In this, Unwin was essentially describing the modern profession of engineering. A post-secondary education, professional accreditation and licensure, and the iron ring in Canada, are designed to foster a sense of responsibility and identity between engineers.

Passage to India

Ernest Hudson departed for India on 29 November 1888 on the special train from Liverpool Street Station in London. They sighted the lighthouse off Colombo, Ceylon, on the evening of Christmas Day 1888. The next day, Hudson breakfasted with E.C. Moreton (CH 1883–7) and Ridout (CH 1885–8) at the Grand Oriental Hotel before driving out to the Bridge of Boats[126] and several of the local villages. John Ridout was supposed to have replaced Hudson when the latter failed his medical examination, but instead joined the Ceylon Survey Department, where he rose to become superintendent.[127] Within eighteen months, Moreton would be dead of tuberculosis at Colombo.[128]

After a hot, rough passage up the east coast of India they reached the mouth of the Hoogley River, leading to Calcutta on 30 December, and the next day they "moored alongside the jetty at Garden Reach at 3 when Henderson came on board to help me with my luggage…. Had to send my gun rifle + revolvers to the Custom House but drove off with the rest of my things to 1 Theatre Road where I managed to get a room. Stopped at the works on my way where I saw Jones who was in charge of the Alipore work shops."[129]

Even during Hudson's short journey to India, the Coopers Hill network was in evidence – Moreton and Ridout greeted him in Ceylon, and Henderson was there to welcome him to Calcutta and show him around. Hudson stayed in Calcutta for a few months, working in the workshops of the Telegraph Department in Alipore (a European suburb of Calcutta) under Col. Henry Archibald Mallock. In April, Hudson was posted to the Telegraph Office in Bombay.

Hudson's transition from Coopers Hill to Indian working life must have been very typical, although the college men usually travelled as a larger group. After he retired, Alfred Cornelius Newcombe (CH 1871–4) recalled his introduction to India as one of the first main batch of Coopers Hill engineers:

> My first voyage to India was in the troopship Malabar, which left Southampton on October 7, 1874. The party to which I belonged consisted of eleven of the first batch of fifty young Civil Engineers just appointed from Coopers Hill College to the Public Works Department of India. The pleasant company of the naval and

military officers and the luxury of a military band twice a-day were what civilian travellers rarely get.[130]

After getting orders at the Public Works Secretariat at Amritsar, my friends and I found ourselves posted in different directions, and we had to part. I was then left alone in Amritsar to work under an "Executive" engineer who had been many years in India, and was now nearing the end of his service. Fortunately, he and his wife were most friendly and hospitable, and many a pleasant evening I had with them at their bungalow in Amritsar and in camp. A start like that, with every assistance in the many little difficulties which arose in adapting myself to the new conditions of life, was most helpful. I was left at Amritsar for several days to collect servants, to buy a horse, and to engage camels and carts for conveyance of my camp-equipage on a journey in the interior along a line of canal which my "Executive" was about to inspect. This by way of introduction to my new duties and to render me familiar with the ways of travelling and managing the natives.[131]

5

An Officer of the Indian Telegraph Department

The Indian Telegraph Department

This chapter focuses on Ernest Hudson's career in the Telegraph service. His diary is a unique record of a young man's transition from being a student at the Royal Indian Engineering College to a working life in the Indian Telegraph Department. As figure 5.1 reveals, this career took him to many parts of British India, including Burma and modern Pakistan. His diary consists mostly of brief entries about his day, the people he met, letters sent and received, and his purchases. He particularly mentioned the people he worked with and provided detailed information and impressions of places and sights when he was away on tours of inspection, in the field constructing telegraph lines, or on hunting trips.

In parallel with his diary, Hudson kept more technical field notes, which he used for personal notes and memoranda as well as details he would need later when preparing his formal reports. Often the entries were accompanied by schematic maps of the telegraph lines, diagrams of the posts on which the lines were carried, and annotations regarding the condition of the line. When he was new to India, Hudson left blanks in his diary entries so he could fill in place names and distances afterwards (or sometimes not).[1] In addition, he sometimes initially wrote names phonetically, before later adopting the accepted spelling,[2] probably because he initially heard the names and routes verbally.

Hudson's diary offers an insight into the practical training he received during his early years in India and his transition to managing his own construction projects. It also reveals many details of his technical, and later managerial, responsibilities and the nature of his day-to-day life and work. This informs a discussion of Hudson's education at Coopers Hill and shows why demands from India necessitated revisions to the college's curriculum at the end of the century. For a superintendent in charge of a division, Hudson's diary also offers an inside view of the pivotal Telegraph Strike of 1908.

At the time Hudson joined the department, the Indian Telegraph Service was well established. From its origins in the experimental work carried out by William O'Shaughnessy, a surgeon and chemist, at Calcutta starting in 1839,[3] a network of several thousand miles of telegraph lines was in place by 1857.[4] These lines were accessible to the public and for private messages, and were adopted by Europeans and Indians alike.[5] Despite technological shortcomings, a poorly conceived network design, and the deliberate strategy by the rebelling sepoys of attacking the telegraph lines, the telegraph was credited by many with allowing the British to suppress the Indian Rebellion of that year by alerting military commanders in time to take preventative measures.[6] In the decade after 1860, the growth rate of the network was relatively modest, although investments were made in reconstructing and improving the existing lines. When Hudson joined in 1889, the system was in its maximum rate of growth, with an average of 1750 miles of line and ninety combined postal/telegraph offices added every year.[7] India was a key node in the global telegraph networks, with links to Europe and Australia.

In common with the other young graduates from Coopers Hill, Ernest Hudson joined the Telegraph Department at the rank of assistant superintendent, class VI, grade 2. This was the bottom rung of the promotion ladder of the "superior establishment" of the department, the group "engaged on the construction and supervision of the various telegraph lines belonging to the Government of India." Two rules governed promotion: the occurrence of "vacancies in the sanctioned establishment" and "selection with due regard to seniority."[8]

A vacancy would typically occur when an officer retired, returned home sick, or died. The rules for retirement could be complex and were changed from time to time, but typically a pension was awarded without a medical certificate after at least twenty years' service. Men over fifty-five years of age could be requested to retire. The value of the pension was determined in rupees even if it was paid in England.[9] The word "sanctioned" was used to indicate authorized approval (for example, funding was sanctioned to build a telegraph line on the basis of an estimated cost), so this allowed a vacancy not to be sanctioned and left unfilled.

In practice, the seniorities of officers were well known and, indeed, played a large part in determining the order of names in the official India List table of members of the Telegraph Department. Passing examinations and the annual confidential personnel reports could also affect the order. A comparison of twenty-five or so names that appear on both the 1888 and 1902 India Office lists[10] for the Telegraph Department shows that the ranking of names remains broadly similar. A few men moved significantly in the sequence between the two dates, often as a result of sickness or secondment to other duties.

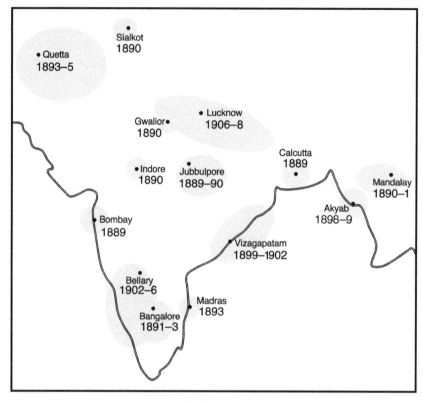

Figure 5.1. Approximate regions in which Ernest Hudson served in India

Learning the Job

It is easy to see from Hudson's experience why the marriage of young officers was, if not expressly forbidden, certainly discouraged.[11] During his early years in India he was transferred to a new office every few months, going to stations that were short-handed as a result of the leaves and transfers of other officers (see figure 5.1).[12] In some cases, Hudson's boxes of possessions were still in transit from his previous posting when he was ordered to the next one. In his first two years Hudson served in six offices – the telegraph workshops in Calcutta; the Bombay office; out on the line between Karkeli and Raigarh in the Nagpur Division; Gwalior in the Indore Division; Sialkot in the Punjab Division in the north-east corner of modern Pakistan; and then to Mandalay in Burma, a distance of some 2100 miles by train and steamer. The experience of Hudson's classmate William Henderson was similar, if not quite as extreme.

These early months formed an essential element of their technical preparations for their careers, taking the largely theoretical knowledge gained at Coopers Hill and adding the practical know-how of their profession.

Most of Hudson's first year was laid out to give him increasing independence, from learning the basics in the workshops to the first trips to inspect telegraph offices. Men appointed before the mid-1880s were sent directly to a division, but from 1885 onwards it became the common practice to start new men off at the Telegraph Workshops in Calcutta.[13] In 1898 just before taking charge of the workshops, Eustace Kenyon (CH 1878–80) commented that there were "at present 3 youths lately out from Coopers Hill all under instruction," but they "will shortly be scattered to the different presidencies."[14] At that time, the workshops employed close to 1000 men on work ranging from instrument assembly to blacksmithing.[15]

Hudson was assigned to do technical drawings by Henry Alexander Kirk, the workshop superintendent. He also worked with Col. Henry Archibald Mallock, a major figure in Indian Telegraph development,[16] on tasks including designing a new telegraph stamp, a proof of payment system analogous to the postage stamp. These designs were good enough to be sent to the secretary of state "for inspection," but the final outcome is unknown. Hudson also recorded that he engaged a "bābu"[17] to give him lessons in Hindustani.[18] Although he does not say so directly, he probably received a thorough grounding in his new profession from the practical men in the workshop.[19]

At its simplest, the telegraph system consisted of a wet electro-chemical battery and switch at a sending station. They were connected to the receiving station by galvanized iron wire supported on modular metal posts by ceramic insulators. When the transmitting switch was closed, a small current flowed out through the wire and back via the earth return to operate an electromagnetic sounder, which reproduced the transmitted code. The challenge was in getting and keeping this simple system working reliably in the field over hundreds of miles of varied terrain and in extreme weather conditions.

After learning the ropes at the workshops, Hudson was sent to the Bombay office, while Henderson was ordered to remain in Calcutta at the signalling office. There, Hudson set to work on drawings for a new battery stand. Over the next eight months he was exposed to a variety of the department's activities. He started going out to inspect telegraph offices in and around Bombay in the company of sub-assistant superintendents, the hands-on technicians who did some of the more complex testing and presumably supported him as needed. This accords well with experience of one former director general of telegraphs, Charles Henry Reynolds, who commented that "when they came out, they had to learn [their work] often from their subordinates."[20]

Visits of inspection to the offices were written up in reports back to the Divisional Office[21] as well as in the visitors books kept at each office. Hudson also started examining probationers for admission into the training classes designed to recruit local people to work on the telegraphs. Two of the senior men, Hutchinson and Dowson, took Hudson to the store yards, where they not only inspected the latest stores recently arrived from England but also familiarized him with some of the older patterns he would encounter in the field. At the Waribund stores in July he also saw a new trolley Hutchinson[22] had designed for laying lead-covered underground wires for connecting the batteries of the Bombay harbour defences; three years later, Hudson would be in charge of a similar installation at Madras.

Hudson's diary entries during this time mainly indicated events of interest, people he met, dining companions, and his frequent visits to the theatre. He noted in his diary letters received and sent, and sometimes made a detailed "ledger" at the back of his diary. For a while he even kept a separate notebook entitled "Places Visited & People Met by E.H. India 1888–1895." Over the years, as people's names appear repeatedly in the diary, a strong impression emerges of the European population drifting around India like a slowly circulating ocean current, passing from station to station as promotions and vacancies arose. The "superior establishment" of the Telegraph Department was quite small, typically around forty, so Hudson met and worked with the same men in a variety of roles over his career.

In November 1899, Gilbert Mahon (CH 1885–6) came in from nearby Kalyan, and together they met ten of the new batch of Coopers Hill men on their arrival from England, as they themselves had been met a year before. They made them comfortable at Watson's Hotel, where they received their orders that same day, going their separate ways to Madras, Rangoon, and Allahabad. Hudson spent the next couple of days showing them the sights of Malabar Hill, the Tower of Silence, and the Elephanta Caves, as well as dining together and seeing them off to their respective destinations. In this way successive batches of Coopers Hill students were welcomed by their fellows into the college's diaspora.

Bengal-Nagpur Railway

The first inkling we have of a change coming for Hudson is a telegram in August from the newly appointed deputy director general, Mr. Brooke, asking when he would be available for construction work. Nothing more is recorded until 29 November, when Hudson received a telegram ordering his transfer to the Nagpur Division, effective in just two days' time. Hudson immediately wired Thompson[23] at Jubbulpore to ask what work he was likely to do there. Thompson's answer is not recorded but he does advise Hudson to get everything he needed at Bombay as "Jub. could only boast of liquor and tinned provisions."[24]

The peregrinations of officers were governed by the rules for "joining time," the allowance for packing up, travelling, and unpacking at the new office during which the officer was still considered to be on duty. For any transfer necessitating a change of residence, Hudson would have been given a base allowance of six days in addition to a number of days based on mileage, up to a maximum of thirty days. Joining time also depended on the mode of transport at a rate of one day per 200 miles by rail, 150 miles by ocean steamer, 80 miles by river steamer, or "mail cart or other public stage conveyance drawn by horses." During the joining time the officer was paid at the lesser of the salaries of his old and new positions.[25] Although Hudson's record shows him joining the Nagpur Division from 1 December, he did not actually leave Bombay until the tenth, travelling the 600-mile journey on the overnight train. This technically put him over the allowance, but it seems to have been with the permission of his superiors.

Each month, Hudson submitted a travel claim for the previous month. He paid for the travelling costs for himself, his horse, and his servants and was reimbursed according to a complex set of rules described in thirty-five pages of the Government of India's Civil Service Regulations, updated every five years. Essentially, the rate was based either on mileage travelled, with different rates according to the mode of transport, or by a daily allowance. The rules were vastly complicated by the need to cover all eventualities over the wide range of possible conveyances, geographic regions, purposes of travel, types of accommodations, and seniorities of officers. For some branches of the service involving frequent travel, such as forestry, officers received a flat daily or monthly allowance instead of a reimbursement. There was even an extra allowance for officers of the Telegraph Department travelling on unopened railway lines (as Hudson did frequently) by trolley, material train, or engine.

For the next seven months, Hudson worked on building a new telegraph line that, like many of his projects, was associated with railway line construction. The Bengal-Nagpur Railway was established in 1887 to construct a line heading roughly west-south-west from Asansol, near Calcutta, to Nagpur. Part of that project was a branch line running in a north-westerly direction from Bilaspur to Katni to join it with the Allahabad-Bombay main line. Both sections were opened in 1891.

Hudson set out from Jubbulpore in mid-December with his cook, "boy" (indoor servant, supposedly from *bhoy*, meaning "bearer"),[26] and two linesmen to join the working party on the telegraph line. He went by train to Umaria and thenceforth down the freshly constructed railway and telegraph lines to Khodri, where the construction work was going on. It took so long to weigh his baggage at Jubbulpore Station (to assess the shipping cost) that it was not complete until fifteen minutes after the train should have left: "To prevent it appearing they were late however they put the clock back ¼ hour."[27]

From Umaria he travelled by railway trolley, foot, or pony, while his possessions were transported on three rather accident-prone camels that broke his new china. Railway push trollies consisted of a bench mounted transversely on a chassis with four railway wheels, pushed from behind by Indian "trolleymen."[28]

They were frequently used by telegraph men and railway engineers to go about their daily business. They seem rather dangerous and uncomfortable contraptions, and over the years Hudson recorded near misses with railway engines (himself and others) and a fatal accident when his men were using a trolley to transport telegraph stores. He described his own numerous freezing journeys through the mountains. To be fair, he also recorded pleasant trips through the jungle, stopping to shoot ducks at the water tanks as they passed, or exhilarating, rattling runs down mountain passes, such as the one shown in figure 5.2. Even during railway construction, trains ran along the rails taking stores, materials, and ballast to the rail head. One assumes the trolliers knew the train schedules and a trolley could simply be lifted off the rails if necessary. While convenient for the passengers, it was tough work for the Indian trolleymen through the same harsh conditions and sometimes over distances as great as forty miles in a day; not surprisingly, turnover seems to have been quite high.

Hudson continued to work with experienced men on the new construction, including Asst. Supdt. John Melville Coode, who arrived in India from Coopers Hill in 1885, and a Mr. Corbett. The general format of the construction process was as follows: the course of the telegraph line was laid out, in this case by Hudson and Corbett; local labourers on contract would dig the post-holes; a working party of Indian telegraph men would erect the posts; and a wiring party would erect the wire(s). The men lived in camps that were shifted periodically as the line progressed. Because the railway was being constructed at the same time, many of the engineering officers had more or less permanent bungalows along the route and Hudson would often stay with them. Once a section of the new telegraph line was constructed, the officer in charge would go back and inspect it. He ordered changes where necessary, drew a map, and wrote up the "line book." The latter seems to have been a fairly broadly defined logbook, not just of the line as constructed but also of other relevant information, such as lodgings, stabling, water sources, camping grounds, and good shooting locations.[29]

Hudson completed his second year in India with two long journeys. The first was a seventy-two-hour train ride 800 miles due north from Indore to Sialkot to fill in for September while G.D. Close was on leave in Srinagar.

Mandalay

Hudson's second journey was from Sialkot to Mandalay, Burma, a distance, as mentioned earlier, of more than two thousand miles. Originally the plan was to construct a telegraph line at Haka, but soon after Hudson's arrival in Mandalay he received a wire ordering the "line to be made right from Pokoku to Kan." This was probably because the military situation in Haka was unstable. The town had been occupied by the British less than a year previously during the Chin Lushai expedition, an attempt to pacify the region by occupation rather

Figure 5.2. "Thirty miles an hour: Trolleying down to New Chaman from the Khojak Tunnel on the Quetta and Khojak Railway." Hudson took this exhilarating trip several times (*Graphic*, 26 May 1894, 620)

than by punitive raids.[30] Guerrilla warfare by the hill fighters, known collectively as the Chin tribes, was still commonplace in the Haka area and elsewhere.

Hudson and F. Mercer (CH 1882–4) travelled west on horseback from Mandalay to Tilin, and then northwards to the village of Kan, on the edge of the Chin Hills just east of Haka. Fever was rampant in the area. A Lt. Warwick was "the 11th commanding officer at Kan who has had to return with fever in as many months," and almost all of the sepoys at nearby Changwa were in hospital. Fighters from the Thetta tribe took advantage and attacked Changwa a few days later, the first in what the British called "a series of outrages," in which an assistant superintendent of police was killed[31] and the telegraph line to Haka was cut again.

No sooner had Hudson arrived in Kan in December 1890 than he too succumbed to the fever and had to be carried by "dooly."[32] He was ordered to return to Mandalay to take over the subdivision, leaving Mercer to finish the work in Kan.[33] On the way back he noticed and repaired a place where, judging from the knife marks on the wire, the line appeared to have been cut deliberately. Hudson reported news and rumours of other attacks on the British, and the recapture of Thetta village by the British. A definite sense comes through in the diaries of the twenty-four-year-old Hudson's excitement at being on the frontier of the British Empire, Lord Curzon's proving ground for the young men of the empire (chapter 2).[34]

Throughout February and March, Hudson was still suffering from acute fevers that left him incapacitated, even to the extent of having to lie down on the road to recover during one journey, only to have it disappear a few hours later. He also suffered periodically from a mysterious malady he called his "old complaint." For a second time, he had to be carried out by a combination of dooly and cart, and "had a miserable time of it what with the heat, dust and jolting." His symptoms persisted until around 20 March, on which day he "sent to the club for fiz and claret to see if they will pull me up."[35] Unlikely as it may seem, they seemed to do the trick and he became healthier from then onwards.

Hudson does not record taking quinine at this time, but he did so in 1898, when he says he took three big doses, only to be bothered by deafness, a side effect of the drug. In his memoirs *Work and Sports in the Old I.C.S.*, William Horne said that "a go of fever" was treated by "quinine, licked out of the hand and not even measured. I have often been deaf as a post from this treatment."[36] From 1896 quinine was sold across India in convenient doses by the post offices, with over seven million doses being sold in 1903–4.[37]

Madras and Bangalore

On 8 April 1891, Hudson left Mandalay for Madras after an eventful five months in Burma. Until this point, he had been working with others, or at most had brief periods in charge of construction. After Mandalay, Hudson would be in charge of his own subdivisions and ultimately of divisions. This therefore marked the culmination of his "apprenticeship," an extended period of systematic exposure to increasingly independent work, coordinated centrally by the Telegraph Department.

Hudson would spend the next two-and-a-half years in the Madras Division in southern India, for most of which he was in charge of the Bangalore Subdivision. He was primarily responsible for the operations the lines from Bangalore east to Jalarpet, south to Podanur, and from there to Ootacamund. Professionally, this was the bread-and-butter of the Indian Telegraph Department.

Perhaps the most interesting aspect of this period was Hudson's friendship with Dr. Ronald Ross. Ross was staff surgeon at Bangalore and was just beginning the work that would lead him to the 1902 Nobel Prize in Medicine for the discovery that the malaria parasite was transmitted by mosquitoes. Hudson met Ross at the end of June 1891 through the Bangalore Golf Club, just two weeks after Hudson's first-ever round. They played together occasionally over the following months and gradually began to dine together towards the end of the year. Many years later in his memoirs Ross described the pleasures of his time at Bangalore, including the time when "a friend (a Mr. Hudson, I think) and I went fishing in the hot weather of 1892 near the Mysore 'keddas',[38] where the wild elephants are trapped by means of a large system of dykes cut through the vast jungles."[39]

Ross had a "forceful, vigorous personality, so full of eagerness and interest, with a resolute, pugnacious mind and fierce yet kindly intelligence."[40] These attributes enabled him to persevere through the difficulties of experimental biology while balancing the needs of work and a young family, but they also led him to be combative and impatient. This negative aspect of his personality emerged during an unsavoury incident during that shooting and fishing trip Ross and Hudson took together. Travelling through the night they had "a lot of trouble" at around midnight changing bullocks for their cart. They were again delayed later in the morning, and Hudson wrote that someone "had to beat the kotwal [police superintendent]" before they could get any fresh bullocks. Clearly this was significant, for on 4 June 1892 Ross stood trial for assaulting the kotwal "in the Mysore court by old Ricketts." Hudson went as a witness giving "a short evidence as to the delay cause by the kotwal in not providing us with bullocks." Three days later Ricketts found Ross guilty and fined him the rather trivial sum of Rs25.

On 14 February 1893 Hudson "played a final round at golf with Ross and made a record drive at the last hole." He also made a round of farewells on the following day and left for Madras by the evening mail train, taking his horse and pony with him.

Baluchistan

The highlight of Hudson's technical career was the three years (December 1893 to October 1896) he spent in the city of Quetta in Baluchistan, close to the

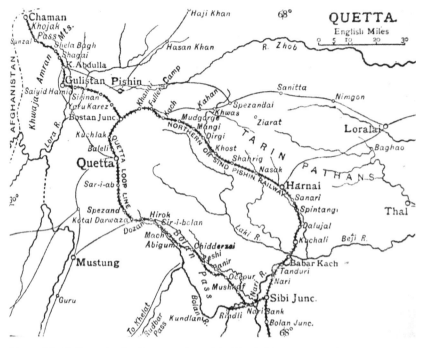

Figure 5.3. Railways in the Quetta region of the North-West Frontier after 1897. Prior to the opening of the Mushkaf-Bolan Railway, the route followed the Bolan River west from Nari Bank, through Rindli to Kundlani, and then turned north towards Hirok (*A Handbook for Travellers in India, Burma and Ceylon* [London: John Murray, 1919], 369)

border with Afghanistan. It was the only place that he expressed regret at leaving when the time came, and indeed he turned down a divisional officer appointment elsewhere to stay at Quetta.[41]

Set in a valley high in the mountains, Quetta was brought under British rule in 1876. It became the administrative capital of Baluchistan with a strong military base protecting the southern section of the North-West Frontier and guarding the strategic Khojak Pass leading to Kandahar on one side and the Bolan Pass on the other. The area was famous for its soft fruit production, and Hudson received requests from his acquaintances for cases of peaches and plums (as well as local crafts). In 1892 Quetta was home to around fifteen thousand people, and while the winters were severe the summers were pleasantly warm. The scenery around Quetta is dramatic, with high mountains, rushing rivers, and precipitous, narrow gorges, in some places only wide enough for four to ride abreast.[42]

Before 1897, Quetta was reached by railway from the junction at Sibi via one of two routes, both with severe difficulties (see figure 5.3). These are discussed in more detail in chapter 7 from the engineering perspective. The Bolan Line ran west from Sibi to Kundilani and then north to Hirok following the Bolan River. It was constructed quickly on the riverbed and was very susceptible to flooding. The alternative railway route (the Sind-Pishin Railway) ran north from Sibi before turning westwards and then curving around to approach Quetta from the north. This celebrated feat of engineering was prone to being swept away by mud slides. The continual flooding of one route and frequent destruction of the other prompted the construction of a railway following a new passage through the Bolan Pass, named the Mushkaf-Bolan Railway. This route is still in use today, and several videos taken from the train are available on the internet that show the dramatic journey through multiple tunnels, over bridges, and along the narrow passes.

Telegraph Construction at Quetta

Hudson arrived in Quetta in December 1893 and set up in a recently renovated bungalow in a compound that also contained the subdivisional office and was opposite the telegraph office (see figure 5.4). On 5 January 1894 he commenced the first of several construction projects in Baluchistan, erecting the telegraph line alongside the southern section of the new Mushkaf-Bolan track, from Nari Bank, through Mushkaf itself and the multiple tunnels at Panir, and across the "dreary, weird, and desolate" landscape to Pishi.[43] The railway then carried on steeply uphill to Kolpur (Kotal Darwaza).

Hudson travelled from Quetta to Sibi by the Sind-Pishin route[44] with his boy, trolleymen, clerk, and *chuprassie* (office messenger). He needed to arrange with the chief railway stores-keeper for a truck to take the telegraph supplies (posts, insulators, wire, etc.) out to Nari Bank and Mushkaf on the material train that supplied the main railway construction. The rails were laid to just short of Ocepur at this time, so Hudson and his men had the convenience of using the new railway line to get around, distribute stores, and obtain provisions from Sibi. A gang of "coolies" (labourers) did the heavy lifting, dug post-holes, and erected the posts. Those Hudson engaged in Sibi were "gradually getting sick of the work and deserting the working party" because of the cold, so he had to hire "very fine" Pathans at eight annas/day to complete the work.

The line men, such as Mahomed Azam and Boota, fitted out the posts and installed the wires, "straining them up" to achieve the correct dip between posts. They were supervised by sub-inspectors such as Shavali Khan and Samander, who worked with Hudson for much of his time at Quetta. Assisting Hudson was a sub-assistant superintendent named E.E. Gunter. As the line progressed, the office at each station was fitted out with instruments by the inspecting telegraph master and connected to the central stores at Sibi.

Figure 5.4. Ernest Hudson's bungalow, Quetta (Hudson Collection, CSAS)

By the end of January, the line was marked out as far as Ocepur, the wires installed to beyond the six tunnels and two river crossings outside Mushkaf, and the offices at Mushkaf and Nari Bank were operational. Once they reached the end of the railway track, the stores were distributed towards Panir by eleven camels at twelve annas per camel per day – which was more expensive than the labourers. The line crossed deep ravines, "some of the spans being very fine being nearly ½ mile long."[45] Hudson met the engineers working on the railway, including Ramsay the engineer-in-chief, and two of the previous year's Coopers Hill men.

Beyond Panir Station was the great Panir Tunnel, 3200 feet long and taking nearly three years to complete. Hudson was able to walk from one side to the other through the "headings," or guide tunnels, that had been completed in 1893. Because trains could not yet run through the tunnel, the engineers had constructed a temporary track from the old Bolan Line to a point just north of the Panir Tunnel for the delivery of supplies.[46] Figure 5.5 shows a general view of the site. The telegraph line was to pass over the hill rather than through the tunnel, but the camel drivers initially refused to take their camels up the hill, it being so steep that Hudson could "quite imagine the camel men funking taking their camels over it."[47]

On 16 February 1894, Hudson saw the new office at Pishi opened, marking the completion of the current phase of the telegraph construction. He paid the

Figure 5.5. Panir Bund from entrance to Panir Tunnel, Mushkaf-Bolan Railway, Baluchistan, ca. 1894 (Hudson Collection, CSAS)

sub-inspector and linemen their bills and labour vouchers for January and dismissed the labourers, "closing the cash account against the Nari Bank – Panir Estimate."[48] Hudson completed the rest of the line from north to south in November and December of that year.

After finishing at Pishi, he returned to Sibi from Mach by trolleying forty-two miles down the Bolan in three and a quarter hours by moonlight: "a grand run down-hill all the way – scenery magnificent and much water in the river through which we dashed throwing up a regular shower bath – arrived for dinner at 24hrs."[49]

Hudson recorded joining the celebration with the engineers at the Quetta Club of the first through-train from Sibi to Quetta on the Mushkaf-Bolan Railway on 29 June 1895. It was opened to the public two years later.

Kotri-Rohri Construction

A construction project of a very different flavour awaited Hudson in the summer of that year. Instead of the dramatic mountain scenery of the Bolan, the flat, dusty Indus plain traversed by the new Kotri-Rohri Line was unexciting. To compensate, the pace of construction was flat out, causing logistical challenges of a different kind. Again, this construction accompanied a new 178-mile[50] stretch of railway being built to connect Kotri, near Hyderabad, to Rohri, on the opposite bank of the Indus River from Sukkur.

Of the country near the start of the line Hudson wrote, "Country very flat and sandy with cotton, millet, Indian corn + c under cultivation where land can be irrigated from canals. Camels employed for drawing up water at all the Persian wheels." The region further north was known for indigo production, "which was cut and pressed in curious pits into which water is run from a small canal ... there is one of these pressing pits behind my tent & the smell is most unpleasant."[51] When the weather was dry, the winds whipped up the dust so it was difficult to see, and on one occasion a train missed the station completely in a dust storm.[52] It could also rain heavily, and flooding made the transport of supplies difficult at times.[53]

When Hudson took over in mid-July 1895, the telegraph was at mile twenty-five and the rail head was at mile seventeen.[54] Hudson's working party completed their section to mile ninety by 2 October, whereas the rail tracks had advanced only to mile sixty-eight by a month later. This meant that the telegraph men were always distributing their stores for the line well beyond the end of the rails, which caused endless delays. Nevertheless, they managed an average of about one mile of completed line per day.

An SAS named Hart had been assigned to start working from the northern end of the line at Rohri, and he and Hudson kept in touch by telegraph.[55] He had Lehma Singh marking out the line, Kaim erecting posts, and Mowla Bux (one of several line men seconded from Quetta) finishing up the work at the stations.[56] In one day in good ground, a labourer could dig four post holes, and the entire gang could complete two and a half miles of post holes. In turn, Kaim and his men could install up to fifty posts per day, which covered about the same distance. In contrast, only about one mile of wire could be erected.[57]

At the beginning, Hudson was scrambling to secure all the bullock carts or camels he could find, supplementing them with the trolleymen. With six carts he was able to distribute thirty posts at a time, and once work was in full swing, he had them doing the distribution overnight, sometimes twenty-four at a time, so the supplies were ready for use in the morning.[58] The district superintendent sanctioned a transportation cost of four and a half *pice* or *paisa* per *maund* per mile.[59] A *pice* was a quarter *anna* and hence one-sixty-fourth of a rupee. A *maund* was roughly equivalent to eighty-two pounds, so a telegraph post weighed about two and a quarter *maunds*. On the basis of this commitment, Hudson was able to sign a contract with a local man, Thal Mal, for distributing all their stores.[60]

From around mid-August Hudson started to comment on delays arising in construction due to the lack of supplies. He had to dismiss some of his working party and had no three-hundred-pound wire or posts suitable for road cross-ings.[61] To add insult to injury, some of the contractor's carts even left to join the railway construction.[62] He called in Thal Mal's agents about this, which clearly had them worried because, when they settled up the Rs800 bill for the work already done, they resorted to attempted bribery: "The contractor's son tried to leave a roll of notes on my table when leaving, evidently as a bribe, although often heard of this sort of thing have never had anyone do it with me before. Seeing his little game ordered him out of my tent and told men not to let him near camp again."[63] Thal Mal seems to have kept the job, perhaps because there was no alternative, but troubles persisted on and off through September. On some days sixty or seventy camels "brought about 50 posts with a lot of wire + insulator boxes," while on others no stores, not even those left next to the line further back, were brought, and the working party sat idle.[64]

By the end of September, Hudson's section of the line was done, although he had some trouble paying his men because the postmaster at Padidan did not have enough cash on hand to honour the service money orders, and "consequently there was much discontent."[65] Then Hudson was ordered to take over Hart's working party and finish the line to his own satisfaction. The work was delayed because they were woefully short of the brackets that supported the wires on the posts – a massive 1260 short, in fact.[66] New supplies were not shipped from Karachi until 22 October, and finally the railway telegraph was connected through on 30 October.

Very Severe Weather

After Hudson returned to Quetta, his younger brother, Herbert, came to visit from England. During his stay Herbert probably wrote a short entry entitled "A Day at Quetta," which was copied into the back of Ernest's *Places Visited & People Met (India 1888–1895)* notebook: "8 o'ck we get up and go into the sitting room for early breakfast by the wood fire. We have a cup of tea, toast, & oranges & bananas, then a hot bath and dress and go for a ride for an hour, then at 10.30 we walk over to the Club for breakfast of chops & omlette [*sic*] etc – with coffee. After which Ernest goes to his office till 4 o'clock, when we have tea, then go for another ride or walk till dark. Read till 8 dress for dinner at the Club. Turn in at 11.30."[67]

This account of a normal day at the office contrasted strongly with the rigours of work in the field. A case in point was recorded in the administration report of the Baluchistan Agency for 1895–6: "Two new lines were added during the year, one from Khanai Station, on the North Western Railway, to Hindubagh, a length of 44 miles, which was erected during the months of January and February 1896 in very severe weather by Mr. E.J.B. Hudson[68] and the other from Loralai to Duki, the head-quarters of the Assistant Political Agent, Thal Chotiali, a length of 20 miles, constructed in March by the same officer."[69]

The work was much the same as usual, involving thirty labourers, sub-inspectors Shevali Khan and Toola, and several dozen camels and/or bullock carts for distribution. The construction took less than four weeks, the last of which saw heavy snow that prevented work and made conditions miserable. These conditions persisted on the way back to Quetta when the line was finished: "Had great difficulty in starting off the men and camels who funked the deep snow on the kotal and bitterly cold wind. Walked all the way [15 miles] in my poshteen [fleeced sheepskin cloak] to keep warm the cold being very intense. Snow about 18 inches deep on road for the first 9 miles."[70] A photograph of Hudson's camel train is reproduced in figure 5.6.

Inserted into Hudson's diary for this period is a cutting from a 1919 newspaper reporting that Hindubagh had been attacked by raiding parties and the telegraph wires – his wires – were cut.

Marriage and the Second Half of Hudson's Career

After 1900, little fundamentally changed in Hudson's working life, although much changed in his personal circumstances.

Hudson met the young woman who would become his wife during his first furlough (October 1896 to April 1898). She was Sophy Ellen Elfrida Alington, known as Elfie, born in Aberdeen on 23 July 1874. During her stay, she and Hudson cycled, sailed, caught eels, and developed photographs together. Hudson records (with a touch of pride?), that after a thirty-two-mile cycle ride one day the party was "all pretty well tired out except Miss Alington." After the visit Hudson sent photographs they had taken and what was perhaps his first gift to her – a fitting one for an engineer – a bicycle wrench.[71]

They were engaged in August 1901 after some hesitation on Elfie's part. It may have been that Elfie was unsure about going out to India, for Ernest mentioned later that on a visit to Sir Alex and Lady Taylor, "Lady Taylor had a long chat with Elfie all about India."[72] Ernest returned to England again in 1902 for "special leave on private affairs," that is to get married. The wedding was a "high festival" in the local community, and the *Lincoln, Rutland and Stamford Mercury* devoted substantial space to its coverage.[73] Frank Cowley-Brown (CH 1889–90) from the Forestry program at Coopers Hill, was the best man. The article concluded with a lengthy list of guests and, as if that was not enough, gave a detailed inventory of the presents given by each guest – 208 of them.

By 1904 Ernest and Elfie had a young son, Thomas Alington Hudson. Fit though Elfie apparently was, she did not have an easy time in India, suffering from a mystery ailment that was variously diagnosed at typhlitis, stomach ulcers, or irritated intestines (or as the doctor jokingly suggested on 1 April 1904, another infant might be on the way). Elfie, Tom, and their *ayah* (nursemaid) accordingly spent time in the sanitorium town of Mahabaleshwar while Ernest

Figure 5.6. Ernest Hudson's luggage setting out from Rindli towards Quetta under armed escort, early 1895 (Hudson Collection, CSAS)

Figure 5.7. Indian Telegraph superintendent's inspection carriage and trusty bicycle, Southern Mahratta Railway, 1906 (Hudson Collection, CSAS)

was based in Bellary. Later, when he had the United Provinces of Agra and Oudh Division, Elfie, Tom, and their second son Adrian, spent as much time as possible in the lovely hill station of Naini Tal,[74] joined by Ernest at weekends and whenever he could get away from his office at Lucknow.

On the occasions when his family was away, Ernest found the bungalow "horribly dull, quiet and lonely ... giving one a feeling of restlessness + longing to be off on tour."[75] Having been promoted to superintendent in 1902, and now in charge of the Bellary Division, his work was less hands-on and more supervisory. He still managed to get out on tours every few months, typically travelling in an inspection carriage that was attached to the end of a convenient train and "cut off" in a siding whenever he needed it (figure 5.7). By 1907 it was equipped with an electric light.[76]

But Ernest did not enjoy the inactivity that inspecting by train entailed: "One soon gets sick of this form of travelling. Monotonous work shut up in a carriage all day and the only exercise possible an evening stroll down the railway line."[77] So, whenever possible, he took his bicycle as well and often cycled thirty or more miles in a day inspecting the line. He mostly relished these rides and chafed at his confinement to the inspection carriage. Even in 1904, after fifteen years in India, Ernest clearly still appreciated the wonder of the country: "Up at 4.30 and started off to Gund[a]lupet as soon as it was light enough. A long ride of 32 miles through the Tippacadu and Gundalupet jungles where there are large herds of wild elephants.... A pretty and interesting ride with changes of scenery from bamboo to teak and other jungle.... Saw many jungle fowl, monkeys and jackals as I rode along in the early morning."[78]

There were bad days too, however, such as the miserable road to Kankowli the following year: "A terribly hot and tiring journey of 26 miles in all. Arrived at 12 o'clock utterly done up ... unable to bicycle more than a few hundred yards at a time along the last 16 miles which were up and down steep hills deep in dust with ruts and boulders of rock – no shade trees and innumerable streams to cross. Mouth parched swollen and bleeding when I arrived."[79]

As divisional superintendent, Ernest Hudson now dealt with more office work, commenting ruefully in his diary on the number of mail bags (one to three) and telegrams (twenty to thirty) awaiting him after a day on the road. He also faced more difficult personnel issues. Ernest had received occasional criticism from his superiors in his younger days, once having been taken to task by the director general for what he considered to be "'neglect of duty' for allowing the battery at the [Coimbatore Combined Office] to fail and so cause interruption in the line for 3 days."[80] Now the roles were reversed, and Ernest had to maintain discipline. Examples of issues coming to his attention were a telegraph master who had taken on his mother as a servant at the department's expense,[81] the painfully slow construction progress made by Sub-Assistant Supervisor Campbell near Kanapur,[82] and investigating insubordination and letters of complaint by Signaller White against Telegraph Master Montgomery in Gorakhpur.[83] Perhaps the most sensitive of these issues was a case in which Mr. Debenham from the Calicut Office had been accused by the local school principal of "irregularities in the conduct of recent examinations" for entry into the Telegraph service – bribery in other words. Hudson was asked to investigate and spent time interviewing Debenham, Father Gonsalves the teacher, Signallers Mannan and Ipe, and most of the probationers. He concluded that simple facts had been "exaggerated into grave offences" and that Debenham was "clear of all blame in the matter."[84] A photograph from this period of men working on the telegraph line is shown in figure 5.8.

Towards the end of 1908, Ernest applied for privileged leave and furlough (he was approved for two years from March 1909). He also sought formal confirmation that he had completed enough years of service to be eligible for a "retiring pension." The examiner wrote back to say that he would have in fact just completed the requisite twenty years' service by the end of his leave and that he would be entitled to retire after that without having to return to India. Hudson did just that in 1911.

Ernest clearly optimized the timing of his retirement to maximize his entitlement to leave while retiring on a pension as soon as possible. He did not explain why, but it is probable that it was mainly for family. Elfie's health was

Figure 5.8. Telegraph line men at work near Almora, 1907 (Hudson Collection, CSAS)

likely to have been one factor in the decision and indeed she died less than ten years after leaving India in her forty-fourth year. Their son Thomas was at the age when he would normally have been sent back to England for school, so retirement avoided dividing the family. It helped that their families were sufficiently wealthy that they did not have to scrape by on Hudson's pension alone.

It is also probable that Ernest was not enjoying his work as much as he had in earlier years. He so obviously enjoyed being out in the countryside on his horse and bicycle, on tour or supervising construction, that the increasingly administrative duties that came with promotion were probably less and less rewarding. This situation would have been exacerbated by the recent unrest and imminent changes in the Telegraph Department.

Telegraph Growth and the Strike of 1908

While the essential tasks of a superintendent had not changed much, telegraph technology had been developing quickly, as had the Indian telegraph network itself. In his senior role with its greater responsibility for finances, Ernest commented several times on the telegraph traffic through various offices. At the Vengurla office in 1905, he remarked on the recent heavy traffic, sometimes representing as much as 90 messages per day per signaller, while down the coast at Nova Goa, traffic was 1000 messages per month higher than in the previous year. [85] Telegraph traffic was also dependent on the season, increasing during coconut harvest time when produce required shipping,[86] or during particular events, such as the trial of rioters in a conflict between village cultivators and planters at Motihari in 1908.[87] During the terrible famine in early 1908, traffic in the major hub at Lucknow where Ernest was based saw an increase from 3000 to over 4000 messages per day.[88]

Over the course of Ernest's career, more than 30,000 miles of telegraph lines were constructed and the number of offices more than doubled, from 940 in 1891 to 2127 in 1904 (almost all combined postal/telegraph offices). The number of messages sent likewise doubled, exceeding 7.3 million in 1904. In addition to the 200 or so people like Hudson responsible for the technical aspects of the telegraphs, the department employed 2300 signallers and telegraph masters and 8000 clerks and subordinates.[89] The telegraph had gone from being the tool of military and government authority to a routine means of communication for the public, businesses, and the press.[90]

By 1906 the volume was over 10 million messages and the Indian Telegraph Department was groaning under the load. "The organisation which was suitable in the infancy of the Department, and has proved itself equal to the care of its vigorous youth, cannot be expected without modification to meet fully the requirements of its maturity."[91] The Government of India decided it was time for a review of the department and established a committee of enquiry to examine all aspects of its operations. Its general goal was to improve service for the same cost to keep telegram rates competitive. In 1907 the committee duly reported its findings and made forty-nine sweeping recommendations. Among these were:[92]

- A reorganization of the administrative structure to devolve responsibilities previously held centrally to eight "circles," corresponding to those used by the Postal Service. The number of divisions would be increased from twenty to thirty-five and they would become the local administrative unit; subdivisions would be abolished.[93]

- The establishment of specialized Traffic and Technical Branches.
- Reorganization of the Signalling Establishment, with better training and recruitment as well as improved pay and conditions, including a reduced frequency of transfers between offices.
- Reorganization of the stores, accounts, and other administrative supports.
- Streamlining of telegraph services and simplifications to the signal office procedures to reduce delays.

One important recommendation that was enacted almost immediately was the long overdue appointment of women as telegraphists (and later as clerks).[94] The government showed a touchingly paternalistic concern for the women's welfare and family responsibilities. Candidates for employment had to be unmarried or widowed and provide an affidavit that they would be living in suitable accommodations, i.e., with parents, guardians, relatives, or institutions such as the Young Women's Christian Association. Moreover, "resignation would be compulsory in the event of marriage." This scheme was open to Indian women as well as Europeans, and the requirements and training would be the same as for the men. The working hours for women were restricted because it was deemed to be inappropriate for young women to be out alone in the evening.[95] Initially, however, the women "were not forthcoming in sufficient numbers to be of use," so the entrance age requirements were relaxed, and the annual salary increments made more attractive.

Despite some concessions, the Indian signalling establishment (mostly European or Eurasian) and the clerks and peons who supported them (mostly Indian) were concerned that many of their submissions to the Telegraph Enquiry had been ignored. Their dissatisfaction had been growing for some years as it became clear that changes were being contemplated by the government. As a result, a Telegraph Association was formed to represent the interests of signallers and clerks.[96]

The eventual implementation of the committee's recommendations was still two years away. In 1909, Eustace Kenyon, then deputy director of construction, was still concerned at the lack of progress, commenting that they had had questions asked in Parliament about the delay: "It was high time we showed the Govt. that we were sick of their slow procedure and wearied of the present state of things."[97]

In the shorter term, though, the uncertainties that resulted from the 1907 report were exacerbated by the appointment of John Newlands, a traffic expert from the British Post Office and an "invaluable advisor and practical reformer," to tour India for two years to investigate the operations of the Telegraph Department and recommend changes.[98] An early lecture by Newlands in

October 1907 was reported in the press, where he seemed anxious to emphasize the reasonableness of his initial findings.[99] Nevertheless, tensions grew in early 1908 as the Telegraph Association headed by a Mr. Barton at Rangoon reported publicly the details of their meetings with officials. The government responded by issuing stern warnings about that breach of confidentiality.[100] Other confrontations followed, including a deputation of 250 members of the signalling staff at Calcutta to the director general on 3 March 1908 demanding revisions to Newlands's proposed scheme of work (particularly on night shifts) and a general 25 per cent pay increase.[101]

As superintendent of an important division, Ernest Hudson was intimately involved with preparations for a possible strike. As early as 1 February 1908 he was monitoring the situation and sending weekly reports to the director general about the mood in the offices of his division: "Held a meeting of the senior men in the Telegraph Office to ascertain feeling of the men in regard to the prevailing unrest and tendency of the signallers to go on strike, also to find out what resolutions were passed at meetings held by the men at Lucknow. Everything seems quiet and although there is undoubted unrest throughout India and Burmah owing to the movement started by Mr Barton a telegraph master at Rangoon, the tendency here is not to implicitly follow his lead but to convene orderly meetings."[102]

Ernest was directly involved in implementing Newlands's changes, and he recorded the events from a unique inside perspective as "the new telegram forms and system of working came into force" on 1 April 1908. He spent an hour in the office before breakfast that day to see how the staff and public were handling the changes, "a few mistakes but things generally working smoothly." This was not the case elsewhere, and the following day he records that traffic was "hopelessly delayed,"[103] with large accumulations at Lucknow and all the big offices as signallers deliberately delayed traffic in protest. The director general broadcast a wire bringing these delays to the attention of the government and the press. Ernest duly took the letter to the manager of the local paper for publication the next day. Meanwhile he made preparations to bring in postal signallers "to replace departmental signallers should the men here give trouble."[104]

An important figure in the crisis was the W.L. Harvey, secretary for the Indian Department of Commerce and Industry, which now administered the telegraphs. He arrived in Lucknow on 5 April and met with Ernest the following day at Government House. He must have made a strong impression, for Ernest was motivated to express a highly unusual personal criticism: "He does not strike me as a particularly pleasant man to work with – a bully & absolutely unprincipalled [sic] in his methods I should say. Probably like the other men in [Commerce and Industry] keen on breaking up the dept. and amalgamating us with the post office."[105]

The next day, 7 April, Director General Trevor Berrington (CH 1876–8) paid an unexpected visit from Simla, spending time with Ernest in his office. They then met Harvey together. "The result of their discussion was a wire to the Supdts at [Bombay, Agra, Calcutta, Rangoon, and Karachi] to summarily dismiss 10% of the staff – a most unjust and disasterous [sic] order to my mind – tending further to wreck the dept, Harvey's ultimate object undoubtedly."[106]

Twenty-five signallers and a telegraph master were in fact dismissed at Bombay,[107] as well as others at Calcutta. As Harvey and Berrington left en route to join Newlands for a meeting with the Chamber of Commerce in Calcutta, Ernest noted that all the signallers at Rangoon were on strike, as well as most of those at Calcutta and Bombay, although "men in offices of my divn. remain steady so far." Ironically, the strike's spread across the entire subcontinent was facilitated by the rapid communications afforded by the very telegraph network at issue.[108]

In response to the strike, reputedly involving 1500 signallers, Berrington ordered Hudson to send twenty of his military signallers to Calcutta, leaving his division "badly crippled" because not enough reservists could be found. Berrington, Harvey, and Newlands met the Chamber of Commerce at Calcutta on 13 April 1908, and rumours of what was going on circulated on the wires. One was that Mr. Barton from the Telegraph Association had been invited to meet with the officials, which Ernest could "hardly believe."[109] Ultimately the strike was resolved when agreement was reached between a large deputation of strikers and the government. The signallers agreed to give Newlands's new system a five-week trial, in return for a pay increase.[110] The strike officially ended on 20 April 1908 when the workers signed a back-to-work declaration.[111]

As a postscript to Ernest's part in the affair, John Newlands visited him in December 1908 to go "into all matters connected with the new system of dealing with traffic."[112] Ernest enjoyed Newlands's company, remarking on his "interesting accounts of his travels through India." They also spent two days sightseeing, or "globetrotting" as Ernest called it.

Ernest's strongly negative reaction to Harvey of the Department of Industry and Commerce, and his impression that the authorities were out to "wreck" the department, are likely to have factored into his decision to retire. He clearly felt that the department as he knew it could not survive much longer. A resistance to such changes was perhaps a natural consequence of the esprit de corps that Coopers Hill fostered.

Education

Ernest Hudson's diary offers a unique insight into the mechanics of constructing and operating the late Victorian telegraph system in India. Of particular interest are the logistics of transporting supplies, laying out the line, erecting the posts, and stringing the wire. Hudson coordinated with other engineers, especially those on the railways, but for much of the time he was responsible for his own affairs, arranging support services as best he could in the circumstances. He worked closely with his team of experienced Indian workers and clearly trusted and respected his senior staff.

Much of Hudson's work was concerned with what might be called "quality control," ensuring that all who were involved with the practical operation of the lines did their jobs properly. One facet was his tours of inspection, reviewing the work of the line men and sub-assistant superintendents in early days, and of the subdivisional officers later on. The other part concerned the administrative, reporting, storekeeping, and accounting functions required to keep a large organization working. In addition, line construction required elementary surveying, cost estimation, attention to detail, and a great deal of logistical planning and professional know-how.

Hudson, like his colleagues, really embodied the qualities needed by the ideal engineer in India, "who will regard an offer of a commission from sub-contractors as a deadly insult; who can keep accounts like a bank clerk, and who, with all that, has a special professional pride."[113]

Despite it taking him to interesting places, each with its particular technical and practical challenges, the basic tasks of telegraph officers did not, at least in the early days, need an especially advanced engineering knowledge. To be sure, they needed experience and skills, but the scope was limited in comparison with the variety of civil engineering projects. This accounted for the relative unpopularity of this service at Coopers Hill that enabled Hudson to obtain his appointment.

Until around the time Hudson received his diploma, telegraph technology was fairly straightforward, and the challenges were generally implementational rather than theoretical. In this context, the education provided by Coopers Hill in physics, mathematics, construction, and elementary electrical engineering was sufficient when supplemented by the practical education he received in India. As the technology grew more complex, however, changes would become necessary in the Coopers Hill curriculum, as discussed later.

Accounting

Engineers in India were in general responsible for keeping their projects on budget, purchasing and managing materials and inventory, paying wages to labourers, working with contractors, supervising staff with budgetary responsibilities, and reporting on financial matters to their superiors. Some years previously, more responsibilities had been passed to engineers to do their own accounting, instead of submitting all requests to higher authorities, in order to eliminate delays and so make the public works system more efficient. The

Indian system could be complex, different from region to region and project dependent, so as Edmund du Cane Smithe (CH 1872–5)[114] – one of the *Oracle's* founding quartet – observed, "Public Works Accounts as they are in India, and the forms of Public Works Accounts, … is very desirable for a man to know something of before he goes out." Frederick Hebbert (CH 1871–4)[115] went still further, saying, "Finance is the soul of engineering, and I have found the teaching of Indian Accounts at the College to be of great value." After many years in India, these two members of the early classes from Coopers Hill chose to highlight the accounting course as one of the most useful aspects of their training.[116]

On the basis of his experiences in the Indian accounting branch, George Chesney introduced a compulsory accounting course into the Coopers Hill curriculum from the very beginning, which he initially taught himself.[117] The course first covered first principles and then mercantile, banking, and government accounts. There was also extensive material related to accounting in the Indian Public Works Department, including topics on contractors, departmental stores keeping, engineers' accounts, and auditing.

When Chesney left the college, James Hurst was appointed as a part-time lecturer to give a series of fifteen accounting lectures on Saturday mornings to students in their final year (the third year for engineers, the second for Telegraph students in Hudson's time). Hurst continued in this capacity until he was dismissed in 1901 at the age of seventy-two. Without Chesney's intimate first-hand experience, the course seems to have become less relevant, even though Hurst originally based his teaching on Chesney's notes.[118] During the re-evaluation of the Coopers Hill curriculum in 1901, Hurst was criticized for focusing too greatly on commercial accounts and not enough on the practices used in India. Hurst believed that his approach was appropriate because only a minority of students in the class went to India, and he argued to no avail that applying their knowledge to PWD accounts if necessary was straightforward.[119] Ernest Hudson's education in accounting, received relatively soon after Hurst's appointment and still bearing Chesney's stamp, was probably reasonably well tailored to his needs in India.

Coopers Hill placed great emphasis on surveying, sketching, and geometrical drawing, so Hudson would have been well prepared in these subjects. He often made sketches of the telegraph lines and sometimes drew the bungalows he stayed in. He also mentioned in his diary helping to prepare a map of the line near Bilaspur and technical drawings in the workshops.

On numerous occasions, Hudson was asked to prepare estimates for telegraph construction or maintenance projects. These were then "sanctioned" by his superior officers and given a serial number and year, by which the entire project would be identified. Hudson did not often record the value of these estimates, but in 1892 he states that he received the sanctioned estimate Rs3050 for construction

Table 5.1. Course of Instruction for Telegraph Students, RIEC Calendar, 1884–5

First Year	Hr/wk	Second Year	Hr/wk
Descriptive Engineering	1½	Applied Mechanics	7
Geometrical Drawing	8	Chemical Laboratory	3
Surveying (partly in the field during 1½ terms)	8	Mathematics (during 1½ terms)	3
Freehand Drawing	2	Applied Mathematics	4
Chemistry	2	Mathematical Physics	3
Chemical Laboratory	3	Telegraphy	7
Physics	1	Telegraph Construction	10
Mathematics	13	Signalling	2
Geology (during 2 terms)	2	Accounts (during 2 terms)	2
Architecture (during 1 term)	2	Physics (lectures during 2 terms)	1
French or German	1½		

of the Patchur-Krishnagiri Line.[120] The only other instance where he attached a value to an estimate was when he needed approval from the divisional superintendent for overrunning by Rs500 the estimate for reconstructing the line from Mandalay to Maymyo.[121] On completing construction in the Bolan up to Panir, he notes with some pride that his estimate was out by just thirty yards in a twenty-three-mile line.[122] Quantity estimation was an important skill for all branches of engineering and was covered in the third year for engineering students; telegraph students covered this material in their course on telegraph construction.

Telegraph Course

Telegraph students shared their first-year courses and some from the second with the engineering students, so their curriculum was as shown in table 5.1.

The contents of the four specialized courses were:

- *Telegraphy:* batteries, earth plates, instruments, lightning dischargers, fault testing, principles of duplex and quadruplex operations[123]
- *Telegraph construction:* materials, soldering, surveying tools, heavy equipment, construction, estimation, land lines and cables, fitting up offices, maintenance
- *Workshop:* disassembling and repairing instruments, constructing and testing a short telegraph line
- *Signalling:* use of Morse ink-writers and sounders

For Hudson's work in India, this appears to have been a useful set of topics. When the Telegraph course of study was introduced in 1878, there was a dedicated technical instructor, Mr. James Duffey.[124] But when he died, he was not replaced and the Telegraph courses were taught by the members of staff responsible for physics and electricity (Prof. William Stocker, and Mr. Thomas Shields, respectively, in 1901), who had no first-hand experience in telegraphy. This may have seemed prudent with only a handful of Telegraph students each year, but as the college moved into the 1890s the equipment and curriculum became increasingly outdated, especially in the newer technologies deployed in India at the turn of the century.[125]

This state of affairs led two sitting directors general of the Indian Telegraph Department to express their concerns about the poor preparation of the Coopers Hill students they received: Charles Reynolds in 1895 and Charles Pitman in 1899. Their communications were included in the 1901 report by the college's Board of Visitors on remodelling the Coopers Hill curriculum. While Reynolds wanted men educated at home to "set a good example in pluck, endurance, and conscientious work," he also wanted those with an interest and education in electrical engineering. This was unlike the current Coopers Hill men in whom there was generally "an absence of anything like enthusiasm in the scientific branch of their work." He went on to say, "Coopers Hill is not a recognized school of repute in connection with advanced telegraph engineering."[126] In other words, technology was changing but Coopers Hill was not.

Pitman agreed and presented detailed suggestions for revising the curriculum, with the assistance of M.E. Nigel-Jones (a colleague of Hudson's), and input from four other Coopers Hill men. Not surprisingly, the Telegraph Construction course came under strong criticism, as it "in no way" applied to India. Likewise, a much stronger training in signalling was called for. Reasonably enough, the consensus was that the Coopers Hill course should employ the posts, wire, tools, instruments, and techniques actually used in India. In addition, a greater emphasis on electrical technology was needed, as well as reference to up-to-date textbooks and the Indian Telegraph Construction Code.[127]

Apparently, the practical part of the Coopers Hill syllabus involved woodworking larch tree trunks to make the supports for a three-post, two-wire telegraph line, which was then adjusted, tested, and analysed. With a knowledge of Hudson's duties, it is easy to see why the experienced officers felt this to be woefully inadequate.

The outcome of the curriculum remodelling for the Telegraph program was that it was made a full three-year program, with enhancements to the existing courses, the addition of a Telephony course, an increased emphasis on electrotechnology, and a specialized Telegraph workshop. The new curriculum included both the theory and practice of insulators used in India, the Indian wire gauge for iron wires, Siemens posts, batteries, and the surveying and construction of lines. A greatly increased emphasis on practical instrumentation was also added.[128] A new assistant instructor in telegraphy, a Mr. Edgar, was hired to supervise laboratories. To address the persistent impression that the telegraph men were inferior to the engineers,[129] their salary was increased to the same level once the new third year came into effect.

Reynolds and Pitman were interviewed again during the 1903 enquiry into closing the college. Reynolds was satisfied with the revised training, saying it was "so much improved that I think it is quite up to our standard." Pitman also felt that all his recommendations had been implemented and that the training was now "thoroughly satisfactory."[130] Unfortunately, these improvements came too late for the college.

Keeping Up to Date

The foregoing describes how Coopers Hill was forced to update its Telegraph curriculum, but how did officers in the field keep up to date on the latest developments? There is no evidence in Hudson's diary to suggest that he attended any structured professional development beyond his initial training, but he, and others in the service, must have learned new technologies and practices in some way.

Until the recommendations of the Telegraph Committee to create a specialized technical branch were put into effect, the main custodian of technical affairs was the workshop in Calcutta. It was responsible for quality assurance on routine supplies, but it also conducted tests and experiments on new technologies. For example, Hudson helped test some "new lead-covered wire which had come over from Germany" at the workshop soon after his arrival in India.[131]

Technical information was formally disseminated to the divisions in three main ways: revisions to the Telegraph Construction Code, technical circulars, and Technical Instructions. Together with the Diagrams of Construction Stores, the Construction Code was the standard to which telegraph lines should conform. Apparently, these were updated frequently because, while everything was still fresh to him, Hudson made the off-hand comment that the day's mail had brought "the usual monthly budget of corrections to the Code."[132]

The department also prepared and circulated "technical data sheets" to assist people when installing and troubleshooting specific instruments. For example, a note in Hudson's 1905 field notebook reminds him to "obtain from office a copy of Circular No. 2749 dated May 2, 1903 on subject of 'Western Electric Pattern Telephone Instruments.'"

The department also periodically issued Technical Instructions. The preface, written by Henry Mallock, instructed officers that all the pamphlets were "to be neatly pasted into a strongly bound file with leather back and stiff mill-board covers" because they represented "a great deal of labour and thought" and "considerable expense in printing." The survival of Hudson's copy of these instructions, dated 1888, is a testimony both to the wisdom of the orders and to his adherence to them. These Technical Instructions contained circuit diagrams in colour, theoretical explanations, practical advice, and any departmental restrictions. Updates to a particular topic were given the same "chapter" number so the reader could readily identify and review a related series of instructions. In many places Hudson had pasted an updated paragraph over the old ones or attached additions in the margins.

Even so, Eustace Kenyon still felt he needed to buy a book called *Examples in Electrical Engineering* when he took over the workshops in 1898. He commented that since he had left the Construction Branch, he was beginning to find himself "most frightfully behind the times in the electrical part of my profession which is a bore when in charge of shops where all the electrical instruments are made up."[133]

We must also expect that a lot of information was transmitted by word of mouth when officers visited each other or when senior officers came on tours of inspection. There is also some suggestion in Hudson's diary that specialists visited the divisions on occasions to provide specific technical advice.[134]

After India

Ernest Hudson's life on the telegraph lines in India is informative, not because it was unusual but because it was typical. In his twenty years' service, he saw a wide range of fascinating places and had wonderful experiences. His diary provides a rare insight into the professional and family lives of the Europeans, and to some extent the Indians, who built and operated this critical and ubiquitous technology. Their way of life has vanished now, and the last Indian telegram was sent in 2013. Diaries like Hudson's are therefore one of the few ways we have of reconnecting with their world.

The last of Ernest's diaries in the public record is from 1908, so information thereafter is sparse. He and Elfie had a further two children, another boy and a girl, and they have descendants alive today on both sides of the Atlantic. For a time, they lived in Cornwall with his brother Herbert before moving into the house in Dorset where Elfie lived until her death in 1918 and Ernest would live for another thirty years after that. A short announcement in the *Times* reported Ernest Hudson's death in 1952:

> HUDSON – On Jan.10, 1952, at Cliveden, Ferndown, Wimborne, Dorset, peacefully while resting after his usual day about the garden, JAMES BONNELL ERNEST HUDSON, aged 85.[135]

6

Jungle Wallahs

Forestry

Indian Forestry

Forests were, and are, a major element of India's landscape and economy. In 1901, approximately 22 per cent of the area of British-ruled India was classified as forest, representing more than 200,000 square miles. The distribution and nature of these forests of course depended on the local latitude, elevation, rainfall, and soil conditions, resulting in a staggering variety of habitats and species.

As is so often the case with natural resources, the broad story of the Indian forests under British rule was one of over-exploitation, belatedly followed by over-regulation. In the first half of the nineteenth century the East India Company, merchant companies, and local rulers effectively had free rein over the exploitation of forests, particularly harvesting the valuable teak trees whose natural oils made the timber highly durable. Teak was particularly sought after for shipbuilding, principally because of its superior lifetime, but also because the British Isles itself was struggling to supply adequate trees.[1]

High demands for teak, combined with ruthless harvesting and inefficient usage (perhaps only 30–40 per cent), soon led to unsustainable felling, several thousand trees per year being cut down in the Cochin region alone at the end of the eighteenth century.[2] In his three-volume history of forestry in India, Edward Stebbing, graduate of Coopers Hill in 1893, described the early British forest industry in India: "The waste in exploitation was appalling, but the material was still in existence within reasonable distance of the markets, and no question of shortness in supply arose between the Home or Indian Governments. Nor did the matter apparently ever cross their minds, the forests being regarded as inexhaustible."[3]

Part of the issue was that, because Britain effectively had no state forests left at home (although there were many in private ownership),[4] there was little government appreciation of the principles of sustainability or scientific management of forests. Stebbing also blamed successive invaders of India before the British for pushing destructive slash-and-burn agricultural practices ever further into the forested regions of the country. Stebbing particularly mentions the deforestation of the Anaimalai region to the immediate north-east of Cochin, where a young surgeon named Hugh Cleghorn recorded the destruction of teak forests due to the *kumri* shifting cultivation system.[5] Cleghorn also noted the damage caused by tree-felling for railways sleepers, estimating that in the Madras Presidency alone 35,000 trees per year were needed for the purpose.[6] To this was added the clearing of forests to make way for increased agricultural land during the early decades of British rule. Whatever the state of the forests before British occupation, their condition worsened substantially during it.

As the nineteenth century progressed, the nature and scale of the unsustainable practices were becoming apparent, but the political will to interfere in the free market remained absent until Lord Dalhousie's administration. In 1852, the province of Pegu in Burma was annexed by the British. This was one of the prime teak-producing regions and had been suffering the same over-exploitation as India. The teak forests in Pegu had been controlled by the royal family, so the British government took this opportunity to claim the same right of ownership for the state. A superintendent of forests, Dr. McClelland, was appointed to review and report on the state of the forests. On the strength of this report, Dalhousie in 1855 affirmed state ownership of the forests and laid the principles for the government, not merchants, to control teak extraction. In India, the appointments of 1847 of Alexander Gibson as conservator of forests in Bombay and of Hugh Cleghorn to the equivalent post in Madras in 1856 also indicated a change of thinking.

The leading proponents of what was termed "scientific forestry" at the time were the Germans and the French. When McClelland resigned, he was replaced in 1856 by a German, Dietrich Brandis, who was the first man to bring a formal scientific background to forestry in British India, albeit as a botanist. Brandis was to spearhead the implementation of Dalhousie's policies, against strong opposition from the established interests: "Brandis' early struggles in Burma are but one instance of the contest against greed, ignorance and short-sightedness, for his efforts at introducing sound principles of organization and protection met with a storm of opposition from all sides, and particularly from mercantile firms engaged in the timber business."[7]

After the uprisings of 1857, the Government of India took direct control of the forests, and in 1864 Brandis was appointed inspector general of forests, with Cleghorn to assist him.[8] The first two appointments to the fledgling

department were also Germans, Wilhelm (William) Schlich and Berthold Ribbentrop. Both men in turn went on to succeed Brandis as inspector general
and, as chapter 3 described, Schlich was later seconded from the Forest Service
to establish the new Forest School at Coopers Hill. There was clear need in
the service for officers with a specialist knowledge of scientific forestry. Accordingly, Brandis obtained the approval of Lord Salisbury, then secretary of
state for India, to educate British men in forestry in Germany and France, and
from 1875 onwards solely at French National School of Forestry in Nancy. This
system remained in place from 1866 to 1886, after which recruitment was exclusively from Coopers Hill.[9]

Forest Legislation

By the time the graduates from Coopers Hill joined the service, the legal framework for the governance of Indian forests was the Forest Act of 1878, which
provided a clear legal framework for forest conservancy and the powers of forest officers. The first such legislation was the Indian Forest Act VII of 1865. It
contained the provisions for establishing "Government Forests," "meaning such
land covered with trees, brushwood or jungle" that local governments decided
should to be protected, "provided that such notification shall not abridge or affect any existing rights of individuals or communities." It empowered the local
governments to make rules on the preservation of designated forests and their
products, the authority of forest officers, and the penalties for infractions. The
definitions in the Act of the permissible scope of local rules and the range of
forest products they covered were broad:

> The preservation of all growing trees, shrubs and plants within Government For-
> ests or of certain kinds only – by prohibiting the marking, girdling,[10] felling and
> lopping thereof, and of all kinds of injury thereto; by prohibiting the kindling of
> fires so as to endanger such trees, shrubs and plants; by prohibiting the collect-
> ing and removing of leaves, fruits, grass, wood-oil, resin, wax, honey, elephants'
> tusks, horns, skins and hides, stones, lime, or any natural product of such Forests;
> by prohibiting the ingress into and the passage through such Forests, except on
> authorised roads and paths; by prohibiting cultivation and the burning of lime or
> charcoal, and the grazing of cattle within such Forests.[11]

An intrinsic flaw in the 1865 Act was that the binary division between open
and protected forests did not contain any grey area in which local customs
and private rights could be balanced. Moreover, the Act did not deal with
privately owned forests, which in many important regions were so interlaced
with state-owned tracts as to render them effectively inseparable. The 1878

Act therefore attempted to combine the assertion of the government's absolute right to the forests with the local villagers' customary, and necessary, usage of forest resources.[12] It therefore introduced the nomenclature of three types of forests:

Reserved Forests: Identifiable forest tracts, close to transportation and suitable for sustainable exploitation, with very restricted rights;
Protected Forests: Government controlled forests of potential commercial value, whose access rights were investigated and recorded, and which could be converted to reserved status;[13]
Village Forests: An allocation designed to protect the rights of villagers to certain forests, but which was not employed in practice due to a cumbersome procedure.[14]

All of this represented a significant curtailment of the traditional ownership, rights, and usage patterns of the Indian population. Colonial authors such as Stebbing, Ribbentrop, Troup, and others regarded these measures as an absolute necessity to preserve the forests and to bring them under a sustainable system of harvesting. Starting in the 1980s, environmental historians took issue with this view, observing that the forests needed protection only because of British commercial extraction, and that colonial revenues were being protected at the expense of the Indian population.[15]

However, the reality was complex. In order to ensure a reliable and predictable supply of forest goods in the long term, the Forest Service focused on sustainability, and managed forests and new plantations.[16] But it was also recognized that forests played other roles, which were captured in 1894 when the Government of India issued what was effectively the country's first forest policy (circular no. 22-F) by defining four broad categories:[17]

Forests the preservation of which is essential on climatic or physical grounds.
Forests which afford a supply of valuable timbers for commercial purposes.
Minor forests, chiefly supplying fuel and fodder or grazing for local consumption.
Pasture lands that were declared as forests under the Act as a convenience in order to settle the rights of the State and private individuals or communities

With regard to villagers' rights, the local implementation of forest policies meant that "most colonial forest interventions were often blunted and reshaped at the ground level by a host of local forces and particular interests." These practices were facilitated by the flexible nature of the 1878 Forest Act and by the pragmatic need to render such policies workable at the local level.[18] Former Coopers Hill student Anthony Wimbush described the working arrangement in the early twentieth century:

> Grazing in the forests was always one of the Forest officer's problems. It was, definitely, bad for the forests but yet the legitimate needs of the villagers had to be met. The browsing and grazing of goats was absolutely prohibited since goats and forestry are quite incompatible. Cattle, however, were necessary for the villagers. Their land outside the forest was all cultivated and there were no grass fields so, from time immemorial, it had been their custom to send their animals into the forest to graze in charge of one or more small boys. In order to regulate this grazing to some extent and in order to make some revenue the forest department used to sell grazing permits for a very small fee per head of cattle. These permits had to be shown to any forest officer on demand so that he could check the number of cattle grazing. The graziers usually carried their permits in a hollow bamboo tube which they always had to take with them to the forest.[19]

With this background in mind, we now turn to the questions of the lives of the forestry graduates from Coopers Hill in India, what work they did, and the role their education played in preparing them for their careers.

Life in the Forests

Forest Writings

Rudyard Kipling is famous for his tales of India, its peoples, animals, and life. He also turned his observant eye on the professions and showed a deep understanding of the spirit, and in many cases the substance, of engineering and other technical occupations in India. In his story "In the Rukh," during which a forest officer met some wolves and the adult Mowgli (in his first appearance), Kipling captured the essence of the Indian forester: "He spends much time in saddle or under canvas – the friend of newly planted trees, the associate of uncouth rangers and hairy trackers – till the woods that show his care in turn set their mark upon him, and he ceases to sing the naughty French songs he learned at Nancy, and grows silent with the silent things of the undergrowth."[20]

The life of a forest officer appealed to a particular breed of men, one more at home in the forests than in the office, that was summed up by their nickname of the "jungle wallahs."

Consider the following description of the *sal* forest: "The general effect is beautiful beyond words, the long valleys, flanked by green slopes, lead the eye over the sea of thickly packed stunted shrubs, the sun shining on the young leaves. Poplar-like sal trees stand as sentinels over it all, some forming little promontories into the valley where slightly higher ground runs from the sides of the hills halfway across the lower ground. The picture of these verdant valleys is framed by dark walls of the high forest along the edge of the hills."[21] And contrast it with the same author's opinion of office work during the rainy season: "In Balaghat I learnt more of the mysteries of office work, which seemed to me to consist mainly in signing letters written by the Babus, checking figures on forms and trying to find suitable answers to objection statements. None of it was thrilling."[22] These observations were written by James W. Best (CH 1901–4) about his work in the early twentieth century.

Several Coopers Hill foresters produced memoirs of their lives in India; Best himself wrote four books, Edward Stebbing (CH 1890–3) the same (not including his monumental *Forests of India*), and Archibald Dunbar-Brander (CH 1896–9) another. A strong vein of romanticism runs through their descriptions of life in the jungles and forests,[23] but a prime motivation for these authors was to describe their adventures on the *shikar*, the hunt for game. The pages of their works are crammed with stories of killing panthers, leopards, tigers, bison, deer, and other forest creatures. To a modern reader, they tend to become repetitive and rather disheartening. It has been argued that the hunting ethos cultivated by members of the Forest Service was an integral part of their identity, serving to distinguish them from members of the civil service through this test of "muscularity, manliness and morality"[24]

Sources used here for understanding the experiences of forest officers include Best's *Forest Life in India*, an interesting and wide-ranging account of his career that talks of his duties in the Forest Service,[25] and Stebbing's *Diary of a Sportsman Naturalist in India*.[26] As the title suggests, the latter is an intriguing mixture of adventure stories and conservationism, activities that Stebbing did not consider to be incompatible. To these published works are added as sources the unpublished memoirs of Anthony Wimbush, written in 1964 when he was seventy-eight,[27] and the diaries of Lionel and Selina Osmaston. Lionel served as a forest officer in the Bombay Presidency from 1891 to 1911.[28]

Lionel's two brothers, Bertram Beresford Osmaston, a contemporary of Ernest Hudson, and Arthur Edward Osmaston also passed through Coopers Hill and joined the Indian Forest Service.[29] Extracts from Bertram's diaries were published after his death by his son;[30] again, these focus heavily on the *shikar*.

Lionel Sherbrooke Osmaston

Lionel "Lil" Osmaston was born into a family of fifteen on 20 October 1870. He was educated at Cheltenham College (as was Bertram) and then at Coopers Hill from 1888 to 1890. Like his brothers, Lionel was a good student, finishing top of the forest students in his first and second years and winning the President's Scholar in Forestry in 1890.[31] He was also a keen rower but was thought to have developed a weak heart while rowing at the college,[32] a condition that was later to send him home from India on sick leave. Together with ten of his classmates, he was appointed to the Indian Forest Department (as it was then) as an assistant conservator, third grade. This entry-level position placed Lionel beneath two further grades of assistant conservator, four grades of deputy conservator, and three of conservator. Collectively, these constituted the so-called controlling establishment.

All "natural-born British subjects" were eligible for appointment to the controlling establishment, provided they remained unmarried before reaching India. In practice, however, very few were Indian, and Coopers Hill supplied none. A notable exception was Framji Rustomji Dasai, the son of a Bombay merchant, who competed successfully for Brandis's European training scheme in 1866,[33] was trained at Nancy, and ultimately rose to be conservator for the Northern Circle of the Bombay Presidency.[34] After 1900, however, an increasing number of Indian names appear in the official gazette entries printed by the *Indian Forester*.

Lionel Osmaston was posted to the Bombay Presidency, which was in 1892 divided into four administrative circles – Northern, Central, Southern, and Sind. Each was in turn divided into divisions led by divisional forest officers. Forest divisions were further subdivided into ranges, managed by forest rangers (or range forest officers), and then into beats that were protected by foresters, assisted by forest- or sometimes beat-guards. These last three types of position constituted the subordinate establishment and were predominantly held by Indians.

Lionel's contract with the secretary of state for India, dated 21 November 1890, stipulated that he would "at all times employ himself wholly, efficiently, and diligently under the orders and instructions of the Government, and of the Officers who may be placed over him, in the improvement, care, and general management of the forests, and in the formation of nurseries and plantations; and he ... will, to the utmost of his skill, instruct and train in forestry and all duties connected with the management of forests, the natives and others who may be placed under him, and as he may be required by the Government." Furthermore, he was required to "diligently study those Native languages which may be prescribed by the Government," and warned that "no rise of pay, or promotion, will take place previous to his passing such examination."[35]

Having already passed the vernacular examination in Marathi in November 1891, Lionel spent the first three months of 1892 preparing for the Higher Standard language exam while on tour, seeking old exam papers and asking a clerk to look over his language exercises after hours. It proved challenging to find time for studying the language, and on more than one occasion he complained that because of office work he was unable to study his Marathi. But a few weeks before the exam Osmaston had the chance to practise his language, because on 15 March 1892 he received "visits from the *Mahalkeri*,[36] Chief Constable and *Talathi*[37] all coming to pay their respects so as they none of them could speak English my Marathi was aired." He travelled to Bombay to sit the exam in April, being tested in Marathi-English and English-Marathi translations on the first day and the *viva voce* on the second. He "did fairly well" and knew he had passed "so now never another exam in my life unless I get transferred to Canara or somewhere,"[38] which would have entailed a test in a new language.

Lionel learned from Shuttleworth,[39] his superior officer, that he had passed the exam with credit, and he received an official certificate to this effect. This paved the way to his promotion to first grade of assistant conservator of forests, approved on 28 June 1892 and backdated to the language exam. His notification noted that "by the conscientious, intelligent and industrious discharge of his duty he has rendered approved service meriting promotion."

L.S. Osmaston's Early Work

Lionel had arrived in India towards the end of 1890, and for his first year he was posted to several locations in both the North and South Thana Divisions. If his experience was similar to others', he spent some of it learning the ropes from a more senior officer. James Best went so far as to write about his early service, "So far in my service I had done little but combine amusement with instruction. Indeed, I was doing so little that was of any use that I felt that I was not earning my pay. This feeling lasted some time. An officer is really very little use until he has learnt the language of the country and the ways of the people in it."[40] Bertram Osmaston was also dismissive of his training in 1888, saying that after "a fortnight's instruction in the method of marking trees for felling" he was left alone with four Indian servants who could speak no English.[41]

This account is broadly consistent with the period in 1896 that Lionel spent supervising Edward Hodgson, a fresh arrival from Coopers Hill. Lionel commented that Hodgson was "to go about with me & learn the business: just like his brother."[42] The brother in question was Lionel's good friend (to whom he occasionally lent quite large sums of money) and Coopers Hill contemporary Charles Hodgson (known as "Hog").

Edward had arrived in India in time to accompany Lionel on tour.[43] Although the regulations were clear about the necessity for proficiency at horse riding, and Coopers Hill took this seriously,[44] Hodgson was not a good horseman. They left a few days later on 4 December, heading south from Poona through the Katraj Tunnel with Hodgson riding Bedouin, one of Lionel's horses. One can sympathize with both men for different reasons when Osmaston wrote in his diary, "Hodgson v. sore & rode awfully slowly so didn't get in till late." In the morning after they inspected a nearby forest, the day went no better: "Hodgson had eno' riding yesterday! So he walked: forgot his topi & so had to go back early."

It is hard to tell whether Lionel was amused or frustrated when the saga continued: "Left Hodgson to come on slowly as he *can't* ride: he let Bedouin lie down in the water while crossing a stream! He can't *bear* riding & has now sent to Poona for his bicycle."[45]

One can only imagine the young man's mortification. He may also have faced administrative difficulties. Best commented that every forest officer "was expected to keep at least one horse and found difficulty in having his travelling allowances passed unless he did."[46] Nevertheless, Hodgson thereafter rode his "byke" from camp to camp by the direct road while Lionel on his horse took shortcuts and stopped to inspect forests along the way. At this point Hodgson received the horse he had bought before leaving Poona, but he still preferred the bicycle. On the positive side, Lionel admired Hodgson's shooting, saying, "He is a grand shot: he shot 12 pigeons out of 13 in eve[ning] all flying shots."[47] Despite this rough start, Edward Hodgson remained in the Forest Service and became chief conservator of the Bombay Forest Department in 1927.[48]

Returning to 1892, Lionel was promoted swiftly to divisional forest officer. In commenting on this progression, Shuttleworth wrote, "Mr. L. Osmaston, who has been in charge of South Thána, is a conscientious and very hard-working officer, and although young in the service, to be selected for such an important administrative position, still having qualified by passing the vernacular and departmental examinations, the Conservator recommended him for the appointment of Divisional Forest Officer, and he has proved a great success, fully justifying his preferment."[49] Unfortunately, Lionel's zeal had taken its toll on his heart, and on 21 July 1892 Lionel was cleared by the Medical Board at Bombay to leave for England the following day.

Working Plans

When Lionel returned from sick leave in late 1893, Shuttleworth gave him responsibility for the preparation of forest working plans for the newly formed Central Circle of Bombay. Working plans were drawn up to provide a systematic management strategy for a particular forest, local conditions, and purpose. James Best gives an example:

> We had to allow the public access to material for house building and firewood, and arrangements were made that the villagers always had some fellings available for their supplies within reasonable distance of their homes. So something like thirty felling series were formed, each being divided up into thirty coupes of roughly the same area or yield, so that every year the same amount of timber and firewood would be on the market in each area served. The constant yield has a wonderful

effect in stimulating a market. Timber merchants know where they are and can make their plans for years ahead. The advantages are obvious.… The main principle of forestry is to fell only as much as grows in a year so as not to encroach on capital. Thus, if a fully stocked wood is renewed in thirty years, then it is safe to cut one-thirtieth of the timber each year.

When the system was fully implemented, the forest therefore comprised "a complete series of trees or crops of all ages from the seedling to the mature tree."[50]

Although many years had passed since the establishment of the Bombay Forest Department, relatively few forests were actually under working plans. At the end of 1897–8 there were working plans for only 17 per cent of Bombay's 14,912 square miles of reserved and protected forests, with a further 24 per cent under development.[51] In Lionel Osmaston's Central Circle when he started in the Working Plan Office, the portion of forest area with a working plan was less than 10 per cent.[52] At least part of the reason this figure remained so low was that an enormous amount of preparatory work was needed before meaningful plans could be drawn up – including designating the forests and settling rights, surveying and demarcation, inspection of the age, state, local conditions, etc. In the Bombay Presidency, work on the demarcation of forests continued into the 1890s. Once the rights to particular forests were settled and they were designated as reserved or protected (or less frequently, delisted), the Forest Survey branch mapped and recorded the boundaries. Boundary markers were erected so that people would know what regulations might be in place. In the Central Circle alone, 28,000 such markers were erected along 1730 miles of boundary in 1892–3. As the work was completed, this number dropped to 13,000 in 1895–6.

The scale of the work was huge, especially in this region of many small forest areas, consisting of hundreds of compartments (433 under Lionel's jurisdiction),[53] each of which may have had numerous felling zones, or *coupes*. James Best again: "The division of my eight hundred square miles into thirty felling series of thirty coupes each was a big job. It meant dividing the forest up into nine hundred different coupes each yielding approximately the same amount of timber and firewood. Natural boundaries had to be made use of so far as possible."[54]

Lionel's work with the Working Plans Division continued through to mid-1896, when he was replaced by Mr. Fry, and the two officers worked together on the plans during the rainy season of that year. The administration report summed up Lionel's activities for 1894: "Mr. Osmaston accompanied the Conservator on tour … inspecting forest blocks and compartments under exploitation, gathering local knowledge, and collecting information and data for the preparation of working plans.… Mr. Osmaston then marched towards Poona inspecting the teak forests of the Ahmednagar and Poona Divisions by the way, and he likewise visited the ever-green forests on the Ghats in the Mival Range of the Poona District."[55]

Early the following year Lionel commented on discussions with Shuttleworth and Hornidge, a fellow deputy conservator, about the working plans under preparation. Lionel continued to be the sole member of the Working Plans Division throughout 1894, with responsibilities for the East Khándesh, Násik, Ahmednagar, and Poona Divisions. He performed his duties with "great industry and zeal."[56]

Lionel then went on a short leave to England, during which he and Selina "Lina" Fanny Plumptre,[57] his first cousin, were engaged. When he left her in London on 12 November 1896, "she couldn't have been braver saying goodbye" whereas he wrote, "More miserable than I have ever felt in my life."[58] Once Lionel returned to India he was, with a few special exceptions, to spend the rest of his career in charge of various divisions, including Poona, Násik, and West Khándesh. Shortly before his retirement in 1911, he was promoted to conservator, based in the Northern Circle. Despite his progress through the ranks, the year-to-year pattern of Lionel's and Selina's lives remained fairly similar.

Forest Inspections

Lionel Osmaston was typically out on tour more or less continuously between mid-November and June. Sometimes he stayed in bungalows and rest houses, but more often he was in camp under canvas. Generally, he moved camp every day, travelling ten to fifteen miles to the next location and performing his inspections along the way. At some strategic locations he might stay for up to two weeks. The daily distance was determined by how far the carts carrying the camp could travel; any greater distance would have necessitated a duplicate set of possessions to be sent on in advance. Each year, Lionel tended to revisit some places he had been before but also new ones. Because of his attachment to the Bombay Presidency, his career did not take him to anything like the diversity of places that Hudson's did (see figure 6.1).

On the basis of the Forest Act, circular 22-F, and the descriptions above, it is clear that what constituted a "forest" in practice was very broad, ranging from teak and *sal* forests proper, to jungles, to open scrubland. This caused Selina to exclaim in disgust soon after her marriage, "Clambered about on a hill inspecting a forest with hardly any trees in it!"[59] Yet there were also majestic stands of teak, "delightfully jungly" camps,[60] and magnificent vistas: "*Lovely* view over Canarese jungles, hill after hill of green jungle as far as one could see with the morning sun lighting them all up on the S. and E. sides & mist like lakes in the valleys."[61]

As the officer responsible for the division, Lionel handled the sales of teak and other major forest produce that took place in February and March each

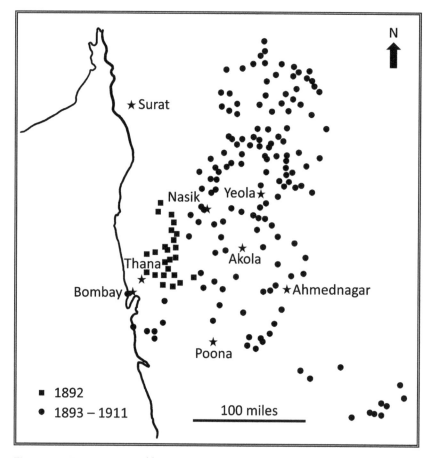

Figure 6.1. Forests inspected by L.S. Osmaston, 1892–1911 (excluding work in Karnataka State in 1910)

year. The scale of the operation was massive. In 1896 Lionel handled 76,000 teak logs: "Karanjali – inspected 23,000 stacked teak trees en route ready for sale at Karwani – a fine sight."[62] At the sales near Akalkua (Akkalkuva) in early March he "sold over 18,000 teak timber for 11¼ thousand rupees: 12% better prices than last years: could hardly help smiling as the bidding mounted up & up!" Not all sales went this well, however, and later the same month he had to halt a sale due to poor bidding. They also involved lots of paperwork: "The conservator's office kept a watchful eye on anything that might be lost, the comptroller did the same and more; he jumped on us if we found anything! When we sold a piece of timber we had to show on our forms whatever sum of money had been spent in preparing it for sale, such as cutting down the tree and hauling the timber. Woe betide us if we sold anything without being able to show that it had cost something to the government!"[63]

The demand for railway sleepers that had done so much damage to the forest and had helped trigger the formation of the Forest Department was still high at the turn of the century for new railway construction and replacement as the old ones rotted. Lionel Osmaston himself mentions working with sleepers only during one period in early 1910 when a Mr. Fowler from the railway came to inspect sleepers at Gunjawati in Karnataka State. Lionel and Fowler met again at Kirivati in February, and finally in April he recorded, "In the morning Lina came out & watched me passing sleeper pieces for a bit."[64] It seems that Lionel and Fowler first discussed the order, then looked at timber in the yard, and finally Fowler inspected the finished pieces with Lionel before delivery.

For his part, James Best heartily disliked the job of selling sleepers to the railway: "The sleeper-passing job took me up to the break of the rains and it was about as dull a job as one could wish for. The railway officer would inspect each sleeper; when he found one not good enough he rejected it. Some had dry rot, others had big cracks, a few had knots in the timber where such things are inconvenient. We had gentle argument over each sleeper with which he found fault. Sometimes, but not often, I persuaded him to change his mind. I suppose that I did some good. Someone had to be there in any case."[65]

On the high hills of the Chakrata Forest Division in Uttarakhand State (near the border with Nepal) it was no simple matter to get the deodar (cedar) sleepers from the forest to the market. Bertram Osmaston described how it was done:

The procedure was to first cross-cut the tree concerned into 10ft or 6ft logs and then to saw these up by hand into broad-gauge or metre-gauge sleepers which then had to be carried down on men's backs for half a mile or more to a point where they were placed, one by one, into a 2 or 3 mile long "wet slide" – a shallow trough-like wooden channel – down which they hurtled in single file to a small rocky stream which then carried them on down for another 5 miles to a depot on the bank of the river Tons. There they remain until September, towards the end of the rains, when all the 100,000 or more sleepers were launched as rapidly as possible [into the river].[66]

One of these wet slides gave the name "Topslip" to a hill in the Anamalai region of Karnataka where Forest Officer Hugo Francis Wood (CH 1890–3)[67] is buried amongst the trees. After retiring to Ooty, "his great pleasure was to entertain his many Forest friends to most sumptuous lunches and dinners."[68] His last wish was to be buried in the small teak plantation which he had created, and it fell to his successors, Diana and Donald Currie, to organize the funeral.[69] The local inhabitants credit Wood with preserving the teak forest in that area of the Anamalai Hills because of his habit, when out walking, of using his walking cane to drill holes and plant seeds. His tombstone is surrounded by the 100-year-old teak trees he planted, and Wood's contribution to forestry is recognized by his modern successors.[70] In 1991 Diana Currie wondered whether the stone they placed over Wood's grave was still there – it does survive, with its inscription borrowed from Sir Christopher Wren's tomb, *Si Monumentum Requiris Circumspice* – if you seek his monument look around.

Despite the undeniably romantic aspect to Wood's final resting place, he was not unusual in his dedication to his forests and interest in their regeneration. Lionel Osmaston maintained at least one nursery for seedlings, tending them carefully until the rains came. He also laid out new forest plantations such as the one he established in 1904 near the attractive hill station of Toranmal at the northern tip of Maharashtra.

Surveying

An important part of Lionel Osmaston's rounds of inspections was to record all relevant local information to develop the working plans. Once the plan was in place, the inspection ensured that the specified schedule of felling, pruning, and cleaning was being performed properly by the local rangers.

The divisional forest officers' inspection included checking the forest boundaries, and Lionel mentions doing so on several occasions. Less than two weeks after getting married, Selina in her lightly humorous way wrote in their diary: "Rode to a forest skirting around a hill to see the boundaries & then as usual went to sleep under a bamboo while Lil strode up the hill through the jungle."[71]

For someone trained at Coopers Hill, the general level of surveying skill required by forest officers should have been fairly elementary. However, not every Forest student applied himself diligently to the course. Wimbush's experience with his survey project on Englefield Green was that because "only one or two people could manipulate and use the instruments at one time" he had been "content to keep in the background."[72]

Sometimes the tasks in India were more demanding, and James Best felt that his education had let him down when it came to surveying the route for a new road for extracting logs: "The hillside was steep and it was difficult for a person with no experience of the work to keep below the maximum gradient between the contours without going away from the direction that one wanted to attain. The corners of the zigzags were an awful problem; it was comparatively easy to keep within the limits of gradient when taking a straight line, but one had to make the road for carts to go along it and be able to turn the corners. A difficulty that I was not taught to overcome at Coopers Hill."[73]

Best's second attempt was an improvement: "The second road was naturally a better piece of work than the first because I had learnt something that I ought to have learnt at Coopers Hill. I believe that the second road was made, but there is no record of the first; the place is probably still roadless jungle as wild and charming as I left it."[74]

In the spring of 1902 Lionel Osmaston also laid out a new road on the hills near Jirwadi (Jirwade),[75] but like Best's there is no confirmation in the diary that it was ever constructed.

Best would have studied surveying in both his first and second years at Coopers Hill, followed by a course of ten lectures on "forest engineering." The need for forest officers to know the basics of engineering was well understood in India, and a Coopers Hill graduate, Charles Rogers (CH 1885–7) was instructed by the Government of India to prepare a manual on the subject for forest officers in the field. The outcome was a three-volume *Manual of Forest Engineering for India*, published between 1900 and 1902, that covers building materials, construction, road making, bridges, transportation methods, and wells.[76]

Fire Conservancy

Much of the office work that Lionel and Selina complained about in their diary came from recording and organizing the information collected during his inspections. In addition to informing the working plans, the Forest Service published an extensive and detailed annual report containing every conceivable statistic and measure of each circle's activities during the year. It was collected into a 300-plus page administration report for the Bombay Presidency as a whole. The work was farmed out to the individual forest officers, so, for example, Lionel commented as early as 1892 that he had prepared the section of that year's report on fire prevention:

> In the two forest divisions of Thána greater attention was paid to fire conservancy; and owing to the energetic co-operation of the Collector, the mámlatdárs [chief civil officer], police and other district officers, have all been enrolled in the cause of the suppression of forest fires…. In the case of fires caused wilfully the main object is shikár, fire being used to remove the undergrowth and leaves, and thus facilitate the pursuit and slaughter of game. The preparation of ash manure in fields, adjoining forest boundaries, has led to the burning of much forest lands in different directions. The damage done by fire in a closed coupe is incalculable, and 15 of the South Thána division fires burnt "closed forest," traversing 755 acres out of a total of 11,112 acres: in one case the fire was caused by smoking rats out of holes, and two of the offenders were caught and punished with rigorous imprisonment for 20 days.[77]

Forest fires were a frequent occurrence and affected huge areas. In 1894–5, for example, more than 200,000 acres were burned, representing about 5 per cent of the forest in charge of the department. This was of course very regionally dependent, ranging from 0.2 per cent in Sholápur to 12.5 per cent in West Khándesh. Lionel remarked on numerous occasions about seeing fires of all sizes between January and April. Near Wassind in April 1892 he remarked on the "frightful views of forest fires over hundreds of acres."[78] An ever-present danger was of village huts catching fire by accident and spreading to the nearby forests. James Best also had to contend with fires in his jurisdiction at Balaghat in Madhya Pradesh state during 1905–6:

> Every day there was some fire or other near my camp and I had to go out with the idea of extinguishing it. It was a hopeless task, with two or three fires raging at the same time we had not enough men to fight them, all we could do was to counter-fire up-wind from one of the fire lines or a river. We could see columns of smoke in almost every direction by day and at night the surrounding hills were marked with long thin lines of flames that flickered and waned in their progress up the hills. From long distances one could hear the explosion of hollow bamboos, and the sudden roar of flames as they reached the denser thickets....
>
> I learnt a lot from those fires. It was hard work putting them out, or trying to, and monotonous writing reports on them.[79]

Valuable forests were protected by the clearing of fire breaks, ironically by controlled burning. James Best, and Lionel and Bertram Osmaston all talk about serving on fire protection duty. To be truly effective the fire-lines had to be monitored from local fire stations, such as the one Lionel set up on a hilltop near Palasner on the boundary between Maharashtra and Madhya Pradesh in February 1909. Best described the idea: "The forests were divided into large sections by means of broad cleared fire lines. At convenient distances along these lines there were fire watchers' nakas – platforms erected high above the ground, roofed in against the weather, from which the lonely fire watcher could look out across the forests for the smoke of approaching fires."[80]

For sportsmen such as Edward Stebbing, wide fire protection lines had other uses, for "in the early morn or just before dark, you might, by quietly walking up the fire-line, come across numbers of sambhar, and with moderate luck bag a good stag."[81]

Forest Offences

With the increased areas of forest coming under the protection of the Forest Act and being demarcated on the ground, dealing with offenders was an important part of the department's business. Forests were patrolled by the beat guards, but officers such as Lionel also came across instances themselves. He mentions several in his diary, particularly in 1892 when it was all still new to the twenty-two-year-old Osmaston: "Inspected the Birwadi coupe on the way, heard chopping in forest & found wild tribesmen cutting huge Hidoo tree, had *awful* chase up precipitous hills till I was nearly dead & came within 25 yards of a woman thief, she began crying & her husband who was mere paces in front chivalrously came back! I then stumbled after him, couldn't catch him & then turned back for wife, she had gone of course so *missed all.*"[82]

Not everyone in the forest was intent on taking wood, however:

> While inspecting forest came on 3 mussulman brahmins[83] living unlawfully in forest & "as they said" making medicine, but the RFO thinks they are alchemists as the compound they were making is said to turn metals into gold: a huge earthern [*sic*] vessel with a grass hood & inside a disgusting sight viz a half-decomposed pig crawling with large maggots: these maggots are made into oil and this oil is the gold making compound, but they said it was being made as a medicine to cure "worms in the head."[84]

Sometimes he gave chase, only to find the woodcutters had a permit or were cutting aloe, not wood, or he was "besieged by petitioners about their being chucked out of reserved forests, they want ground for cultivation."[85]

Serious cases, such as those involving contractors or fake permits, were taken to the magistrates' court. In 1891–2, 1262 cases were prosecuted in the Northern Circle, of which nearly 80 per cent were convicted. For more minor offences, forest officers could "compound" cases by assessing a fine on the spot, or as Lionel put it, "judgement given in the jungle."[86] Five years later, the Central Circle reported 1610 cases before the magistrates and 2,091 cases having been compounded by divisional forest officers. In addition, an astounding 101,192 animals were impounded for grazing in "open forests" without permits, as well as 138,770 for "trespassing" in "closed forests." This enormous total was 82,000 fewer than the previous year.[87]

These numbers highlight the impact that forest regulation had on the local populations. While the official policies claimed that villagers' needs would be met, the British restrictions on previously open lands seemed inequitable, arbitrary, and punitive. But that former openness had permitted the exploitation and deforestation perpetrated by others, an example of "the tragedy of the commons."

Famine Relief

Agriculture in India is very dependent on the monsoon rains, and failures of the monsoon have resulted in terrible famines, leading to tens of millions of deaths. Policies implemented under British rule have been criticized for exacerbating the suffering during famines by promoting the growing by villagers of non-food crops (such as cotton), the continued exporting of food even during famine years, a mistaken belief in the effectiveness of market forces, and the lack of timely responses. Rarely were famines the result of an absolute shortage of food but "complex economic crises induced by the market impacts of drought and crop failure."[88] The issue was often the cost of food, not the lack of it.[89]

Writing about the 1907 famine, Frederick Canning (CH 1900–3) described the response in his division on the Nepal border:

> The famine of those years marked an epoch in provincial history. The suffering was very great, one in ten in our district were on the roll of relief. But the famine arrangements were rather wonderful, works were arranged all over the affected areas and everyone went at it all out and red tape disappeared.
>
> The chief thing the Forest Department did was in addition to the hitherto usual throwing open of areas for forest grazing, the first supply of forest fodder grass, sent by rail to distant districts. A special train load went off every day and during the moonlight periods the continuous clank of the grass hand presses day and night is something one still remembers....
>
> The sal forests suffered terribly. 10,000 acres were killed and 55,000 acres very seriously affected.[90]

During the widespread 1896–7 famine, the Bombay Forest Department responded with measures that had, regrettably, become standard practice. To provide a source of income for the affected villagers, they stepped up the construction of boundary markers, and in the Central Circle a whopping 66,000 were built for a cost of almost 17,000 rupees (expenditures of this kind was part of the Circle's accounts but charged to a "famine fund" to track costs). To help feed cattle in famine-stricken regions, grass was cut, pressed, weighed, and transported by rail, as Canning described.

Lionel Osmaston mentioned the famine throughout the first half of 1897, sometimes in the context of inspecting the "damage done by [the] relief work Johnnies to [the] forest,"[91] or monitoring the grass cutting. Selina wrote about riding to "one of the Forest Relief works just started – building forest cairns – saw all the workers having their names called out & beginning work, about 200 of them."[92] She also recorded seeing over 2000 women "sitting in lines breaking up stones for road making, they all looked fat and pretty festive."[93]

The dry weather had a number of other effects, including the necessity for increased fire prevention measures. There were also issues of cattle straying into sensitive areas, destroying saplings by eating or trampling them. Labour for routine operations was hard to find because people did not want to travel far from home for work.[94]

Lionel was assigned to famine fodder work from the end of October 1904 to the end of May 1905, being in charge of the grass cutting around Sindkheda in northern Maharashtra. Selina took and developed several photographs of the grass pressing work, one of which is reproduced in figure 6.2.

Figure 6.2. Selina Osmaston's photograph of pressing grass cattle fodder for famine relief, taken near Sindkheda, 25 November 1904. It shows pressing the grass in the foreground, weighing the bales in the middle distance, and loading them into railway trucks in the distance. This work was performed under Abdul Rasul of the Bombay Forest Service (Osmaston collection, CSAS)

As an interesting insight behind the scenes, Lionel afterwards wrote to Mr. Fry, the conservator, observing that he had undertaken this work in addition to his West Khándesh Division, and requesting either additional privilege leave, as had been granted in the 1900–1 famine, or additional remuneration. The "outturn" of his fodder production was 183 lakhs of pounds, or nearly 8,200 imperial tons, which he stated exceeded that of any fodder section in 1900–1. While Fry was supportive, the undersecretary to government responded bureaucratically that "while Government acknowledge the zeal and promptitude with which Mr. Osmaston's work was carried and the consequent success of the grass operation," the 1900 rules were no longer in effect and a "special allowance or honorarium" was contrary to the principles laid down in several government resolutions.[95]

Camp

Come November each year, the forestry officers were eager to leave behind the stuffiness and routine of the station and return to the forests: "Station life becomes dull and irksome after three months of office routine by day, followed by the usual bridge or snooker-pool and dinner-parties at night. The time comes when social amusements lose their attraction and one longs for the freshness of camp life, changes of scenery and the company of the men of the fields and the jungles."[96]

After the bustle of a trip to Bombay to see Lord and Lady Curzon arrive in India, Selina wrote thankfully of her return to camp at Chalisgaon: "Back again in our peaceful camp, a blessing to have done with the racket of the hotel."[97]

A forest officer was in camp for months at a stretch, as James Best described: "Normal routine work meant much marching and touring throughout the open season from November to June. During that time I very rarely saw another white man. One would start out with cook, butler, cook's mate, *saises* [grooms], sweeper, laundry-man, orderlies, tent-pitcher and his assistant, and a number of cartmen or camelmen."[98]

But the outdoor life had its disadvantages too. Lionel described in his diary periods of loneliness, and other writers also commented on periods of intense isolation.[99] Writing to his mother from camp in 1910, another Coopers Hill man, Jack Wildeblood (CH 1887–90), confessed that he lost track of the days of the week while travelling and concluded, "I have no news for you from here as I don't see anybody.... This camping life is very healthy, but it is rather lonely."[100] Eustace Kenyon shared this difficulty: "The novelty of having nothing but one's own society, seeing nothing but interminable forest or a vast flat stupid paddy plain, eating nothing but moorgie,[101] & doing nothing but cooli drive, wears off very quickly, & jungle life palls upon one."[102]

In addition to an officer's own work, the task of managing the large group of servants and monitoring the range forest officers he was inspecting could be challenging. Throughout Lionel's first months on tour he contended with an irritating string of minor disobedience, theft, injuries, breakage, and squabbles that tested his patience.

After they were married, Selina joined Lionel on tour, later taking the children as well. Indeed, Selina was out on tour very soon after her marriage.[103] On Tuesday, 23 February 1897, Lionel arrived in Bombay from the countryside to receive a wire from Aden: "Try & arrange wedding Friday." Accordingly, the next day saw Lionel "Rushing around changing wedding to Friday & c." Thursday was spent the same way and, although Lionel does not say so, it seems to have been the day Selina first landed in India. At the bottom of that day's entry he writes. *"Lina will do the diary job from today!"* Friday's entry is indeed in her handwriting – her first word was *"Married!"*[104] The newlyweds spent Saturday morning shopping and packing, and Selina notes, "Got to our lovely first camp between 6 & 7" in Lanouli – she had landed, married, and travelled to camp in three days.[105]

Clerks and Logistics

A few days later Selina wrote, "The 7 clerks all came up in a row to pay their respects to me – isn't it awful!" Those clerks were, however, essential for keeping the official side of the operation running. When Lionel was without clerks for a time in April 1894 it was greatly inconvenient, one entry reading, for example, "Finished the awful Nandgaon Range report 7.5 sheets (office copy and fair copy as no clerk) & sent it off."[106] The eventual arrival of a head clerk in June was greeted with cheers.

Best gave an idea of the range of other tasks the clerks performed: "The most invaluable of all the retinue was the camp clerk; he was responsible for informing the range officer in advance where we expected to go so that arrangements could be made for our comfort; he controlled the postal runners and the arrangements for sending the post to and from headquarters in our own special bags; he read out the vernacular correspondence, copied urgent letters in camp, and did many little things to make every-one's life more comfortable."[107]

The levels of detail and sheer quantities of numerical data compiled manually for the administration reports of the various agencies in India are quite astonishing. Alfred Newcombe explained the working of the Public Works Department that facilitated such detailed record keeping and accountability:

> The office work went quite smoothly. This was partly because the accountant was well trained, and both he and the clerk were intelligent, steady workers; and partly because the public works system is arranged so as to keep the records of transactions and the compilation of accounts in clearly defined grooves. At each stage of a work some special list or report or account has to be written, the earlier ones being the bases on which the later ones depend. The final records showing work done to date and expenses incurred are sent forward monthly to the examiner of accounts at headquarters, supported by copies or originals of the initial and intermediate accounts made up as the work progresses; and, after scrutiny by him, if satisfactory, they are passed.[108]

Each of the Osmastons' diaries lists that year's servants, when they joined, and how much they were paid. In 1901, for example, they had eleven servants, including the ayah to look after the children, butler, cook, two grooms (*syces*),

Figure 6.3. Selina Osmaston's place at Sharanpur during the rains, 1904, while Lionel was at Dhulia, Khandesh (Osmaston Collection, CSAS)

a water carrier (*bhistie*), a clothes-washer (*dhobie*), a house boy (*chokra*), a sweeper, and a gardener (*mali*). Most of the functions and many of the individuals would have accompanied the family on tour.

With tents, furniture, housewares, provisions, clothes, guns, and myriad everyday items for several months in camp, as well as all the office accoutrements, even a relatively modest family needed a procession to transport their chattels. Usually the "kit" was carried by waggons, which caused endless trouble: breaking down, being unavailable, not turning up, dropping things, or slow to arrive. Occasionally, where the roads were unsuitable for waggons, labourers were employed to carry the goods on their heads. In one case Lionel described using "28 private coolies & 17 office coolies" to transport their possessions.[109]

On official business, the tents used were "large and well furnished with carpets, chairs and tables,"[110] so people used to life in camp "contrive very luxurious, or at any rate comfortable, makeshifts, which are not to be called hardships."[111] Ernest Hudson called these "swiss cottage" type tents, an example of which can be seen in figure 6.3. Lionel also had an office tent.

For the first few months of 1892, Lionel kept track of his purchases of stores in the front of his diary. On the lists were essentials such as flour, butter, oatmeal, grain, potatoes, salt, sugar, and kerosene. To drink, he took tea, mocha coffee, French coffee, cocoa, tonic, and soda water, while milk was bought from villagers. Bread was baked in camp, so Lionel packed an assortment of jams and marmalade. Meat was supplied by chickens, goats, and sheep bought along the way (on several occasions these were snatched by panthers), and supplemented with potted beef, brawn (half of which was eaten by a "pi" wild dog), and tinned "ham & chicken." Then, of course, there were the other essentials: Scotch, Irish whiskey (one stolen), champagne, port, and cheroots.

On Christmas Day 1893, Lionel, Mr. Hodgson senior, and another forester Mr. Millett sat down to dinner of a fine tinned plum pudding, tinned salmon, and chicken "arrangements," before singing round the campfire to the music of Lionel's banjo. Their meal was washed down with a bottle of champagne, one of those itemized above, "which had followed [him] all over Thana district in 1891–92." One sympathizes with the servants who toted the bottle from place to place; stories like this fuel the slight ridicule that often accompanies the image of the British officer's baggage train wending through the countryside. Considering they were transporting everything needed for a house, an office, and a staff of perhaps twenty for many weeks, the operation was inevitably on a large scale.

Despite a superficial similarity, the Osmastons' working lives were quite different from the Hudsons'. Although both went on tour, Hudson's rarely lasted more than a few weeks, in contrast to Osmaston's of many months. Even then, Hudson was usually close to habitation because the telegraph ran between settlements or alongside railway lines. Hudson frequently stayed in a dâk, PWD, or military works bungalow, or with other Europeans working in the area, and so he was rarely short of company. By necessity, the forest officers' work took them into the wilderness and away from their English communities, so they had to be much more self-sufficient, both materially and socially.

Memsahibs

This book is mainly about men. Women did not join the technical services. Nor, during the time of Coopers Hill, did they earn engineering credentials. Nevertheless, European women in India played a vital role in the Raj, although it was typically "less flamboyant" and more concerned with the "details of everyday life."[112] The stereotypical image is of the station socialite, freed of responsibilities for the household and child care by her Indian servants, indulging in garden parties and gossip, and overseeing the "absurd punctiliousness about the official order in which one goes to table and who sits on the right hand and who on the left." During the hot weather, the stereotypical memsahib headed for the cooler climate of the hills. But not all: "Jill is not always wafted hillward by the first whiff of hot air from the dread furnace to come. She does, on occasion, stand by her husband, through bitter and sweet, through fire and frost; and what such a standard of wifehood costs the brave women who live up to it, only the wives of India know. For these unrecorded heroines are a nation without a history."[113]

Elfie Hudson and Selina Osmaston come across in the diaries as this second type of memsahib. To be sure, they participated in the normal round of badminton mornings, tennis tournaments, golf games, and other station pursuits, but after all, what choice did they have? They also went to Naini Tal or Mahabaleshwar for vacation or when sick. But they were also resourceful and resilient women, fashioning the best life possible for themselves and their families in their circumstances. These were no shrinking violets; Elfie could cycle thirty miles with ease and Selina could ride and shoot, and even went to ladies' pig-sticking.[114] They were both very much "part of the team," even if their roles were distinct from their husbands'.[115]

An impression of how overwhelming the arrival in India could be was given by Rosamund Lawrence, who even twenty years later recalled her amazement at her first sight of the streets of Bombay: "And there on the quay alongside, customs officials, friends, servants, scallywags of all kinds. I had never seen so many people; a mixture of brown faces, and dirty white garments and spotless uniforms, and helmets, mixed up with oxen, mangy dogs, crows, and beggars, and driving through narrow streets between tall colour-washed houses, with vivid trees jammed between them, jingling victorias [horse-drawn carriages] and bullock carts round you, and parrots shooting across the road over your head, black crows squawking. People. People. People. And your frock stuck to your shoulders."[116]

Often the period of engagement was "dreary waste of a year and four thousand miles that lay between [her] and memsahibship."[117] Ernest and Elfie travelled to India together. After her twelve-month wait, Selina was whisked through her marriage and off to camp within seventy-two hours, so perhaps she did not have a chance to remain overwhelmed. At her second camp there was "no noise here at all & it's a lovely camping ground in the very midst of steep high hills."[118] She seemed to relish the experience, and it may have given her the opportunity to love the countryside before dealing with the city.

While we do not know whether Selina read it, there was substantial advice available to the Englishwoman going out to camp. First among these would have been the iconic *Complete Indian Housekeeper & Cook*, by Flora Annie Steel and Grace Gardiner. Dedicated to "the English Girls to Whom Fate May Assign the Task of Being House-Mothers in Our Eastern Empire," this guide had gone through multiple editions since its publication in 1888 and enjoyed a similar status to *The Joy of Cooking* in modern North American households. Amongst chapters on the duties of the mistress and servants, accounting, livestock, gardening, and raising children, was one called "Hints on Camp Life."[119]

"The first axiom for camp is not to do without comfort, if it does not entail discomfort by increasing the trouble." Accordingly, the advice given in that chapter covered subdividing tents to make them more comfortable, how and when to move the milk cow between camps, types of bed, cooking arrangements, and the convenience of multi-use packing boxes. The chapter also talked about the difficulty of purchasing supplies from villagers since they were generally reluctant to sell produce:

> It is easy to understand how fearful a weapon for oppression that appalling necessity of camp life, the táhseel chuprassi, or táhseel office orderly,[120] may become.
>
> That he is necessary in the present state of civilisation few will deny; if they feel inclined to doubt it, let them go into a village with a large camp, and see for themselves. They will be exceptionally fortunate if they can get even grass for their horses. As long, therefore, as the present system lasts, one of the chief duties of the mistress in camp is to see, as far as in her lies, that no oppression is committed in her name.[121]

In keeping with the first axiom, the concluding advice of the chapter was: "In regard to stores, it is well to take as few as possible, especially tinned provisions, but do not make yourself uncomfortable for want of things to which you are accustomed. That is the great secret of camp life."[122]

For the other half of Selina's time, and much more of Elfie's, they lived in rented bungalows. There were times for both women when they seemed forever packing and unpacking, hanging pictures and hunting trophies, acquiring or selling furniture. For the Osmastons this happened twice a year, of course, but there was also transfers, times spent at hill stations of Mahabaleshwar or Naini Tal, and vacations at hunting camps. In the words of Margaret MacMillan, "Like ants, they constructed nest after nest."[123] Furnishing was a mixture of rented and owned, and sales of house contents were commonplace as people were transferred or returned to England. Cumbersome items such as pianos were also rented.

Other than the club and possibly the gymkhana, the important place for entertaining was in the home. Acquaintances from the station frequently dined together at someone's bungalow, but these dinners were often with other officers from the host's department, both junior and senior to the husband. It was also common for people in the same service to stay with one another if they were visiting from elsewhere, such as on a tour of inspection. So the quality of the meal and the comfort of the surroundings were important for the family both socially and professionally, and in this way the home became to some extent "branch offices in the business of empire."[124]

In her *English Bride in India*, Mrs. C. Lang (writing as "Chota Mem," or junior memsahib) strongly urged "young inexperienced English girls starting housekeeping in India" to keep lists of all the household possessions, specially silverware, china, glass, linen, and saddlery, as well as rented furniture.[125] This would enable her to hold the *kitmagar* (butler) responsible for them and allow her to keep track of anything broken or lost. Chota Mem's little book was full of practical advice on food, which servants to employ, and on what salary (both of which are consistent with the Osmastons' experiences), daily routines, and useful household hints.

The wives of Coopers Hill men were, like themselves, from fairly affluent families. For all the many differences from home, life in India had its similarities too; the social hierarchy, outdoor pursuits, domesticities, entertainments, and employment of servants would not have been totally alien. Chota Mem advised, "Be patient with your servants and treat them more or less like children, remembering they love praise, and don't treat them as if they were machines."[126] This single sentence captures so much about the attitude of the British to their Indian staff – exasperated, condescending, and yet affectionate. Although the context was Indian, the behaviour was rooted in English society that employed thousands of lower-class, menial domestic servants, and exploited thousands more in the mines, mills, and factories of the English industrial cities. Nevertheless, the relationship between employer and servant could be long; Anthony Wimbush's "boy," Parthasarathy Pillai, served him and his family for twenty-eight years.[127]

Children

"If marriage complicates a woman's life in India, the advent of children – blessed advent though it be, – scarcely tends to simplify matters."[128] For women with children, the ayah was therefore a servant of especial importance. She was often regarded as an ally in household management.[129]

> Her duties vary very much with her situation, as of course if there are children and no English nurse kept she naturally looks after them, otherwise she does duty of lady's maid with some duties of housemaid combined.... The ayah is a most useful servant and if she is willing and clever will be a tremendous help to you, and you must own it is nice to have one woman in the house. It is such a comfort when you come in hot and tired to have her to take your shoes and clothes off, and put out what you want to wear, to brush, and fold up your things, and generally look after them.[130]

However, as Steel and Gardiner observed, "A good *ayah*, however, is diffi-cult to get."[131] This was certainly the case for Selina's first child, Erica, born 5 September 1898. Before the birth, "'Margaret' the ayah arrived & I like her looks which is a blessing."[132] However it appears that Margaret did not want to go on tour because just as the Osmastons were starting to pack,[133] Selina "had the great blow of finding that Margaret Ayah wants to leave" and had to find another on short notice. They took on "Fatimah Ayah" on the day before they left for camp, but she only lasted until January 1899, when Selina had to "send off poor wretched Fatima ayah & take on the new 'Angelina' let's hope she *will* be angelic!" But she was not, and got "worse and worse," so eventually they "left the Ayah behind chiefly because of a fracas over her chota hazri [little breakfast]!" Somehow they managed to replace her, but her successor was again short-lived, and she was fired on 21 March 1899.

In the first months of 1899, baby Erica went through several periods of poor health. She eventually died on the evening of 22 March in the bungalow at Nan-dur in north-west Maharashtra. Knowing what is to come, it is heart-breaking to read in the diary of the parents' anxieties, and their optimism each time Erica appeared to improve. They described their final tender care of the dying baby and efforts to get medical assistance from Nasik thirty-two miles away (the doctor came the next day, having been out of town when the summons arrived). They made a coffin for her from Lionel's gun box and buried Erica on the hillside at sunset. Later they built a wall around the grave and made sketches of the site. On their last day in that camp, Selina wrote, "Our last day here; we went down to our little grave: we are not leaving little Erica there we know, & yet it is hard to go away & we don't know whether we shall ever come here again."[134] Sadly, Lionel's and Selina's loss would be repeated twice more, as three of their first four children died in infancy (Osmund and Judith were bur-ied side-by-side in Nasik cemetery). More happily, they also had five children who survived to adulthood, including Fitzwalter Camplyon Osmaston, who followed his father into the Indian Forest Service.

The Hudsons also used an ayah when the children, Tom and Adrian, were young. Later they followed Steel and Gardiner's advice that "for children out of arms, a good, well-principled English nurse was essential. However good native servants may be, they have not the same up-bringing and nice ways, knowl-edge, and trust worthiness of a well-trained English nurse."[135]

Like many children born in India to English parents, "Bobbie" and "Campie" Osmaston (figure 6.4) attended school in England from ages four and six, respectively, for "few will deny that the sooner after the fifth year a child can leave India, the better for its future welfare."[136] Elfie and Ernest Hudson returned to England while their children were still young, perhaps to avoid that eventuality.

Figure 6.4. Lionel Osmaston with sons Robert Lionel and Fitzwalter Camplyon, outside their hut at Toranmal, 1904 (Osmaston Collection, CSAS)

Plenty of Tigers to Spare

As we have seen, hunting and shooting were a major part of life in India for many of the British residents, so an understanding of the lives of Coopers Hill graduates in India would not be complete without a brief consideration of the *shikar*. College graduates who published books on game hunting include Bertram Osmaston, Edward Stebbing, James Best, W. Hogarth Todd, and A. Dunbar Brander.[137] Ernest Hudson was a keen snipe shooter and the Osmaston brothers killed their share of large cats. This was perfectly acceptable because, after all, there were "plenty of tigers to spare."[138] They also routinely, usually daily when on tour, shot for food; Hudson described distributing the spoils of his snipe- and duck-hunting expeditions to others in his station.

In the first chapter of *Jungle Byways in India*, Edward Stebbing painted a wonderful picture of two English *shikaris* "tanned and burned brick red by the days of exposure," unloading their "battered collection of kit," "black-bearded and fierce-moustached" servants, and hunting trophies from a train after their return from leave:

The joy and pride on their faces is for the bundle of horns which they carefully see lifted out of their carriage, together with several old battered leather rifle and gun-cases. A couple of leathern trunks and two rolls of bedding, water-bottles, a wooden store box, and an old shikar topi or two.... On looking on this battered collection of kit, your eye will run critically over the horns. Not much, perhaps, will be your verdict. Nothing big. No, there may be no record heads. But to their proud possessors, ... that little pile of horns represents hours of patient toil and tramp, hours of discomfort through the long hot day when the flies nearly drove one mad, and the heat temperature went up and up and up, until you felt as if you would never and could never get cool again.[139]

Stebbing contrasted the dishevelled but happy young shikaris with the "beautifully tailored and outfitted sportsmen" to be seen departing from London by train for grouse shooting in Scotland, with their "chaotic mass of smart leathern trunks, dressing-cases, kit-bags, immaculate gun-cases and, last but not least, dogs." "We can imagine the sniff of contempt of our immaculate home sportsmen! And yet the experienced know that with that kit and by means of it those youngsters have had far finer sport than money could purchase in the Old Country, whilst that brawny servant with a couple of stones and a hole in the ground will and often in the past has turned out a dinner which, to a hungry sportsman, be he prince or subaltern, is all that can be desired."[140]

A fixture in the annual social calendar in India was the Christmas camp, where large groups of friends and family convened at some suitable hunting ground to celebrate the holidays with a shoot. The Hudsons, Osmastons, Best, Todd, and Stebbing all attended these gatherings. The style of the hunt depended on the preferences of those involved and the choice of location, but the most dramatic involved the elephant line, such as that shown in figure 6.5 during Frederick Canning's Christmas hunt in 1906.

Figure 6.5. "The Line" near Patihan, Uttar Pradesh, Christmas 1906 (Canning Collection, CSAS)

In *Wildlife and Adventures in the Indian Forest* it is estimated that Bertram Osmaston shot twenty each of leopards, panthers, and tigers during his career, hastening to add that these animals were so numerous and troublesome to villagers that they were regarded as vermin. Judging from his diary, Lionel Osmaston probably shot around twenty in total, many of which were tracked down after taking farm animals from his camp.

The ultimate instance of the pest tiger was the iconic "man-eater." Both Lionel and Bertram attained some notoriety for disposing of man-eaters. Bertram's exploit occurred less than five months after his arrival in India when he had "never seen a tiger outside a zoo," and made the front page of the *Pioneer* newspaper on 17 May 1889. The spot was marked with a signboard that read, "Near this spot in May 1889 a man-eating tiger, the last known tiger to have inhabited the district, was shot by Mr. B.B. Osmaston I.F.S. while it was in the act of mauling Mr Hanserd [*sic*] a forest student."[141] Anthony Wimbush later commented on camping near this sign in around 1920.[142]

In his diaries, Bertram credits Hansard's survival on the combination of the "thick woolen muffler around his neck" and the tigress's worn and decayed teeth.[143] Hansard was mauled badly and spent several months in hospital but was eventually discharged and went on to marry his nurse. Bertram learned later that Hansard ultimately died from the "after effects of that terrible encounter."[144]

Probably some animals were deemed man-eaters in order to enhance the reputation of the hunter. Nevertheless, forest officers did serve as de facto unofficial gamekeepers for their forests: "The Forest Officer has had the duties of gamekeeper added to his other arduous ones in the forest. He issues the permits for shooting; allocates the blocks between the various permit-holders, possibly finding when this distribution has been made that there will be but a small area left in which he may fire a rifle himself."[145]

So, when over a period of months, a panther had been attacking people in his division, it was out of duty as well as sport that Lionel went after it. The first mention of the panther was on 9 February 1902 when it killed one woman and injured two more at the village of Mulher; Lionel went off at once, tracked the panther, and organized two beats through the jungle, but to no avail. The next we hear of the panther is on 13 March, when Lionel received notification that it was again killing people in Mulher. He went off alone to track the panther and as Selina was writing the diary, there was no news for eight days until she wrote, "The wretched panther still lives to triumph again!" Finally, in November, Lionel heard that the panther had attacked again. Leaving Selina and their one-year-old son, Camplyon, again, he rushed off to Mulher, but it was not until 5 December that Selina's diary entry gave the brief but triumphant news of Lionel's return: "To our delight Lil appeared in the tonga [a small, two-wheeled carriage] about 3pm *with* the dead dreadful man-eater! It *is* grand – he shot it yesterday eve, sitting over the remains of a poor youth the panther had killed the day before."[146] A narrative of the hunt was formally published in the *Journal of the Bombay Natural History Society* in 1903.[147] Lionel speculated that the panther had taken to eating human flesh during the 1900–1 famine when, hungry itself, it would have come across weak or dead people in the jungle. He reckoned that the panther had killed thirty people in the Mulher region alone and had attacked eleven more who had survived.

The Sportsman Naturalist

Stebbing, Best, and the Osmastons were ardent naturalists and conservationists, which seems at odds with their shooting habits. They write almost lovingly about the animals they hunted, and poetically about the jungle plants and creatures. So, for example, Stebbing commented on the magnificence of a bison he had tracked down: "Eighteen hands – and this specimen stood well over that – of coal-black beauty shining like satin on the back and sides, where the light filtering through the branches struck him, with four clean white stockings from the knee downwards." This did not stop him from shooting it, however: "I let the barrels sink slowly down till the sights made a bead on the shoulder and immediately pressed the trigger."[148] The tale continued with the tracking of the wounded bison into the jungle and its ultimate demise.

Stebbing, in particular, went to some lengths to reconcile these two facets of his life: "It is an experience common to many true sportsmen, I believe, that they soon grow tired of the mere slaughter of the animals they go out to seek. Gradually the fascination of the jungle lays its hold upon them, and of the jungle-loving denizens. It becomes a pastime of absorbing interest to watch the life of the jungle in its daily round from early morn to dewy eve, and again in the solemn watches of the night."[149]

By the time *Diary of a Sportsman Naturalist in India* was published in 1920, Stebbing was concerned about the falling numbers of game animals in India, which he believed were caused by better, cheaper firearms and the reduction due to agriculture of jungle tracts capable of supporting the larger animals.[150] Stebbing's observation that more, better-equipped hunters had contributed to the reduction of animal numbers rings true, but there is more than a little hypocrisy in this statement: the slaughter was fine while it was confined to the privileged few, but not when it was within reach of the masses.

Because Stebbing also believed that the "nefarious practices" of an "inhuman class of slayers," namely Indian poachers operating outside the government regulations, were also responsible for the reduction of the animal population, he also explored the notion of game reserves: "The idea of the Game Sanctuary was a natural outcome of the indiscriminate slaughter to which wild animals have at all times and in all countries been subjected by man. So long as it was man imperfectly armed against the animal with its natural sagacity or fierceness to protect it, conditions were equal, or in favour of the animal, and there was no reason for intervention."[151] In these sanctuaries all construction, forestry and hunting should, he argued, be forbidden and police should be responsible for enforcement.[152] There was much common sense in Stebbing's proposals, especially given the context of the times. To conclude his book, he quoted extensively and supportively the opinions of Peter Chalmers Mitchell, secretary of the Zoological Society of London, including the very modern-sounding words written in 1912: "I do not admit the right of the present generation to careless indifference or to wanton destruction. Each generation is the guardian of the existing resources of the world; it has come into a great inheritance, but only as a trustee."[153]

Education

Of the three subject branches at Coopers Hill, the Forestry Branch probably best prepared its students for work in India. It had a number of factors in its favour: a focused academic program, designed and led by acknowledged experts in the field; professors with extensive Indian experience; purpose-built new buildings; and a discipline of critical economic importance to India. It was also in the privileged position of being the only forest school in Britain.

When he established the Forestry School, William Schlich stated that Coopers Hill was not his preferred location but that it was mandated for political and economic reasons. With its physically separate facilities and different subject matter, the Forest School remained rather aloof from the rest of the college. President John Ottley was very clear about this during the 1903 enquiry into the college's future when being questioned by Charles Crosthwaite, chairman of the committee:

> Q: They [the forestry students] are really not an integral part of the College? A: No, not at all.
> Q: They are sort of an excrescence on the College? A: Quite so.
> Q: Do you think they are of use to the College beyond paying so many fees? A: They are of financial use, but that is the only use they are to the College.[154]

The students themselves seemed to have integrated more fully because social and sporting activities brought them together, enabling telegraph student Ernest Hudson to be friends with foresters Edward Oliver and Bertram Osmaston.

There were probably several related reasons behind Ottley's antipathy towards the foresters at a time when one might have thought that maintaining a united front would have been preferable for the future of the college. Four of these will be discussed further: the different selection processes for engineers and foresters, increased expenditures, specialized curriculum needs, and the developing idea of moving forestry out of Coopers Hill to a university.

Selection Process

Unlike the rest of the Coopers Hill students, the selection of forest students remained by competitive examination under the auspices of the Civil Service Commissioners, a system inherited from the time when foresters for India were trained in in France and Germany. Initially, the exam was tailored towards forestry and included physics, chemistry, and botany. In 1890, however, the secretary of state for India, echoing Chesney's original admissions policy, removed these subjects so as to cater to the backgrounds of students from the country's private schools. After much lobbying by Schlich and others, elements of science were eventually reintroduced in 1903.

By this selection process, the number of students admitted to the Forestry course was equal to the number of available appointments in India. The situation was therefore very like the early days of engineering, whereby a student was guaranteed a position unless he did very poorly in his studies. Ottley thought that their presence was a disadvantage to the college because forestry students were "not at all pre-eminently known as workers," in contrast to the engineering students who were competing for a limited number of appointments.

Increased Expenditure

Over the years since the introduction of the Forestry course at Coopers Hill, Schlich and his colleagues had used their position of strength to progressively enhance the program. Starting with Schlich as professor of forestry and Harry Marshall Ward as professor of botany, William Fisher was recruited as a second professor of forestry in November 1890. By 1895, the Forestry Branch comprised an expert group of professors, including Walter F.H. Blandford as lecturer in entomology,[155] Arthur H. Church as lecturer in chemistry of soils and vegetation, and Baden H. Baden-Powell teaching forest law, all of whom were paid by fee. For some years Dietrich Brandis himself led the practical tours to the Continent that the students undertook in their final year. When Ward left to go to Cambridge in 1895, he was replaced briefly by Charles A. Barber, and then permanently by Percy Groom in 1898.[156]

Between 1889 and 1896, the number of Indian Forestry appointments per year hovered around ten or eleven but then fell to around half that in subsequent years.[157] Despite its uniqueness, therefore, the Forest Branch contributed to rather than resolved the financial insecurity of the college.

Curriculum

The driver for the increased staff complement had been the desire to increase the specialization and relevance of the academic curriculum to meet the needs of Indian forestry.

Initially the course lasted two years. General subjects taken with the other Coopers Hill students included engineering, surveying, mathematics, geometrical and freehand drawing, and accounting. These subjects are clearly relevant to the foresters' work in the field such as surveying boundaries, mapping and demarcation, road-building, timber sales, and administrative reporting.

William Schlich covered the technical aspects of forestry. In year one, this included the environmental aspects of tree growth, managing forests, and protecting from damage from people and insects. In the second year, students studied the commercial management of forests, such as working plans, forest inspections, forest produce, felling and transport of timber, and technical qualities of woods. Again, the direct relevance to the duties of Lionel Osmaston and other foresters in India is apparent.

Under Marshall Ward, Forestry students also learned about the identification, anatomy, physiology, reproduction, and diseases of plants. This work included field trips to local woodlands, an extended visit to a British forest, and excursions to Kew Gardens where they could see Indian species first-hand. The course culminated in a tour of three or four months to Germany and France to "examine the systems pursued in the large and more systematically managed forests of those countries."[158]

In 1888 Schlich had made "three proposals in order to afford a more complete training to officers sent out for appointments in the Imperial Forest Service of India." These were to increase the course duration, appoint a second professor of forestry, and relieve himself of the responsibility for taking students abroad.[159] His proposals were approved, so when Fisher joined Coopers Hill as the second professor of forestry in 1890 the length of the course was increased from six terms to nine (twenty-four to thirty-four months). Within a few years, the last two of those terms were spent at the forests abroad, with students working in pairs under forest officers in Prussia.

Anthony Wimbush's time in Germany made a great impression. He and his friend Thomas Whitehead (CH 1904–6) were sent to the small town of Lauenau near Hannover to live with Herr Forstmeister Schulze and his family. With his uniform, Drillinger gun and back-pack of butterbrod, Schulze was the object of "the greatest admiration" by Wimbush and Whitehead. They also learned a great deal:

One of the main operations which required great skill, was the marking of trees for thinning in woods of different ages. After initial instruction the Forstmeister would hand over an area to Whitehead and me and tell us to complete the marking. This might keep us employed for a number of days and I have no doubt that the Forstmeister inspected the area by himself, from time to time, to see how we were getting on. He would see very easily whether we understood what we were doing and, if mistakes were being made he would put us right.[160]

They also worked with the forest's charcoal burners and roamed the forest in search of deer, which, Wimbush said, "enabled us to learn a great deal of wood-craft which was all part of our training and stood us in good stead in later years in India." After their time in Lauenau, the two students visited the Black Forest and Bavaria to study the spruce and oak forests there.[161]

After Schlich's changes, the program therefore included two additional academic terms, which were used to provide greater coverage of practical forest management and forest law. At the same time, the Forest Engineering already discussed replaced the Descriptive Engineering course taken by the engineering students. B.H. Baden-Powell's *Forest Law*, the book associated with his lecture series at Coopers Hill, is a lengthy compendium of twenty-seven lectures, ranging from the basics of legal rights, through the legal classification of Indian forests under the 1878 Act, to the legal powers and responsibilities of forest officers.[162]

The course at Coopers Hill did not include language training. This may at first seem a surprising deficiency, given Best's assertion that "an officer is really very little use until he has learnt the language of the country and the ways of the people in it."[163] However, the geographically defined region of the forest officer's duties in India meant that he needed knowledge of the relevant local language. As required by his contract, therefore, Lionel Osmaston passed his vernacular exam, as well as the higher standard exam, which were in his case both taken in Marathi. Ernest Hudson learned Hindustani (chapter 5), as did irrigation engineer William Sangster.[164] An officer in Burma, for example, would have needed a different language, hence Lionel's joke about it being his final exam unless he was transferred to Canara. From a practical standpoint, the college could not offer instruction in this range of languages.

Although Schlich would not have chosen to set up the Forest School at Coopers Hill, he quickly set about making the best of it, establishing a nursery to educate students in planting and nursery management. The nursery supplied 60,000 young pine trees each year to reforest an 800-acre wasteland area that the college leased for the decade from 1890 at £300 per year.[165] Once this was accomplished, the Crown resumed management of the area and the college's young trees were sold instead. Schlich also developed a series of quarter-acre model plantations on college property for scientific research and teaching.[166]

Oxford

The fourth potential reason for John Ottley's negativity towards the Forest School was that he may have already foreseen its departure from Coopers Hill. From the outset the foresters had been distinct, with their own buildings and curriculum, and the original vision shared by Schlich and Brandis included a school to educate foresters for Britain as well as India (the Government of India, which was paying, disagreed). The issue was still clearly on Schlich's mind in 1897 when he gave a lecture to the Royal Scottish Arboricultural Society, musing about the role of Edinburgh University in educating foresters for Scotland.[167]

In 1902, a committee was established by the British president of the Board of Agriculture to "report as to the present position and future prospects of Forestry" in the country. William Schlich was appointed as member. When it reported later the same year, it recommended that a forest school similar to that at Nancy could be established relatively cheaply in Britain by transferring the school at Coopers Hill to one of the universities.

So by the time the 1903 enquiry commenced, Schlich's position was already clear. Moreover, owing to the school's almost standalone structure, it was conceptually relatively straightforward to shift it in toto to a university (as indeed happened in 1905). Despite his protestations to the contrary, it sounded from his evidence to the committee as if Schlich had already explored the idea with Cambridge University. If so, his former colleague Harry Marshall Ward would have made a natural contact at that institution. However, Oxford was ultimately considered to be more suitable and the forest branch of Coopers Hill moved there (see chapter 8).

Practising in India

Indian Institutions

Three institutions had been established in the early days of organized forestry in India that served to educate, coordinate, and inform the widely dispersed community of forest officers. The first was the regular Forest Conference that brought together senior officers to discuss topics of common interest. The meeting held in Allahabad in 1874 that led to the 1878 Forest Act was one example, and Lionel Osmaston attended another with a focus on forest fires in 1909 at Poona.

Another outcome of the 1874 conference was the establishment of a quarterly *Forest Magazine* with Dr. Schlich, then conservator of Bengal Forests, as the honorary editor. The first edition came out a year later. Two issues were published in 1875, and four the following year under the title *Indian Forester*:

A Quarterly Magazine of Forestry. In 1883 the frequency was increased to monthly and the title changed to *Indian Forester: A Monthly Magazine of Forestry, Agriculture, Shikar & Travel.* Schlich served as editor for the first three volumes, succeeded first by James Sykes Gamble, and then in 1882 by another familiar name, W.R. Fisher. The journal is still in production today, one of the oldest forestry publications in the world.

The *Indian Forester* served as a general communication method between the foresters covering a wide range of topics and locations. It included technical articles and correspondence, reviews, official papers and announcements, extracts from the official gazettes of appointments, and updates on the timber market. Through the magazine's pages we can follow the career progress of Lionel and Bertram Osmaston and their colleagues as they were promoted, gave or received charge of their divisions, and asked questions on technical matters.

The third institution for educating foresters in India was the Forest School at Dehra Dun, located in the Himalayan foothills of Uttarakhand State between Simla and Naini Tal. The school had been established in 1878 by Brandis, who served as its first director, with a view to training Indians for the subordinate branches of the Forest Service. Within a few years it was offering two programs to forty-six students, an eighteen-month course in English for the Forest Rangers' Certificate and a twelve-month one in Hindi for the Foresters' Certificate. Prior to entering the school, each candidate was required to pass an exam and to have worked on probation in a forest for at least a year, to demonstrate his suitability for the work. The ranger students studied the theoretical aspects of forestry, practical surveying and silviculture in the forests, working plans and forest management, and culminated with further practical experience. The Forester Certificate focused on their work in the forest, such as "felling, pruning, thinning, natural regeneration, protection against fire, and making lime and charcoal, as well as the measurement of timber and the construction of forest roads and simple buildings."[168]

In 1906 the Government of India issued a resolution to rename the school to the Imperial Forest Research Institute and College and added "a staff of experts who will be in a position to devote a large proportion of their time to the prosecution of scientific research connected with forest produce, as well as to give the best available training to candidates for the Forest Services."[169] This staff was to be seconded from their home Forest Service to do research and teach in the college. Bertram Osmaston served as president during the First World War.

Dehra Dun did (and still does) form the spiritual home of the Indian Forest Department, serving as the repository, producer, and disseminator of technical information and training. In this it performed a function similar to that of the Telegraph Workshops in Calcutta, although the latter relied on Indian colleges such as Roorkee for training. The combination of Dehra Dun, *Indian Forester*, and the Forest Conferences served to keep the technical knowledge of India's forest officers up to date.

Scientific Limitations

In his memoir published by the *Coopers Hill Magazine*, Ernest Hudson quoted a witty remark by a forest officer friend to the effect that men in the Telegraphs, unlike foresters, did everything in "such a desperate hurry" because "trees take a long time to grow."[170] The unfortunate corollary of this long timescale was that any mistakes or limitations of scientific understanding took a long time to become apparent. During that time, the world did not stand still – an officer could not wait for twenty years to see how a particular forest management technique worked before deciding whether or not to use it again. Forests still required management, commercial extraction continued, and deforested areas still needed remediation.

It was therefore almost inevitable that some approaches to managing forests would prove to be unsuccessful. To really evaluate a forest system took several "rotations" through the entire planting-felling cycle, and there had been no time for any Indian forest to even complete a single rotation. But by the 1920s, some of the deficiencies in earlier schemes were becoming apparent. One was the widely used "coppice with standards" method in which some trees were coppiced (cut down to the ground to encourage new growth) but others (the standards) were allowed to continue growing. Wimbush described the process: "There was often considerable competition amongst the local contractors and usually the highest tender received would be accepted. The successful contractor had to sign an agreement undertaking, amongst other things, to save and protect the standards and to finish the felling and removal of the firewood before a specified date. It was the responsibility of the subordinate-forest staff to supervise the work of the contractor and to see that he kept to the conditions of the agreement."[171]

However, the coppicing with standards approach proved to be unsuccessful in many areas, and in 1922 Stebbing concluded, "Although this system is still in use in most parts of India ... modern practice has condemned its application on the large scale to replace the Selection System.... A decade to a decade and a half was sufficient to demonstrate that this form of management applied to large areas of forest was faulty."[172]

In valuable forests, the temptation was to leave too many standards, potentially causing the forest to gradually decline, leading to reduced forest coverage and a risk of soil erosion. Other misguided local policies included efforts to increase the yield of valuable timber products from a forest by changing the proportions of tree species, tending to convert mixed forests into monocultures, often of fast-growing species that could be felled sooner.[173]

Even if the science was correct, the forests did not exist in isolation but were part of a complex environmental and socio-political system.[174] So even a sound

working plan could fail if circumstances changed over the several-decade duration of the plan. Unusual weather, poor local implementation, short-term
exigency, greed, or political policy could all disrupt the forest officers' carefully
thought-out schemes.[175] The need for timber during the First World War was
an extreme example.

As captured by the 1894 circular 22-F, the overarching philosophy of the
Forest Service was that proper conservation and management could achieve
both environmental and commercial benefits. There is some suggestion that
the focus of the service came to favour exploitation over conservation.[176] Certainly, the total out-turn of timber and fuel increased from approximately
150 to 350 million cubic feet between 1900 and 1925. Stebbing considered
this to be evidence for the success of the forest management strategy, stating,
"The general introduction of better Working Plans and more intensive management, necessitating the provision of adequate funds, will yield far higher
results."[177]

Despite Stebbing's claim that the forests could sustainably provide this out-
turn and more, India's forest cover declined progressively during the century
between 1880 and 1980.[178] However, this deforestation was also mirrored by an
increase in the area devoted to agricultural land, suggesting that deforestation
was at least partly (depending on locality) the result of agricultural pressures
caused by a 600 per cent increase in population. Since 1980, however, the re-
forestation policies of the Indian government have led to a measurable increase
in forest area.

Return to England

Lionel Osmaston officially retired from the Indian Forest Service at the
relatively early date of December 1912; Bertram, in contrast, remained in India
until 1923.[179] On his return to England, Lionel served as an advisor to Lord
Robinson in the Forest of Dean, Gloucestershire, and transferred to the Forestry Commission when it was created in 1923. He eventually retired in 1931.

In his obituary in 1969, one of Lionel's colleagues in England commented on
his punctiliousness:

L.S. Osmaston was a man of strict and deeply religious[180] principles. Right was
right and wrong was wrong, but never the twain could be brought near together
by compromise.

In the 1920s I was helping Osmaston to draw up a Working Plan for the New
Forest. At 9 a.m. each morning I had to appear before him to state what I had done
in field work on the previous day and to tell him of my proposals for the coming
day. 8.55 would be too early, 9.05 would be too late.

… When I was working on this Working Plan job, Osmaston cannot have been in charge of the New Forest for more than 18 months, but I was amazed at his intimate and detailed knowledge of its every nook and corner. If his working day was strictly limited to 9 till 5 with a half-hour's luncheon break, he certainly never wasted a moment in idle chatter or study of anything outside the actual job.[181]

Selina and Lionel retired to Kent, just a few miles from Goodnestone, where she was raised. Selina died in 1949 in her eighty-second year. Lionel, Bertram, and Arthur Osmaston each lived to a ripe old age, reaching ninety-eight, ninety-three, and eighty-seven, respectively.

7
Engineers in India

Engineering at Coopers Hill

There is ... no country in the world in which the engineer has such an extraordinary and inexhaustible field for his labours as in India.[1]

Nearly half the graduates from Coopers Hill joined the Indian Public Works Department. In some ways their working lives were very similar to those of their Telegraph and Forest colleagues, especially in their junior years. But whereas the work of one telegraph or forest officer was generally similar to that of another, there was a great variety of engineering specialties in India. Likewise, while a senior engineer might ultimately be in charge of large construction projects, the telegraph and forest work did not typically change fundamentally as men progressed through their careers.

The engineers typically started their careers by being responsible for the inspection, maintenance, and operation of a local region, such as a stretch of road, railway, or canal. But the goal for most of them was to work on, and ultimately lead, high-profile construction projects, to leave their marks on the face of India. If they were successful in this objective, the engineers' later careers could be quite different from their early years. Moreover, the types of work undertaken – bridge, railway, canal, town sanitation – could vary widely from one engineer to another.

So while this chapter again asks what Coopers Hill engineers did in India and how their education prepared them, this question is approached through a selection of large-scale projects involving college graduates rather than using an individual's diary. In addition, the engineers were considerably less inclined to publish their personal recollections than were the forest officers. Amongst the few who did, Alfred Cornelius Newcombe's (CH 1871–4) informative *Village, Town and Jungle Life in India* stands out. William Hogarth Todd (CH 1897–1900) also published tales of tiger hunting in *Tiger, Tiger* and a memoir containing some of his engineering experiences in *Work, Sport and Play*.[2]

Instead, the engineers published detailed technical accounts of several construction projects in the *Minutes of Proceedings of the Institution of Civil Engineers*, which will be used here as the basis for understanding their work. Coopers Hill men also published technical books on their fields. These sources will be supplemented by the diary of canal engineer William Peter Sangster (CH 1891–4) and personal recollections of others from the *Coopers Hill Magazine*.

Coopers Hill Curriculum

About designing the Coopers Hill curriculum, George Chesney wrote, "Of course, in framing a scheme of education, one is met at once by the ever recurring difficulty where to draw the line between thoroughness and superficialness; between selecting to pursue one branch of science or attempting to survey briefly the circle of the sciences, and the difficulty is especially felt in a technical college of this kind in deciding between the conflicting claims of technical training and education *per se*."[3]

Chesney's "recurrent difficulty" still applies. There is never any shortage of technical material to cover, so what is the right balance between breadth and depth of content? Engineering educators are also rightly conscious of the need for students to understand the economic, environmental, business, global, and social aspects of their profession, as well as to appreciate the humanities and other areas of human expression. Since 1871, however, the emergence of multiple fields of engineering and the wide variety of institutions now offering an engineering education mean that current students have a wide choice of technical specialty and contextual framework.

When Coopers Hill opened, the curriculum consisted of nine academic terms, two of which were spent on practical training, so time was constrained. In contemplating his sweeping revisions to the academic operations of Coopers Hill, the newly installed President John Ottley wrote in 1900 that the completion of the whole engineering curriculum in ninety weeks (nine terms of ten weeks each, the Practical Course having been ceased) was challenging: "This period is all too short for all that we profess to teach, and the only logical conclusion is that we must adhere to essentials, and rigorously discard all luxuries, however tempting they may appear to be."[4]

Three incarnations of the Coopers Hill engineering curriculum can be identified, corresponding roughly to the 1870s, the 1880s and 1890s, and the 1900s (summarized in table 7.1). Except for transitional years between 1882 and 1884 as the New Scheme and new workshops were being fully implemented, the college's curriculum remained divided into four branches, as originally established by Chesney (see chapter 3). Material covered in the four branches was generally spread out over multiple academic years. To graduate, a student had to both exceed a minimum aggregate mark in each branch and attain a specified overall total mark (which could include any optional courses).[5]

Table 7.1. Evolution of the Coopers Hill Engineering Curriculum, 1874–1902

	1874–5	1895–6	1901–2
Branch 1	**Engineering**	**Engineering**	**Engineering**
	Descriptive Engineering	Descriptive Engineering	Engineering
	Surveying	Surveying	Engineering Designs
	Architecture	Elements of Architecture	Surveying
	Geometrical Drawing & Estimating	Geometrical Drawing	Building Construction
	Accounts	Estimating	Building Construction Designs
	Freehand Drawing	Accounts	Geometric Drawing
	Notes & Reports	Project (survey and estimate)	Freehand Drawing
	Project	Engineering Design	Estimating
		Mechanical Design	Accounts
		Workshop	Workshops
		Reports	
		Mechanical Laboratory	
		Architectural Design, or Alternative subjects (see below)	
Branch 2	**Mathematics**	**Applied Mechanics**	**Applied Mechanics**
	Pure & Applied Mathematics	Construction	Strength & Elasticity of Materials
	Applied Mechanics	Hydraulics & Mechanism	Hydraulics & Pneumatics
			Mechanical Laboratory
			Theory of Structures
			Mechanism & Heat Engines
Branch 3	**Natural Science**	**Mathematics**	**Mathematics**
	Experimental Science, Geology, & Minerology	Pure Mathematics	Pure Mathematics
		Applied Mathematics	Applied Mathematics
Branch 4	**Languages**	**Natural Science**	**Natural Science**
	Hindustani and History & Geography of India	Chemistry	General Physics, Sound, Heat, & Light
	Choice of two: Latin, Greek, French, German	Physics	Electricity & Magnetism
		Geology & Mineralogy	Electrotechnology
		Geological Excursions & Practice	Telegraphy & Telephony
		Chemistry Laboratory	Physical Laboratory
		Physical Laboratory	Electrotechnical Laboratory
			Chemistry
			Chemical Laboratory
			Geology

(Continued)

Table 7.1. Evolution of the Coopers Hill Engineering Curriculum, 1874–1902 *(continued)*

1874–5	1895–6	1901–2
Alternative subjects	One of: Architectural Design, Advanced Chemical Laboratory, or Advanced Physical Laboratory One of: Freehand Drawing, French, or German	

Examinations were given in each of the autumn and Easter terms. Results of the annual examinations held at the end of the summer term were printed in the college calendar, not only for the most recent session but for all previous years. For better or for worse, a student's academic record was therefore available for all to see, even thirty years later. For the second- and third-year examinations, separate rankings were given for each branch, and student performances were divided into first, second, and third classes (later reduced to two).[6] Students achieving a first-class ranking in one or more branches were awarded a higher diploma, instead of an ordinary diploma, and were thereby exempted from the entrance examination of the Institution of Civil Engineers.[7]

A portion of the assessment for all branches was performed by external examiners (36 per cent for Engineering, 27 per cent for Applied Mechanics, 28 per cent for Mathematics, and 50 per cent for Science).[8] Several Coopers Hill staff (Minchin, McLeod, Ward) performed this service for other institutions, and it was a common method by which professors could both supplement their incomes and demonstrate eminence in their fields. In 1871, fifteen external examiners were employed for a fee ranging up to £35 and costing the college £328 in total.[9]

According to John Ottley, these examiners were formally selected by the president, typically on the basis of suggestions from the individual instructors. Reading between the lines, Ottley was concerned that the relationship between the professors and the external examiners may have been too close to permit true independence. Certainly some examiners did the job annually for many years.

Students passing out of the college "with special distinction" were appointed "Honorary Fellows of Coopers Hill," abbreviated FCH. Typically, two or three fellowships were awarded each year, with the annual number ranging from zero in 1875 to six in 1900. Eighty-seven recipients were listed in the final 1902–3 calendar in the prestigious location immediately following the list of college staff. Anthony Wimbush wrote in his memoir that there was an additional benefit to being a FCH on the voyage to India: "As a concession to these men

Government ordered that in addition to free passages they should be intitled to have their drink bills on the voyage paid by the Government." Wimbush continued: "When the first contingent went on board ship for India a batch of them were standing talking on deck when an officer came up and said 'Are you Coopers Hill fellows?' Of course they answered that they were, whereupon the officer told them that they were entitled to free drinks throughout the voyage. This only happened once!"[10]

As discussed in chapter 3, the engineering curriculum included the Practical Course. In addition to the academic subjects, students participated in drill and gymnastics, for which a small number of marks were awarded. Under President Ottley, "physical drill and gymnastics," "drill and musketry," and "conduct and discipline" were assessed elements of the curriculum and required "optional" subjects for men seeking appointments in India.

1870s

In keeping with the changing times, therefore, the breadth of engineering content in the Coopers Hill curriculum progressively increased. However, in the 1870s much of that content is "hidden" under the applied science rubric, and the following courses contained significant engineering subject matter.

Descriptive Engineering (Branch 1)
A wide-ranging, non-mathematical introduction to engineering, including
    construction techniques; materials; canals; irrigation; steam engines;
    locomotives; bridges; tunnels; railways (see below)
Pure and Applied Mathematics (Branch 2)
Principles of statics, dynamics, kinetics, and hydrostatics
Applied Mechanics (Branch 2)
Construction: Stress, strain; mechanics of materials and structures;
    bending; torsion; frames and trusses; girders; joints
Hydraulics: Movement of water in orifices, pipes, canals, and rivers
Mechanism: Machine movement, pulleys, gears, hydraulics, work of
    machines, friction

Students could also opt to take advanced versions of several courses, including Descriptive Engineering, Architecture, Applied Mechanics, and Natural Science.

In these early years, Branch 3 was Natural Science, consisting of geology and mineralogy applicable to construction, as well as chemistry and physics under the umbrella of "experimental science." Branch 4 consisted of Hindustani and Indian history, together with modern European and Classical languages.

1880s and 1890s

From 1884–5 onwards, Branch 3 was retitled Applied Mechanics and comprised courses in Construction and Hydraulics & Mechanism. Branch 3 became Mathematics, consisting of Pure Mathematics and Applied Mathematics. These remained essentially unchanged until the late 1890s.

Branch 4 became an expanded section on Natural Science, with an increased emphasis on Chemistry and Physics, with their associated work in the laboratory (figure 7.1). Modern languages (the old Branch 4) could still be taken as an alternative to Architecture (Branch 1), but Hindustani and Indian history and geography disappeared.

At that time, there were several additions to the Engineering Branch of the curriculum, all taken in the third year:

Engineering Design
Design for a bridge or large roof, including notebooks, calculations, and
    drawings "sufficient for its actual execution"
Mechanical and Hydraulic Design
Design for a construction project (such as canal locks, aqueduct, or reservoir)
    or machine (steam engine, turbine, pump), with notes and drawings as for
    Engineering Design
Mechanical Laboratory
Using the Mechanical Laboratory to test iron and steel specimens and
    instruction in precision measurements.

The descriptions of the two new courses in Construction and Hydraulics & Mechanism (both split across years 2 and 3) were in fact almost identical to the two sections of the previous Applied Mechanics course. Similarly, the old Pure and Applied Mathematics course was separated into its constituent parts with little change in the list of topics and taken in first and second years.

The new mechanical testing laboratory of Coopers Hill had been installed in 1883. The main testing machine (figure 7.2) was capable of applying a load of 100 tons on a metal sample and was used to measure the sample's extension or compression under load. It was based on a commercial machine, but embodied "many improvements suggested by Professor Unwin." For teaching, it enabled students "to realize how far and under what loads iron and steel or other metals can be bent, crushed, or extended."[11] It was also used for commercial testing services, carried out by the superintendent of the workshops, mainly on "axles and tyres" destined for the Indian railways under contracts to the Government of India.[12] An analytical testing laboratory was also set up along similar lines to perform chemical testing.

Figure 7.1. Messrs. Atfield, Lewis, and Stotherd in the Coopers Hill chemistry laboratory, the "Stinks Lab," 1891 (© British Library Board, photo 448/1)

Under Natural Science (Branch 4) came fundamental classical physics (Years 1 and 2) and chemistry (Year 1), each with a laboratory component. An advanced third-year laboratory in either could be taken instead of the Architectural Design course from Branch 1. Geology and Mineralogy was split between first and second years.

1900s

By the 1901–2 academic year, a number of changes had been introduced into the engineering syllabus. The Applied Mechanics Branch (2) was reorganized to separate the various subjects into clearly identified courses, although the total emphasis on Branch 2 was in fact reduced slightly (because of the expansion of electrical technology).

Both changes resulted from Ottley's curricular and financial reforms (see chapter 4). The new construction curriculum was intended to make that branch more efficient by improving the alignment of the courses with instructors. Ottley also proposed the addition of mandatory material in electrical engineering, noting, "Electrical work is becoming daily more and more important to Engineers, and no man at the present day can be said to be properly equipped if he

Figure 7.2. The 100-ton testing machine in the Coopers Hill Mechanical Workshop, ca. 1900 (© British Library Board, IOR MSS EUR F239/82)

has not been taught at least the rudiments of the science, and something of its numerous applications in the profession."[13]

The Board of Visitors concurred, and the changes were brought into effect in the 1901–2 syllabus, necessitating the departures of Stocker and Shields. As will be described in chapter 8, Ottley also proposed many other adjustments to reduce costs and increase relevance, which touched on almost every aspect of the Coopers Hill curriculum. They also allowed the college to respond to the complaints from the Indian Telegraph Department about the college's coverage of electrical engineering (chapter 5).

Interestingly, the fraction of the curriculum devoted to what can be identified as "engineering" content did not change very greatly throughout the college's history (as determined by the number of marks allocated to the various subjects). As figure 7.3 shows, however, the combined fraction of engineering and natural science did become steadily greater. From 1877 to 1878, the academic program lasted the full three years as students destined for India started to take their practical training after the completion of, rather than as part of, their third year. The additional laboratory courses, resulting in the significant jump in combined percentage from the 1870s to the 1880s,

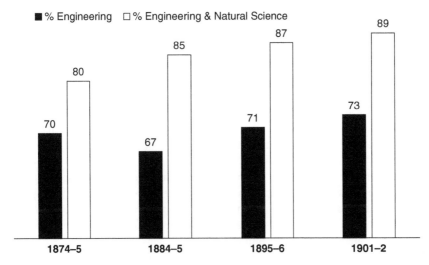

Figure 7.3. Fraction of the overall Coopers Hill course marks assigned to engineering subjects and to the combination of engineering and natural science in various academic years (Data from the college calendar for each year)

and the extra time meant that the absolute number of hours also increased. Later increases were from the inclusion of the theory and applications of electricity.[14]

Descriptive Engineering

Many of the Coopers Hill engineering courses were highly technical and would not be out of place in the first two years of a modern engineering program. These included the mathematical analysis of structures, basic materials testing, calculations for trusses and beams, and classical hydraulics and fluid flow. The empirical understanding of chemistry and classical physics were quite sophisticated, but of course the theory in both fields would undergo a revolution in the early 1900s with the advent of quantum mechanics. Likewise, electrical engineering was in its infancy. In some areas, however, such as engineering drawing, surveying, and workshops, the levels of skill demanded at Coopers Hill were considerably higher than in today's undergraduate programs. In the context of its time, the quantity and depth of technical expertise and practical know-how at Coopers Hill was immense.

In order to provide context for students' studies, most modern engineering programs include first-year courses on topics such as the engineering profession, engineering design projects, and introductions to the areas of engineering specialty. The equivalent at Coopers Hill was the Descriptive Engineering course, which covered "so much of the subject as can be dealt with without mathematics."[15]

The context in this particular case was that of engineering practice in India, so when the course was first introduced in 1871 the journal *Engineering* remarked, rather inelegantly, that in the course "special attention is paid to the peculiar requirements of engineering practice in India, and a desire is evinced to make the student partially acquainted with materials and the modes of dealing with them, that he will find most useful in the probable field of his ultimate usefulness."[16]

However, thirty years after having taken Descriptive Engineering himself, Robert Egerton was able to recall only a small part of the course that dealt specifically with India: "We were taught in particular ... the methods peculiar to India of brick making, brick burning, lime burning, and details of the limited materials available in India for building purposes."[17]

Considering that most construction undertaken by Coopers Hill engineers in India used stone, bricks, mortar, cement, timber, and other building materials sourced locally, this was in fact highly relevant content. Frederick Hebbert considered the main differences of practising engineering in India to be the cheap and plentiful labour, the widely varying conditions at different times of the year, and the adjustments needed in construction work to adapt to the monsoons.[18]

The full content of Descriptive Engineering in 1884–5 is shown in table 7.2. This course outline follows quite closely the contents of the textbook for the course, which was (ironically, given the inferior status accorded to Thomason College) the *Roorkee Treatise on Civil Engineering in India*, in two volumes.[19] The first edition had been assembled in the late 1860s by Major J.G. Medley of the Royal Engineers from teaching and other materials used at Roorkee, with an emphasis on engineering practice in India. Expanded later editions of these volumes were brought out by A.M. Lang, Medley's successor as principal of Thomason College, and the series was subsequently expanded into other areas (e.g., drawing, earthworks, railways).

Ernest Hudson's notes survive for part of the Descriptive Engineering course, dated 30 September 1885 at Coopers Hill, his first term at the college. This hardbacked notebook contains carefully written notes on the course material, along with beautiful coloured diagrams. True to its name, the course described general techniques for engineering construction, ranging from the correct proportions of ingredients for concrete to the organization of tramways for excavating cuttings.

Topics covered in Hudson's notebook are indicated by an asterisk in table 7.2, suggesting that this was one of several notebooks for the year, and that the content was grouped according to the expertise of the instructor rather than strictly following the sequence of the syllabus. The various elements of

Table 7.2. Topics Covered in Descriptive Engineering Course, 1884–5

Year 1		Years 2 and 3
Construction	**Design & Execution of Structures**	**Hydraulic Engineering**
Natural stone*	Foundations	Measurement & storage of water
Qualities of building stone*	Bridges	River engineering
Quarrying stone*	Tunnels*	Artificial canals
Bricks*	Roads	Water supply of towns
Brickmaking near London	Railways	Sanitary engineering
Cementing materials*		Irrigation in India
Wood varieties		
Felling & treating timber		**Mechanical Engineering**
Varieties of iron		Elementary pieces of machines
Production of pig iron		Hydraulic prime movers (e.g., waterwheels)
Production of wrought iron		Steam engines
Production of steel		Locomotives
Preservation of iron		Machinery for raising water (i.e., pumps)
Masonry*		
Bricklaying*		
Carpentry		
Use of metals in engineering & building		
Earthworks*		

Note: Topics marked with an asterisk are covered by Ernest Hudson's surviving course notebook.

the course were normally taught by Callcott Reilly (professor of construction), Thomas Hearson (professor of hydraulic engineering and mechanism), and Arthur Heath (assistant professor of engineering). Although the calendar listed all three names, Heath was considered to be primarily responsible for the course and was probably Hudson's instructor.[20]

Heath, who was formerly an engineer in the Great Northern Railway in Britain and who joined Coopers Hill in 1875, was popular with the students because of his extensive participation in the sports and clubs of the college and the "public spirit and high ideals of life and conduct."[21] Reilly and Hearson presumably gave the lectures in their own areas of expertise. Surviving examination papers from the early 1870s also show that the course at that time also started with the topics in Hudson's notes.[22]

At the time of its introduction, the Descriptive Engineering course was unusual in its breadth of coverage and integration of material. It was an innovative way of familiarizing students with the terminology and techniques of their new

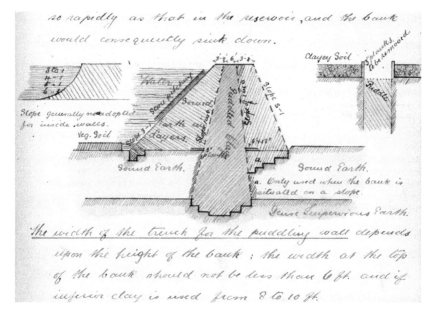

Figure 7.4. Diagram of earthen embankment from Ernest Hudson's "Descriptive Engineering" notebook, 1885 (Hudson Collection, CSAS)

profession, providing a broad general knowledge of the different branches of the subject, and for generating a sense of purpose for the engineers about to go to India. Other institutions, such as Owens College (forerunner of Manchester University) and Trinity College Dublin, adopted elements of the course and similar titles over time. In later years, it was apt to be looked down on by practising engineers as being insufficiently technical.

Descriptive Engineering was renamed simply "Engineering" for the last few years of the college's existence, and the content was updated. For example, the section on foundations was modernized to reflect the techniques then in use for bridge construction in India, as described later in this chapter. The first-year component was still essentially descriptive, but the course progressed to the "scientific principles underlying the practice of engineering" and the "practical application of these principles" to the construction of railways, canals, and other civil works.[23]

Tanks and Bunds

Figure 7.4 shows Hudson's diagram for an earthen embankment from his Descriptive Engineering notes. The "puddled" clay (made from a mixture of sand, clay, and water) forms the waterproof barrier, while the earth on either side provided strength and stability. Hogarth Todd described a similar structure used to create a small "tank" or reservoir in 1908–9: "Once the puddle-trench, as it is called, has been properly keyed along the whole length of the dam into sound earth or rock the construction of the headworks is plain sailing. The whole base of the embankment has first of all to be 'benched,' that is, cut into steps before the work is started; the earthwork has to be well watered and rolled,[24] and in this way the embankment of the 'tank' is rendered watertight for all time."[25] This embankment, or "bund," was built across a watercourse in the dry season and was equipped with a sluice to draw off water for irrigation and a channel to distribute the water to the fields, as well as a "waste weir" to divert flood water around the dam when the tank was full.

Todd went on to describe the construction of the similar, but much larger, Tendula (Tandula) embankments across the Tandula River between 1910 and 1921 (which are still in operation).[26] The work involved "thousands of coolies, helped by a large number of miniature trains." Since the work lasted several years, one of the measures specific to India was, as mentioned by Hebbert, the need to prevent each summer's floods from cutting away the unfinished embankment "like butter." This was achieved by building temporary waste weirs that were some four feet lower than the height of the partly finished dam. The weir needed sufficient capacity to handle the full flood of the river; the one Todd described was 600 feet long and designed for a depth of more than seven feet.

When that year's rains broke, the water quickly reached the maximum depth over the weir, and the escaping water roared "in a turgid stream down the rocky hillside to meet the riverbed again below the embankment." All sorts of driftwood, snakes alive and dead, and even the corpse of a fully equipped horse washed down to the dam. "Towards midnight the water ha[d] only another six inches to rise before it top[ped] the bund," so with the rain still falling they ordered the evacuation of the villages downstream in case the dam burst. But the rain eventually stopped, the water level fell little by little, and the dam was safe. They notified the villagers that it was safe to return home, only to find that none of them had actually evacuated.[27]

Learning about India

One of the main justifications for the continuation of Coopers Hill in the face of its financial challenges and the rise of other universities was that it alone provided specialized education for Indian service. When challenged to explain where in the curriculum this occurred, supporters of the college typically pointed to Descriptive Engineering, surveying, and accounting. They also observed that because of the "college traditions" a new recruit "imbibed unconsciously, more or less, a great deal of knowledge which it is very difficult to put down on paper, or to impart to him in any other way." The prospect of serving in India set the whole tone of the institution so that a young engineer taking up his appointment had a clear idea of what lay ahead and was, in the words of President John Ottey, "getting saturated with Indian ideas." Ottley cited as examples the presence of students of British descent who were raised in India, four or five "old Public Works servants" serving as tutors, previous graduates visiting from India, and letters from recently graduated friends. With some gall, Ottley also considered that part of the college's "Indian atmosphere" came from the presence of Indian students themselves.[28]

For the years they were published, the *Oracle* and *Coopers Hill Magazine* played important roles in this extracurricular dissemination of information, printing accounts of engineering works in India, hints for those about to go to India, stories of *shikar*, and articles about life in the country. With stirring stories of capturing elephants, tigers, and panthers, transformative engineering works, and adventures in foreign countries, the magazines would have fuelled excitement amongst the students working towards their diplomas and a sense of unlimited opportunities throughout the empire.

The college library subscribed to *Indian Engineering* as well as British publications such as *Engineering* and the *Engineer*. There were informal channels of communication as well. We know from Hudson's diary that he corresponded from India with his friends still attending the college, doubtless telling tales of his adventures. It was also customary for men on leave from India to call into Coopers Hill to visit their former professors and share their experiences. Lastly, many students already knew about India, having been born there or by following family members into the college. In 1891, for example, eighteen students in residence were born in India (15 per cent of the total), including Hugo Wood. A decade later, more than twenty students were born in India with a similar number from another dozen countries around the world.[29]

Practical Course

Engineering students who had achieved positions in India were appointed as assistant engineers from around the beginning of the October after their graduation. For the twenty years between 1878 and 1898, the expectation was that the young engineers would spend the next year gaining practical experience with an engineer in England. The Government of India paid the pupilage fees in addition to a salary of £150. Two half-yearly "premiums" of £10–25 each could be awarded "according to the degree of proficiency and diligence shown" by each student, as judged from their notebooks and plans.[30] The total cost per student could therefore be up to £200, as well as the pupilage fee. If a man withdrew from the Indian service, he was required to pay back the pupilage fees, and if he "did not make sufficiently good use" of the Practical Course, he could be required to repeat it with no pay, or in extreme cases his appointment could be cancelled.

The wording in the college calendar was that "every Assistant Engineer *may* be required to go through a course of practical engineering" (emphasis added). The initial uptake of the fourth-year Practical Course was fairly low, perhaps because there was still a demand for new engineers in India. Of the forty-three men appointed to the PWD in 1879, for example, three did not take up their appointments, fifteen went directly to India, and twenty-five (58 per cent) went on the Practical Course.[31] In the early 1880s, the participation rate varied between 56 and 71 per cent, but by later that decade the majority of PWD appointees took the practical experience year. The annual costs were tabulated in the report of the 1903 enquiry and comprised the salaries of the participants, pupilage fees to the engineers, and costs for supervising the work experience.[32] This supervisor was typically an officer on leave from India who received £150 per year and a travelling allowance of one guinea per day.[33] For the years 1879 to 1884, the average annual cost per participant was £301,[34] significantly higher than the £200 estimated by Chesney when he proposed the scheme.[35] In 1879, twelve engineers joined companies working on civil engineering (iron works, water works, docks, and canals), while thirteen joined railways and locomotive works.

The number dropped to zero in 1899, the year after the course was officially stopped. In cancelling the Practical Course, the Government of India canvassed the opinions of administrators and engineers in India on the view that "the knowledge of engineering methods gained by the students during their practical course was incommensurate with the time spent thereon and with the attendant expense to the state, and that such experience could be as readily acquired by the young engineer on works in India without any preliminary practical course in England."[36] The Practical Course was "condemned by a large majority of those consulted" and it was therefore cancelled. This was a sign of how circumstances had changed since the late 1870s when Chesney believed that the level of engineering activity in India was insufficient to guarantee meaningful practical training.

While some students lamented the abolition,[37] a leading article in the *Coopers Hill Magazine* explained further. Probably representing the official line, it suggested that on the Practical Course in England a student was merely a "visitor" with no responsibilities and that he served under an engineer whose interest in their education was uncertain. The only professional control of the experience was the college's supervisor of the Practical Course, whose visits were "like those of the angels, 'few and far between.'" On the other hand, the magazine opined immodestly that the year would be better spent in India as "a member of an organised service, the proudest and most efficient that the world has ever seen." There he would have real responsibilities under the guidance of an engineer with similar motivations and a personal interest in their success, for "without responsibility there is no such thing as 'practical training.'"[38]

After 1898, a few students (up to two per year for a maximum of two years) could be recommended by the president for additional practical training for particular areas of specialization,[39] and the expenditures indicated that one or two students per year were indeed selected.

Coopers Hill Engineers in India

Public Works Department

The rules of the PWD were laid out in its code, published approximately every five years. Unlike a modern building code, however, it was not a technical manual but an administrative one. It outlined the rules of the department on a wide range of issues such as duties, appointments and promotions, personal matters, and accounting. In principle, the administrative structure for public works in British-ruled India between 1871 and 1906 was fairly straightforward: The department was under the overall authority of the Public Works secretary to the Government of India, who was assisted by deputy secretaries for each branch (Buildings and Roads and Irrigation, later combined into Civil Works, Railways, and Accounts). One or two chief engineers were allocated to each of the twelve provinces, depending on the size.[40] The reality was more complex because of a variety of exceptions – special rules often applied to the presidencies of Madras and Bombay, state railways were managed separately, civil works in areas of strategic military importance could be under Military Works, and princely states managed their own affairs.

A larger province might have two chief engineers, one for Buildings and Roads and one for Irrigation. Each was in overall charge of that branch and reported to the local government. Reporting to the chief engineer(s) were superintending engineers, who were in charge of engineering works for a geographic area known as a circle. Within each circle were divisions, each led by an executive engineer.[41]

Within each division, assistant engineers were employed to work on specific projects or areas under the orders of the executive engineer.[42] Coopers Hill men were typically appointed as assistant engineers, second grade at a salary of Rs4200 per year.[43] Their duties as described in the PWD Code included making surveys; collecting information concerning – and drawing up – projects, designs, and estimates; and superintending the construction of works.[44]

These ranks together constituted the Engineer Establishment of the PWD, drawn from Coopers Hill, Indian engineering colleges, the Royal Engineers, and other qualified army officers. There was also an Upper Subordinate

Establishment (sub-engineers, supervisors, and overseers) and a Lower Sub-ordinate Establishment (sub-overseers), which were both staffed mostly by Indians.

With respect to initial training, the code was vague and brief: "Assistant Engineers, on being first posted, should, as a rule, be kept at the Head-Quarters of the Division, until they have made themselves thoroughly acquainted with the office work and forms of account, and with the general routine of duty, as also for the purpose of studying the vernacular."[45]

The Assistant Engineer

For example, the first few months of 1896 saw William Sangster (CH 1891–4) patrolling the canal around Madhopur, inspecting the canal bed, repairs to the rapids,[46] and silt clearance operations during a closure of the canal for maintenance. Two decades earlier than Sangster, on the other hand, Alfred Newcombe had taken the same trip from Bombay to the Punjab, also to work on the Upper Bari Doab Canal. While his Coopers Hill classmates were dispersed to their postings, Newcombe's practical training commenced by accompanying the executive engineer and his wife on a tour of inspection. The routine was much like Hudson's or Osmaston's: in the morning they travelled twelve to fifteen miles between bungalows along the canal "inspecting the earthwork, bridges and other works in progress," and in the afternoon doing office work. As the next step in his training, Newcombe was set to surveying minor irrigation channels used for distributing the water from the main canal into the fields, which involved supervising a team of workers in addition to his own servants.

Newcombe had obviously performed his initial tasks satisfactorily because he was next given a thirty-five-mile-long subdivision of the canal to look after, based at a bungalow at Gurdaspur next to the canal between Amritsar and Madhopur. The bungalow was a comfortable base from which to ride out and inspect the canal construction. But, in common with the experience of murdered railway engineer Edmund Elliot, the location was very isolated, with the nearest European company about ten miles away.

The Executive Engineer

William Sangster worked on a number of impressive irrigation projects, but his name was particularly associated with the construction of the most challenging part of the Upper Swat River Canal (opened 1914). As executive engineer, he was responsible for construction works in his division, including driving a 2¼-mile canal tunnel through the hard granite of the Malakand range, the longest of its kind in India (see below). In 1913, as construction was in full swing, Sangster's day followed a fairly stable routine.[47] He was typically up at around

6:15 a.m. and went straight to the office to sign official *dâk*, such as work ab-
stracts, salaries, and registers of works. He sometimes used this time to catch
up on technical reading, prepare papers, or write his home letter. At 8:15 or
so, he would go home to his wife for breakfast and "glance at" the newspapers.
Afterwards, he would open the day's post and send it over to the office. He
might discuss work at home with one of the European engineers, return to the
office to deal with other urgent matters, or go out to inspect the construction
works. After lunch at home at about 2 p.m., Sangster would return to the office
for more work or go on inspection until late afternoon. He would then change
for tennis or for his particular passion, polo. At around 7.30 p.m. he would
return home to change for dinner, often dining with colleagues or visitors and
chatting afterwards. In the absence of guests, he would read magazines such as
the *Strand* or the *Tatler*, a non-fiction book, or a novel, and then retire to bed at
around 11.30 p.m. While he travelled almost daily in his local region, the local-
ized nature of the construction work meant that Sangster did not undertake the
lengthy travels of Lionel Osmaston or Ernest Hudson.

Newcombe on the Weather

Alfred Newcombe painted particularly vivid pictures of the tribulations of the
Punjab climate – intense, dry heat, dust storms, and the insects one might encounter on the dinner table. He particularly described the isolation faced by the
assistant engineer:

> For those whose headquarters are in isolated parts, and for days at a time obliged
> to remain mostly indoors, the solitude occasionally becomes oppressive. One can-
> not read and write much for pleasure, especially after some hours of office work or
> preparation of plans.... Sometimes I held imaginary conversations with imaginary
> people, and at other times I would shout out loud to break the intolerable silence.
> One's nerves get out of order, and such trifles as the chirruping of the squirrels
> outside on the verandah roof or in the trees, or the too rich scent of the lime-trees,
> may irritate them.[48]

The hot season also brought challenges at work, especially after a disturbed
night:

> At [the] office many sets of papers are on the table, and have frequently to be re-
> ferred to. To keep them from being blown about by the punkah [swinging ceiling
> fan] it is necessary to use a paper-weight on each set of loose papers, and to keep
> one hand on the paper one is reading or writing on. The other may be used to raise
> a weight to get at some other document for reference. It is then that one finds that
> two hands are not enough, for, as soon as the weight is raised, a wave of the punkah

may cause half a dozen loose sheets to sail away across the room. When a man has had a bad night and a few whisky-and-soda pegs, and comes to office in a not very placid state of mind, such occurrences do not improve his temper.[49]

In the light of these comments, it is perhaps easier to empathize with Robert Hawkins, the murderer of Edmund Elliot in 1893. Reading back over the newspaper accounts of Hawkins's trial, it is hard not to sense Hawkins's mounting paranoia.

Hawkins shot Elliot next to the railway line as they met to hand over some measurement books in advance of Hawkins's termination (chapter 2). His explanation was that Elliot's final comment, "I have done for you now," coming on top of a "long course of insults and taunts so deprived him of control" that he acted on impulse. Significantly, the executive engineer believed that the trouble between the two men started in the rains of 1892. Increasingly intemperate and melodramatic communications, on Hawkins's part at least, developed in the communications book. Sensitive to his social position and the criticisms of a more fortunate contemporary, Hawkins felt the situation was "as much as any mortal could bear."[50]

The defence's argument of temporary insanity was unsuccessful because Hawkins apparently "knew perfectly well, before and after, what he was going to do, and what he had done," and because he had not shown any trace of aberration of the mind after his arrest.[51] However, in the light of Newcombe's descriptions above, one could easily understand that Hawkins, feeling sick and lethargic and driven to distraction by the accumulation of minor annoyances, could very well have "brooded over wrongs and nursed his wrath on such trifling grounds."[52] He may not have met the definition of insanity in the court's eyes but he was clearly unwell.

Coopers Hill Engineers in Action

According to the Coopers Hill calendar, the first person to be appointed from the college to India was William Patrick Brodie in 1872, just a year after the college opened. He was followed the next year by Robert Grieg Kennedy, Charles William Hodson, and John Benton (of the latter two, more later). Their appointments were expedited because they had already had practical training and, in Benton's case, had studied engineering in Edinburgh before entering Coopers Hill.[53]

Most Coopers Hill graduates had retired from service under the superannuation rules of the PWD before 1940. The last Coopers Hill men on active service may have been Arthur Brokenshaw (CH ~1903–6, who as a young telegraph officer worked with Ernest Hudson in 1907) and Francis Farquharson (CH 1904–6, chief engineer, Punjab irrigation), both of whom retired in 1941.

Muhammad Ahsan (CH 1900–3), who as Nawab Yar Jung Bahadur Ahsan served the nizam's government in Hyderabad as chief engineer, probably also retired at around this time.[54]

Hence, there were Coopers Hill engineers working in India for almost seventy years and they dominated the PWD well after the college's closure. In 1912, nine of the thirteen chief engineers (69 per cent) and forty-five of the seventy-one superintending engineers (64 per cent) were from Coopers Hill. A decade later, despite the goal of the newly formed Indian Service of Engineers to recruit more Indians, these numbers were still 62 and 53 per cent, respectively.[55]

From the hundreds of projects, large and small, carried out by Coopers Hill engineers, a few are selected for discussion here. They cover railway construction, tunnelling, bridge building, and canals. These examples are chosen because detailed technical descriptions were published, first-hand accounts or photographs survive, and a connection can be made to the Coopers Hill curriculum.

Railways

The Kandahar Railway

For much of the existence of Coopers Hill, one pressing political concern was the threat of Russia attacking the north-western borders of India through Afghanistan. At the end of the Second Anglo-Afghan War (1878–80), Afghanistan was vacated by the British. The country was left under the authority of Abdur Raman Khan, under British protection and control of foreign policy. The cantonment at Quetta was therefore strategically critical for guarding the Bolan Pass, the passage between the Punjab and Kandahar. The town also became the base for thousands of British troops poised to rush to the defence of Kandahar if necessary. With the extreme difficulties of maintaining the supply route to Kandahar top of mind, the construction of a railway that notionally joined the Indus Valley State Railway to Kandahar – the Kandahar Railway – was therefore officially sanctioned.

Such was Quetta's importance that considerable funds were spent to build this line with the "utmost rapidity." A few years later, similarly large sums were expended to push through and maintain the highly problematic Bolan and Sind-Pishin Lines to Quetta (figure 5.3). There was also a military road following the Bolan route and an extension line to Chaman on the Afghan boundary. Once the line was through to Chaman in 1892, enough railway supplies were stockpiled there to extend the last seventy-five miles to Kandahar in an emergency. In 1896, Ernest Hudson was responsible for moving the telegraph military reserve stores from his compound at Quetta to Chaman so that communications could be extended equally quickly to Kandahar.

Each of these railway lines provided employment for Coopers Hill engineers. George Moyle (CH 1872–5) worked on the Kandahar line in 1879 and a profile in *Indian Engineering* commented that, despite his youthful appearance, "he was a man of action, vigorous and effective, a doer of things, rather than a talker about them."[56] Because of the urgency to build the line to support the ongoing engagement in Afghanistan, materials and supplies were requisitioned from far and wide, with the result that a great variety of rail shapes and fittings had to be assembled into a working line. Moyle was at pains to point out that the circumstances prevented "economy being studied as a primary consideration."[57] Transport of those materials was difficult because this section of the railway was separated by water from the rest of the Indian rail system – the Indus had not yet been bridged,[58] so supplies had to be ferried across the river or delivered by sea to Karachi. Despite these challenges, 133½ miles of railway were opened to military traffic after just 101 days (6 October 1879 to 14 January 1880).

In his 1890 *Notes on Permanent-Way Material, Platelaying and Points and Crossings*, William Henry Cole (CH 1873–6) wrote that the plate-laying on the Kandahar Railway was "perhaps the smartest piece of work of the kind on record."[59] The arduous business of plate-laying, as the railway track construction is known, was described by Moyle himself:

> The organisation of the platelaying gangs was as follows:– The work was divided into three departments, each under the charge of a picked European subordinate. These were – 1. Material, six hundred men; 2. Laying, six hundred; and, 3. Lifting and rough packing, four hundred; altogether, sixteen hundred men. The material gangs had to unload all permanent way from the trains, load it into carts, carry it to the tip and unload it, distributing rails, sleepers, and fastenings exactly as they were required; also to clear the line behind of all surplus material.... Next in order the laying gangs picked up the material from alongside the line, placing it so as to give as little trouble as possible to the linkers; these were in turn followed in order by keymen, gaugers, borers, spikers, pick-up trollies, rough straighteners and lifters, each distinct operation being looked after by a native platelaying inspector. The third party of men lifted, packed, and straightened the line sufficient to allow of trains passing over it safely at a speed of from 15 to 20 miles an hour, and also carefully tested and perfected all bolting, keying, gauging, and spiking.[60]

Moyle went on to clarify that these 1600 men were only those employed on the actual laying; the camp contained 3500 men in total, with the others being used for earthworks, bridging, water distribution (20,000 gallons per day), and sanitary. There were also cart-men to carry the 600 cartloads of materials that were required for each mile of track (mostly for the rails and sleepers). Assembling these workers and training them on the seven different rail systems for which they had supplies proved to be challenging. Moreover, it was difficult to convince the local labourers to work in the desert between Jacobabad and Sibi. Indeed, for a five-day period in early November, the workforce went on a general strike until it was "terminated by a judicious removal of dissatisfied men, and by explaining carefully to the remainder what arrangements had been made to ensure their safety from privation in crossing the desert."[61]

The Descriptive Engineering course at Coopers Hill contained a reasonable overview of the principles that Moyle would have needed to know when starting on the railways: "Permanent way of railways. Gauge of railways. Ballast. Timber sleepers. Rails. Chairs. Rail joints, fish joints. Cast iron sleepers. Wrought iron sleepers. Cant of rails. Elevation of outer rail on curves. Sidings. Switches and crossings. Turntables."[62] Assuming the *Roorkee Treatise*, as textbook for the course, is a good guide to what was discussed at Coopers Hill, the students learned quite a detailed set of terminologies, rules of thumb, dimensions and weights, specifications, calculations of stresses, and examples from Indian railways connected with laying the permanent way of a railway. This would have been quite adequate for students to understand Molye's and Bell's papers on the construction of the Kandahar Railway, if not all the practical know-how necessary to lay the line with such speed.

The Sind-Pishin Railway

With the end of the Second Anglo-Afghan War in September 1880 and the election of Gladstone as British prime minister on a platform of peace, the urgency to build the next instalment of the railway to Kandahar temporarily faded. In fact, extensive preliminary surveys had been made of the two potentially viable routes from Sibi to Quetta by Sir Richard Temple, and it had been decided to follow the Harnai[63] route because the gradients were less steep. This was the Sind-Pishin Railway.[64]

Gladstone's government reversed its decision in 1883 after another war scare with Russia, and the project was restarted under the code name of the "Harnai Road Improvement Scheme" (supposedly to save the prime minister's face). In due course "a large and enthusiastic staff of military and civil engineers"[65] gathered at Sibi including Coopers Hill men: Charles Hodson, Ernest Shadbolt, Frank "The Baron" Fowler, William Johns, George Rose, Henry Savory, Richard Woods, Francis Pope, Edmund "Friar" Tuck, and Charles Cole. Several of these men would go on to work on other Baluchistan railways.

The Harnai Line was put under the command of Col. James "Buster" Browne R.E., one of the larger-than-life military figures popularly associated with British India: "Distinguished alike as a brave soldier, a scientific and able engineer and an accomplished linguist, he was above all one who ruled over men in the fear of God and won the warm affection of all who served under him."[66] However, two very different views emerged of the roles of the Royal Engineers and civil engineers, providing a graphic illustration of the tensions still present between the two groups twenty years after the establishment of Coopers Hill.

Browne's personal assistant, G.K. Scott-Moncrieff R.E., a self-proclaimed planet around Browne's sun, minimized the role of the civil engineers: "A large number of officers, chiefly civil engineers, arrived on the works. Many of these were first-rate men, as good as one could wish to find, but there were also some very sore trials. One man came to the work and remained for five weeks doing absolutely nothing, bewildered, I suppose, with the rush and energy all round, and then he put in a medical certificate and departed. Another pair were continually quarrelling and reporting each other to headquarters."[67]

Alternatively, in its profile of Ernest Shadbolt, who had been in charge of the difficult Nari Division, *Indian Engineering* later wrote about Browne's leadership: "There was no competent designing, no method, no co-ordination; the whole thing was a nightmare of ineptitude; and on this amateur performance the young Coopers Hill executives and assistants, posted to it, opened their eyes in wonder. It seemed to them that all that was desired was to push the construction on at speed regardless of expense, accuracy of design, or of the quality of the work."[68]

Even Moncrieff admitted that "with two or three exceptions, the Royal Engineer officers sent up to the works knew nothing, practically, about the technicalities of railway engineering," and even the more experienced civil engineers still had something to learn from the "peculiar circumstances of the country."[69] Moncrieff went on to describe the need for and challenges of precise surveying in laying out the railway route, in which the Coopers Hill men would have been well versed.

Amidst heavy rains and cholera, the route of the line was resurveyed in 1883, and construction started using several thousand sappers under Browne's command and an Indian labour force of some 15,000. The first stretch of railway from Sibi through the gorge of the Nari River with its "turmoil of waters" during times of flood was challenging for its steep cliffs and numerous bridges and heavy cuttings. The narrow, winding ravine near Kochali was similarly difficult with tunnels and difficult cuttings, some 100 feet deep and held by heavy revetments.

From Spintangi the route was straightforward until the "extraordinary freak of nature" known as the Chappar Rift was reached.[70] This fissure, crossed by the dramatic Louise Margaret Bridge, was the key to the Harnai route because it provided a way to get up to the next, higher plateau with an acceptable track incline. The last particularly difficult stretch of the line traversed "mud gorge" with its treacherous loose rock that turned to mud with any rainfall. Here the line was carried in cut-and-cover tunnels, protected by banks and channels, to minimize the chance of disruption. Despite these precautions, traffic was interrupted several times per year, often for many days, by washouts, landslips, and flooding near Mudgorge station. This was clearly not desirable for a critical military supply line.[71] Hudson commented in March 1894 that after three days of heavy rain the Sind-Pishin Line was breached in fourteen places. Later he walked to Mudgorge station and "saw the places where the hill side had slipped down onto the R[ailwa]y at 'Puddle Hollow' last month."[72]

After furlough in England and a period as quartermaster-general for India, James Browne was appointed governor-general's agent and chief commissioner to Baluchistan, based at Quetta.[73] There, on 28 April 1894, Ernest Hudson dined with Sir James and Lady Browne at "a very sociable little dinner party." Hudson also recorded Browne's serious illness due to gout during late May and early June 1896, followed by his death on 13 June: "Sir James Brown [sic] died at 6.30 a.m. and was buried in the evening at about 8 o'clock. We all assembled for the funeral procession at the Residency and then marched up to the cemetery. All the troops turned out to line the road and do escort duty."[74]

Bolan Line and Chaman Extension

The Sind-Pishin Railway was opened at Chappar Rift on 27 March 1887 by the Duchess of Connaught (the eponymous Louise Margaret, at whose wedding the Coopers Hill Volunteer Corps had served as honour guard, chapter 4). But even before the completion of that line, military exigency necessitated a much quicker way of establishing a supply route to Quetta, and the idea of rapidly constructing a temporary route from Sibi to Quetta through the Bolan Pass was conceived. The fifty-seven-mile line was constructed between May 1885 and August 1886. Referring to figure 5.3, the route (although not shown) first went west from Nari Bank to Kundilani, with its 800-foot cliffs, and then north to Hirok following the Bolan River. The rails ran along the riverbed on what Hudson called an "Irish causeway." In times of rain, the river rose extremely quickly, tossing down huge boulders, so the route was frequently flooded and often needed reconstruction. On one occasion Hudson wrote that the river was "coming down like a mill race & rails 10–12ft under water." A month later his train was nearly derailed as the engine's wake in the flood water caused a bank to crumble.[75]

The construction of the Bolan Line was initially under the command of Col. Lindsay (from the Kandahar Railway). But when he was injured by falling from a trolley, the line was taken over by another larger-than-life character, Francis Langford O'Callaghan, and a team of civil engineers.[76] This line was a cheap, quick fix, with an inconvenient change to a narrower gauge at Hirok to assist with the steep gradient. It was completed in time to help supply the upper reaches of the Sind-Peshin construction. While it achieved its purpose, the Bolan Line was very susceptible to flooding, as Hudson's experiences travelling by train and trolley over the submerged rails demonstrated.

O'Callaghan was also given responsibility for constructing the challenging Chaman Extension Line that led north-west towards Chaman on the Afghan border from the junction of the Bolan and Sind-Pishin Lines at Bostan.[77] From about 5200 feet above sea level at Bostan the line climbs to 6400 feet at Shelabagh, before descending again to Chaman at 4300 feet. This route features the famous Khojak Tunnel, which passes beneath the Kwaja Amran mountain range. At nearly two and a half miles in length it was the longest in India at the time. The tunnel was surveyed by O'Callaghan in late 1887 and the first train ran through it on 1 January 1892.

With its usual urgency, the army could not wait for completion of the tunnel to commence rail traffic to Chaman. Instead, O'Callaghan assigned a Coopers Hill engineer, Walter Weightman (CH 1878–81),[78] to develop a temporary method for connecting the two stretches of permanent way on either side of the hills. Weightman's solution was a "rope-incline" that could lift railway waggons up one side of the mountain, roll them along the ridge, and then lower them down the other (figure 7.5). As well as supplying Chaman, it enabled the transport of machinery and supplies to the west end of the tunnel during its construction.[79] Weightman noted that he designed and constructed the ropeway on O'Callaghan's instructions and that his solution was approved by the consulting engineer to the government, Sir Guilford Molesworth.

After being drawn up to the start of the incline at Wallers Camp by powerful locomotives, the ordinary locomotives and waggons were shunted onto a special vehicle, shown in figure 7.6, that kept them level as they were winched up the incline. Weightman described the workings of the rope-way over the summit:

No. 1 incline, worked by a stationary engine and wire rope, rises with a gradient of 1 in 2¾ till it reaches the summit of the range, a height of 7,260 feet above sea-level. Next follows a short length of single broad-gauge line (5 furlongs) along the ridge, with a ruling gradient of 1 in 40, worked by small tank-engines; and then commences the descent of nearly 2,000 feet on the west side. This is achieved by three separate rope inclines, the first having a gradient of 1 in 2½, the next of between 1 in 7 and 1 in 10, and the third of between 1 in 8 and 1 in 13.[80]

Inclines 1 and 2 were worked with one car ascending and one descending. The tracks were only one waggon wide but had a passing place midway. To avoid the complexity of switching between them, Weightman used a clever system of parallel rails placed six inches apart that divided at the passing point. During its three years of operation from late 1888, the rope-way carried more than 95,000 tons of cargo over the Kwaja Amran range.

A photographic record of the incline rope-way construction was kept by another Coopers Hill engineer, Richard Douglas Perceval (CH 1877–80). His four albums contain informative photographs of Wallers Camp, the scenery

Figure 7.5. Eastern section of the Khojak Rope-Incline (Incline 1), built by Walter
Weightman in 1888, with trucks at the passing place (© British Library Board, photo 481/3)

around the Khojak Tunnel, and the railway construction. A group photograph
(presumably from 1887) shows engineers on the Kwaja Amran Railway Survey
including O'Callaghan and Coopers Hill men R.J. Woods (CH 1875–8), S. de
Brath (CH 1874–7), Perceval himself, and E. Napier (CH 1880–3). Weightman
was there too, sporting his monocle, floppy moustache, and Coopers Hill tie.

Figure 7.6. Covered railway wagon on an incline truck at the top of the eastern incline of the Khojak Rope-Incline (© British Library Board, photo 481/2)

Because of the unsettled situation in the region, the Khojak Tunnel was protected by watchtowers, turrets, and block houses. In April 1896, Hudson described running the telegraph cable into some of these defensive positions and installing lightning dischargers. This was part of the larger installation of cables through the tunnel to avoid the more vulnerable route over the summit. Communications were switched to the new cable in July 1896.

Tunnels

The Panir Tunnel on the Mushkaf-Bolan Railway

Repeated breaches in both the Bolan and Sind-Pishin Lines, and the high cost of repairing them, soon made it clear that a more reliable alternative route to Quetta was needed. That new line, the Mushkaf-Bolan Railway, was introduced in chapter 5 because of its place in Hudson's career; here the focus is on the route's many tunnels.

James Ramsay, the chief engineer of the line, was not from Coopers Hill, but most of the other engineers were. Charles Hodson replaced Ramsay when he reached retirement age in July 1894.[81] In charge of divisions were Charles Cole (responsible for the Panir Tunnel and Lower Bolan), William Johns (Mushkaf Valley) – both veterans of the Sind-Pishin Line – and Thomas Curry (Upper Bolan). While the line was under construction, the young engineers Horace Walton and Edmund Lister lived in the fortified station buildings at Pishi and Panir stations, respectively.[82] A close community developed amongst the engineering leadership along the line, and Hudson mentioned in his diary dining, shooting, or conferring on technical matters with all these Coopers Hill colleagues.

In his overview of the Mushkaf-Bolan Railway,[83] James Ramsay listed sixteen tunnels, several of which contain more than one section. Some are named after their location (Mushkaf, Panir, Sir-i-Bolan) or given descriptive names (Rift, Red Clay). Others, however, are more fanciful – Cocked Hat, Bella Vista, Cascade, Windy Corner, and Mary Jane. This last is said to have been named for the wife of Francis O'Callaghan, evoking images of a lonely engineer far from his family.[84] This attribution does not work, because O'Callaghan's wife was Anna Maria Mary.[85] However, Thomas Curry's obituary explained that he named the tunnel Mary Jane as a joke on the name of the Indian contractor working on its construction, Mera Jan.[86]

Descriptive Engineering on Tunnels

This section first summarizes Ernest Hudson's notes for Descriptive Engineering to understand what first-year Coopers Hill students were taught about tunnel construction.[87] It then compares that to the construction techniques used on the Mushkaf-Bolan Tunnels by Johns and Cole and the Malakand Canal Tunnel by Sangster.[88]

The Descriptive Engineering notes stated, "The tunnel is commenced by driving a *heading* or *drift way* – a narrow passage 5 ft. high, 3 to 4 ft. wide – completely through the hill at or near the level of the formation surface." To do so, careful surveys were made of the hill's profile along the route of the tunnel, and fixed reference points were established, from which the entire path of the tunnel could be seen. Typically, shafts were then sunk along that path from the surface to the level of the heading, with the desired depth being determined from the profile of the hillside and measured down the shaft using ten-foot surveying rods joined together. Inside the heading (or, later, the tunnel), reference points were established by driving iron spikes into the timber supports. Candles suspended from the spikes were used to survey the route. Once the heading was complete, the centre line of the tunnel was marked at intervals between the shafts to ensure the excavation of the main tunnel was accurate.

Figure 7.7. Entrance to Panir Tunnel during construction, ca. 1894 (Hudson Collection, CSAS)

The major tunnel on the Mushkaf-Bolan Railway was about twenty-five miles from Sibi where the railway passed beneath the hills dividing the Muskaf and Bolan Valleys. This was the 3218-foot-long Panir Tunnel, running almost due east-west (figure 7.7). In this case, the surface was too far above the tunnel for shafts to be practical. None of the other tunnels on the Mushkaf-Bolan Railway used shafts either; most were too short. The difficulty of constructing the line came from the sheer number of tunnels and bridges. Some were cut-and-cover, used, for example, to protect the railway from landslides where it ran along the edge of a steep hillside.

As indicated in figures 7.8 and 7.9, the headings could be placed towards the top of the final tunnel or near the bottom. Johns commented that he used a top heading on the relatively short tunnels of his division because the ventilation was better, "an important consideration ... when the thermometer stands at 120° in the shade."[89] For the Panir Tunnel itself, a top heading (eight feet square) was used from the east in soft ground, and both top and bottom headings were dug from the west in harder rock. In the western half, the removal of debris from the upper heading was facilitated by shovelling it through holes in the floor into the waggons of a tramway built along the lower heading. The rate of progress in both headings averaged thirteen feet per day, and they met approximately midway on 21 August 1893, when "the horizontal error on the junction of the headings was 3 inches."[90]

The engineer's relief when the headings joined successfully was expressed by Stephen Martin-Leake (Hudson's friend from 1890) about his construction of the Bhanwar Tonk Tunnel on the Bengal Nagpur Railway: "The headings were started from both ends and met after just a year's work. The meeting was somewhat exciting.... The final shot was fired from the south side, while we awaited the result on the north. On going up to the face, we found a small hole, just large enough to shake hands through! For several days the question had been: 'Shall we meet?' It was a great relief to be able to walk through the heading instead of over the top."[91]

Once the headings were completed, the rest of the rock was removed. This was not done indiscriminately but in a precise sequence designed to support the roof and simplify the extraction of debris. First the base of the heading was lowered, and then the excavation was extended sideways across the whole width to allow timbering to support the roof as the tunnel was excavated. The spoil was tipped down into waggons running on the foundation level as it was removed a little behind the construction of the arch brickwork. The arch was underpinned on walls as the rock at the lower sides was finally removed.

Errors were still possible at this stage, as Martin-Leake found: "An error in alignment ... crept into the first few lengths of widening and lining on the south side and the remaining curved portion had to be 'faked in.' ... We did not talk about it at the time!"[92]

Bricks

The Panir arch was constructed of five rings of bricks made locally, with any gap between the arch and the rock above filled with brickbats and mortar. Ramsay

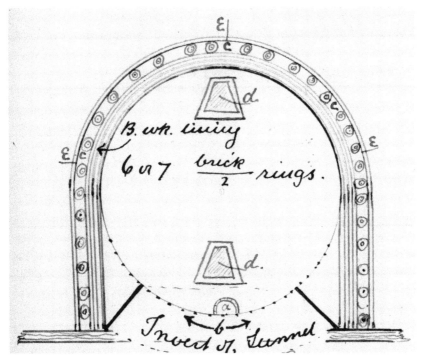

Figure 7.8. Ernest Hudson's Descriptive Engineering course notes on tunnel construction. Cross section showing: a. Culvert; b. Invert; c. Timbering; d. Heading; e. Two feet allowed beyond the timbering (Hudson Collection, CSAS)

estimated that the construction of the entire line used 29 million bricks made near Sibi and a further 11 million from Jacobabad. Egerton's comment about Indian engineers needing to learn brickmaking makes sense in this context, and Hudson devoted more than eight pages of notes for Descriptive Engineering to the composition and types of bricks.

These bricks were set in mortar made with Portland cement, imported from England – 194,000 barrels in all for the line, and 15,000 for the Panir Tunnel alone. This mortar was a key determinant in the strength and durability of the structure, so the use of a cement whose properties were consistent and well known was worth the expense. Portland cement was also known to perform well in wet conditions and did not degrade during transport. Moreover, the engineers' technical papers took pains to report the ratios of sand to cement they used for the various applications, so Portland cement formed a de facto technical standard. The cement was mixed with local sand to make it cheaper

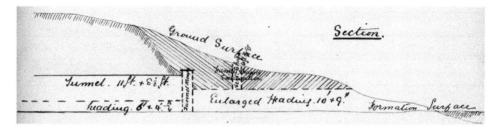

Figure 7.9. Ernest Hudson's Descriptive Engineering course notes on tunnel construction –
longitudinal section. The area to the right of the tunnel mouth formed a cutting on
completion, which shortened the tunnel and reduced expenses (Hudson Collection, CSAS)

(although Cole comments on the expense of transporting the sand) and harden
better. However, too much sand made the mortar weaker under tension.[93] Hudson's notes recommended a sand-to-cement ratio of no higher than 3:1. The
ratio used for the Panir Tunnel arch was 4:1, which was acceptable only because
the masonry in the arch and walls was always under compression.

Completion

Many lives were claimed by accidents during the construction of the railways
to Quetta, although these would have been far fewer than those lost to disease. In the great cholera outbreak of June 1885, for example, 2000 labourers
on the Sind-Pishin Line died of the sickness out of a workforce of 20,000.[94]
Ernest Hudson's diary mentions some of the accidents that occurred while he
was there. In one of what must have been many similar accidents, Hudson described how a block of stone fell onto an Indian labourer from a cutting near
Mushkaf and "smashing his head in."[95] Another common type of accident involved engine derailments, one of which blocked the line near Mushkaf for
three days and one where a temporary bridge of railway sleepers gave way while
the train was halfway across. More serious was an incident in which five ballast
trucks became detached at Shelabagh, ran down the hill, and into back of a mail
train, killing an Indian and an English guard.

To reduce this type of risk on the new railway, five "catch sidings" were installed above stations on the steepest sections. Points installed on the main
track diverted trains into these sidings by default; to continue its journey, a
train had to come to a halt so the operator could switch over the points.[96]

Not all points were of this fail-safe type, however. Walter Weightman recalled
a time when, as he and a local assistant were walking towards the Khojak Tunnel, they heard the "ominous rumble and rattling" of breakaway spoil trucks
from the excavation. Seeing that the carts were heading for a crew working on

Figure 7.10. James Ramsay (*rear*) and guests at the Summit Tunnel, Mushkaf-Bolan Railway (probably late 1894) (Hudson Collection, CSAS)

a bridge, Weightman's companion Abdulla threw down his instruments and dashed for the lever that would divert the trucks into a catch siding. He managed to turn over the lever in time but was hit and killed by the waggons.[97]

Finally, on Saturday, 29 June 1895, Ernest Hudson was able to write in his diary, "They ran the first through train from Sibi up to Qta on the M.B.Ry. in about 6½ hours. Johns, Cole, and the other Engineers came up by her and I met them all at the club in the evening."[98] A photograph of engineer James Ramsay and his guests is reproduced in figure 7.10.

This track now forms the only rail link between Sibi and Quetta. Stations on the Bolan Line were already being closed and the line dismantled in 1894 during the Mushkaf-Bolan construction, and Hudson was realigning the telegraph line appropriately. Traffic on the Sind-Pishin Railway eventually came to a halt on the night of 11 July 1942 when the hillside supporting the track leading northwards out of the Chappar Rift was washed away by flood waters rushing down the chasm. By that date, traffic on the line was light and it being war time, the decision was taken to close the route beyond the coal mines at Khost.[99] The metal spans of the Louise Margaret and other bridges were recovered in 1944 by Percy Berridge.[100]

Malakand Canal Tunnel

By comparison, William Sangster's construction of the Malakand Tunnel on the Upper Swat River Canal followed the classic approach taught in Descriptive Engineering. When the tunnel was opened, it was named in honour of the canal's designer, John Benton.

The Upper Swat Canal took water from the Swat River and transported it south under the Malakand Range through the two and a quarter-mile tunnel (comparable with the Khojak Tunnel). It was then distributed to the fertile plain around Mardan by two canal branches that split to go eastwards and westwards, hugging the surrounding hills. The tunnel passed 775 feet below the highest hill and, unlike the Panir Tunnel, did not require a lining because the rock was hard granite. In order to maintain the required flow rate, the tunnel was designed to achieve a "very high [water] velocity, and a minimum of rock-blasting." The tunnel bed was therefore sloped at a gradient of 1 in 215, representing a height difference of about fifty feet between the head and tail of the tunnel. This was unusual in a tunnel of this length and caused problems due to the groundwater draining downhill and flooding the heading working face. Moreover, the cross section was much smaller than a railway tunnel, making work and spoil extraction more difficult.

Electric and compressed air rock drills, both powered by hydroelectricity from the Swat River, were used to bore holes for the explosives. As a result of the hardness of the rock and the limited working space, progress was much slower than for the Panir Tunnel, just four to five feet per day, representing the blast of one set of thirty dynamite-filled holes. Despite the use of electric fans to ventilate the headings, the atmosphere inside was hot and "almost fetid."[101] Overall, the construction took three and a half years and was, in John Benton's words, "an exceedingly difficult and heavy task."[102]

The path of the tunnel was marked over the hillside by a line of pillars, and permanent measuring stations containing theodolites were installed at each end to facilitate surveying.[103] Three shafts were sunk to the tunnel bed, two for working and one for ventilation. These were a challenging undertaking because of their depths (from north to south: 104, 246, and 297 feet), the hard rock, and water seepage. Although excavation of the headings was carried out from the four faces of the two working shafts, they ultimately represented a small fraction of the total length.

The meeting of the first two headings (from the tunnel head in the north and Shaft 2) took place on 26 June 1913. The following day, Sangster's diary entry simply stated, "Went down shaft 2 and in to Tunnel H[ea]d thro' the hole. Quite a big hole it was."[104]

Headings from Shaft 3 (or "main shaft") and the tunnel tail were drilled through on the evening of 24 September 1913 and blasted through the following morning. "Went down main shaft with Obidullah & on into the T. Tail Heading *through the gap*. Then went on into the other heading where 3 Holman drills were working."[105]

Finally, on Saturday 4 October 1913, a bright red entry in the diary recorded, "Tunnel holed thro' this evening."[106] Sangster sent wires to his superiors with the good news, including his immediate superior, Richard Tickell.[107] He received a flurry of congratulatory telegrams back, including from the chief commissioner of the North-West Frontier Province, John Donald. They also arranged a celebratory feast for the tunnel workers.

The Upper Swat River Canal officially opened six months later on 14 April 1914, although it was not then fully completed. It was found that the actual water flow through the tunnel was not as great as predicted, mainly as a result of the roughness of the walls, so the tunnel section was later smoothed and enlarged to increase the flow.[108]

Bridges

A 2018 document from the Indian Standing Committee on Railways[109] reported that 37,689 railway bridges in the country were more than 100 years old (from a total of 147,000 bridges), and it is probable that the number of bridges built under British rule greatly exceeded 100,000. Many, or even most, of these would have been constructed under the supervision of Coopers Hill engineers. The vast Indian rail network would therefore have been impossible without the engineering capacity and expertise to build reliable, economic bridges. These ranged from humble culverts to the massive structures of the Lansdowne and Hardinge Bridges. This section will consider two of them.

Constructing the Rupnarayan and Hardinge Bridges

Two detailed accounts of building such bridges from the ICE will be discussed here – the relatively standard Rupnarayan Bridge,[110] constructed by the Martin-Leake brothers and their team (figure 7.11) between 1895 and 1900, and the enormous Hardinge Bridge,[111] constructed by Sir Robert Gales (CH 1883–6) from late 1908 to 1915.

Ernest Hudson first mentions meeting the brothers on the last day of 1889. He dined at their bungalow near Khodri on the Bilaspur-Katni Branch of the Bengal Nagpur Railway, along with "Johnny" Coode (CH 1883–5), who was so sick he had to be carried there. They hunted and inspected along the line together in the early months of 1890, including one memorable trolley ride through hail storms during which the poor "trolly men could hardly get along at all."[112]

The Martin-Leakes came from a family of six boys and two girls.[113] The eldest brother, Stephen (1861–1940), entered Coopers Hill in 1880. He was not on the list of students admitted that year by competitive examination,[114] so he seems to have been an "extra student" admitted under the New Scheme. His brother Richard ("Dick," 1867–1949) was educated at King's College London.

Stephen Martin-Leake (known at Coopers Hill as S.M. Leake) did not in fact complete his diploma, leaving during his third year.[115] Sir Trevredyn Wynne (CH 1871–4), his superior on the Bengal-Nagpur Railway, recalled, "Mr. Leake when at Coopers Hill was always last in his year and as he could not possibly pass out he took an appointment on the Bengal and North-Western Railway."[116] This raises a question about whether the railway was hiring substandard engineers or whether the Coopers Hill environment did not bring out Martin-Leake's best qualities. It seems to have been the latter, for Wynne continued, "Mr. Leake very soon acquired a high reputation, and when this Company was formed I was glad to get him as one of our District Engineers and put him in charge of almost the heaviest district on the line."[117]

The Rupnarayan Bridge crossed the river of that name at Kola (Kolaghat) and enabled the Bengal Nagpur Railway to run directly into Howrah Station. The river flows into the Hooghly just above Diamond Harbour, where the passenger steamers docked, and was 1800 feet wide at low water. It was an important route for river traffic, so the bridge needed to maintain a twenty-five-foot clearance to allow steamers to pass. Unusually for India, the river's flow did not vary substantially with the season but instead was tidal, with a twelve-foot rise and fall.

Like many Indian bridges, it consisted of multiple "spans" supported on a series of piers built across the river. The word "span" was used to mean the complete structural unit of the bridge resting between the piers, as well as the distance between the piers. Each span therefore consisted of the "girders" that provided the strength, usually constructed from trusses, plus the flooring, sleepers, rails, and everything else necessary for the working structure. Bridges were designed after systematic surveys of the river's characteristics and the geography of the location, and an estimate was prepared for governmental approval. In sanctioning the Rupnarayan Bridge, the Government of India specified the number (seven) and length (314.5 feet) of the spans and determined that the pier foundations should be constructed using the same method as the nearby Jubilee Bridge. As was usual for the time, the span trusses and other critical metalwork were manufactured in England, and the pieces shipped to India for assembly. Each Rupnarayan span weighed a nominal 546 tons, but the heaviest single piece was just over 5 tons. Martin-Leake did not say who designed the trusses, but in similar cases such as the Hardinge Bridge, they were designed in the London offices of Sir Alexander Meadows Rendel, consulting engineer to the secretary of state for India. Figure 7.12 shows the bridge during construction.

Figure 7.11. "The Bridge Builders," engineers in charge of building Rupnarayan Bridge, 1900. On the basis of personnel listed in Stephen Martin-Leake's ICE paper, the men are tentatively identified as (*left to right*) Stephen Martin-Leake (superintending engineer), unknown, Dick (Richard) Martin-Leake (district engineer), K.W. Digby, E.V. Bowman, unknown, unknown, G.L. Molesworth, D. Leslie (Hudson Collection, CSAS)

Instruction in construction at Coopers Hill was under Callcott "Pomph" Reilly, from 1871 until he was required to retire in 1897. His early work on the analysis of bridge trusses, with reference to the actual built structures, established him as an expert in the field. As immortalized in the poem "The Great Professor," "Strength he could measure, stress and strain declare / Bridges design, and with most wond'rous care."[118]

Reilly was ultimately replaced by the team of Richard Woods (assistant professor of engineering) and Arthur Brightmore (professor of engineering). Woods "had the advantage of both high theoretical attainments … and fifteen years of practical experience in survey, construction, and project work of a valuable kind."[119] He was a Coopers Hill alumnus with an excellent academic record and had worked on the challenging Sind-Pishin and Chaman Extension Railways. Brightmore had experience on British aqueducts and waterworks before joining Coopers Hill and had interests in steam-powered motor vehicles. Volume 1 of the *Roorkee Treatise on Civil Engineering in India*, textbook for

Descriptive Engineering, contained an extensive chapter on analysing stresses in roof trusses and bridges. Later, Woods's 1903 book for Coopers Hill engineers, *Strength and Elasticity of Structural Members*, also covered strength of materials, bending of beams, and the analysis of trusses; similar material would be in the second or third year of a modern civil engineering degree. Men like Stephen Martin-Leake would therefore have been familiar with the mathematics, analysis, and design principles of their spans, even if the design work was done elsewhere.

The work in India was of a very practical, implementation kind, consisting of a great deal of problem solving, application of hard-earned know-how, and logistics. In the early years, there is some truth to the criticisms of grizzled engineers in India that the Coopers Hill men knew too much theory and not enough practice. However, the profession was changing. It was becoming more complex and less centralized so that each engineering project would be designed and constructed by a team of people with diverse specialties, including designers as well as constructors. Engineers of the new century would need to know theory as well as practice.

Foundations

I now come to a very interesting subject to the Indian engineer, which involves several specialities, and applies more or less to all the constructions.... I mean the subject of Foundations.[120]

Foundations were a critical issue for engineers constructing railway bridges over the wide, silty, seasonal Indian rivers. Descriptive Engineering discussed several topics relevant to bridge builders in India, including soil testing for foundations, how to build foundations, bridges, and river engineering (measuring flow, dredging, embankments, etc.). In common with many Indian rivers, the bottom of the Rupnarayan consisted of silt "which was constantly on the move, and varied in depth according to the state of the tides."[121] Beneath that were other unstable strata measured by test bore holes and extending to a considerable depth. Foundations therefore had to be deep, in order to reach a suitable load-bearing material[122] and to withstand the scouring effect of the river in flood, which could quickly erode holes tens of feet deep in the "shadow" of the piers, causing subsidence.[123] At 80 feet below mean sea level, the foundations met with sand that was able to support the piers, so the foundations were excavated a few feet lower to 88 feet. This was not a particularly unusual depth; those of the Hardinge Bridge went down to 150 feet.

The basic technique for sinking deep foundations was adapted by the British from techniques used historically by Indian well-builders and was well established by the time of Coopers Hill:[124]

Figure 7.12. Construction of Rupnarayan Bridge, January 1900 (Hudson Collection, CSAS)

Nearly all the ancient bridges of Upper India, and even in the more remote regions beyond its boundary, were founded on brick cylinders sunk into the sandy beds of the rivers.... In these wells the brickwork is commenced on timber, or more usually on wrought-iron, curbs of great strength.... The sand or other material met with in sinking is gradually withdrawn through the central hollow of the well ... and after the sinking has been carried to a certain depth – usually about one to two diameters – the wells are loaded with a heavy weight, such as iron rails or bricks stacked upon them, in order to overcome the side friction of the cylinders. These foundation-wells are frequently forced down to great depths, such as 80 to 100 feet, and reaching even 140 feet.[125] When the sinking to the desired depth is complete, the interior space (or spaces) is sealed at the bottom with a certain thickness of cement concrete deposited in the water; the latter is then removed, and the remaining hollow portion is plugged with ordinary concrete or brickwork.[126]

Because the Rupnarayan River was tidal, it was decided to use steel caissons to form and protect the foundations of the piers, similar to those shown in figure 7.13. These were essentially large metal boxes sunk into the riverbed in the same way as the wells above, and filled with a masonry, concrete, or sand. Those used for this bridge were each sixty-three by twenty-two feet in area with semi-circular ends and ultimately eighty-eight feet in depth, weighing 385 tons complete. They were divided into three vertical internal dredging chambers, which could also be used to adjust the buoyancy of those caissons lowered through the water. As the caissons sank, new rings were rivetted to the top. Of the eight caissons, three were lowered into the river from floating stages, three were constructed from temporary islands, and two were on dry land.

A number of difficulties were experienced getting No. 4 caisson to sink vertically as a result of scouring of the riverbed. Although it is not mentioned in the official publication, Hudson remarked in his diary that he "went to visit No. 4 pier which had been giving a lot of trouble and saw them explode a charge of four dynamite cartridges with the object of giving it a start down it having refused to move for several days."[127] About four hours later the pier suddenly fell three and a half feet to a depth of about ninety feet.

Each caisson was sunk so its top was at mean sea level and then filled with concrete at the bottom and masonry above. These then formed the foundations for the construction of the eighteen foot, visible masonry piers of the bridge; the weights of the various elements of each complete pier are itemized in table 7.3. The spans were constructed from parts imported from England directly on pontoons moored by the shore. Guided by multiple hawsers, these pontoons were then floated into position at high tide, so each span was over the piers it was to rest on. When the tide turned, the pontoon naturally fell away from beneath the span, leaving it resting on the piers. They were then jacked into their final positions and rivetted in place. Each span took about a month to assemble and two days to be floated onto the piers.

Figure 7.13. Caissons for a pier of Godaveri Bridge, 1898 (Hudson Collection, CSAS)

It took four and a half years between arrival of the staff on site and the first train running across the Rupnarayan Bridge on 19 April 1900. Although the piers were built for two tracks, the girders were erected for only one. Thirty years later Stephen Martin-Leake revisited his bridge: "As far as we could see and hear, there had been no movement of the piers and the brickwork looked as sound as the day it was built. It is another question with the girders; one road is no longer sufficient for the traffic and the existing girders are not equal to

Table 7.3. Weights of Elements of Each Complete Pier of Rupnaryan Bridge

Component	Weight (tons)
Filled caisson	6,025
Pier	1,722
Bridge span	546
Live load (train)	314.5
Total	8,607.5

present loads. The girders for the second road have been ordered and a scheme drawn up to get them into position."[128]

Martin-Leake emphasized that because the seasonal volume of water flowing in the Rupnarayan River was roughly constant, they were able to work night and day year-round with the works illuminated by electric arc lamps. In his description of the much larger Hardinge Bridge (with its fifteen 345-foot spans and 150-foot foundations), Robert Gales gave some idea of the logistics supporting the construction efforts, which included one of the first examples of the use of electrical power throughout:

> The early part of 1909 was occupied in preliminary surveys and investigations; prospecting for the supply of stone; arranging for locomotives and rolling stock of both standard and metre gauge, as well as standard- and metre-gauge track[129] for service-lines at the quarries and at the bridge; chartering steamers and flats to handle the large amount of pitching-stone[130] and other material required; acquisition of land; starting brick-fields, and the provision of steamers, launches and barges. Thereafter building quarters for staff, hospitals, workshops, power-houses, water-supply and other service works on both banks of the river, and medical and sanitary arrangements, were put in hand. There were eventually in use on the operations 81 miles of standard, metre and 2-foot 6-inch gauge service track, and twenty-four locomotives, twenty-eight brake-vans and 830 trucks.[131]

In addition to his being a "damnably competent" engineer, Gales excelled at this kind of organization with "the mastery of detail, and the confidence with which he accepted his responsibilities and faced complicated problems that gave the impression he saw his way very clearly."[132]

As well as the main paper on the Rupnarayan Bridge that Martin-Leake presented to the ICE on 16 December 1902, the *Minutes of Proceedings* also include text of the discussion that occurred after his talk. He was put under detailed and rather aggressive questioning by some of the most prominent bridge builders of the day, including Frederick Robert Upcott (PWD secretary for railways) and Frederick Ewart Robertson (in charge of constructing the Lansdowne Bridge at Sukkur). The wide range of the issues discussed emphasizes the complexity of even a "relatively standard" project such as the Rupnarayan Bridge. As well as the myriad financial, administrative, and logistical matters, the job entailed practical issues of construction, civil engineering theory, and increasingly elements of mechanical and electrical engineering.

Efficiency

It has been said that an engineer is a man who can do for ten shillings what any fool can do for a pound; if that be so, we were certainly engineers.

Nevil Shute, *Slide Rule*[133]

Public Works Department engineers in India were obsessed with costs, partly because they operated in a public works system that required a painstaking accountability at all levels, from the largest project to a pot of jam. But it was also a point of professional pride – finding that elegant solution that did the job for ten shillings instead of the pound. It is noteworthy that the first item on *Indian Engineering*'s list of objections to James Browne's military engineers on the Sind-Pishin Line was that they pushed on with the construction "regardless of expense." At the 1903 enquiry about whether to close Coopers Hill, Edmund du Cane Smithe was asked, "Is the student taught at Coopers Hill to make a bridge in the cheapest way?" He responded, "You get him with the qualifications, which, when he has seen a little of India, will enable him to do it in the cheapest way."[134]

The drive for economy permeated the PWD at all levels, and the need for "good management and economy" was expressed explicitly in the duties of both superintending and executive engineers.[135] These rules were enforced by a small army of office clerks and accountants who insisted "strictly on what may or may not be done; and no deviation from the Code rules is, in their opinion, even to be thought of without at least a Government order from authority higher than that of the officer under whom they are serving."[136]

For the railways in particular, the PWD's focus on efficiency after 1871 was in part a response to the early days of Indian railway construction (1853–71), during which guaranteed returns to investors and poor oversight had allowed construction costs to balloon. Some of these costs were legitimately incurred because the terrain was difficult, but examples of overbuilding and incompetent engineering were frequent. With the era of state railways (1871–9) the government attempted to bring these issues under control, and the establishment of Coopers Hill was part of this general strategy. India's railway construction was returned to private hands after 1880 (with the exception of strategic military lines, such as those in Baluchistan). Under this second guarantee period, private construction proceeded apace and India's largest rivers were crossed by a sequence of massive, and to some excessively engineered, bridges (Dufferin,

Lansdowne, Jubilee). A combination of greater experience, better materials and design, and improvements in the skills and management of the Indian labour force managed to offset some of these additional costs.[137]

To demonstrate their economy, defend apparent excess, and provide useful data for their colleagues, engineers included in their technical papers a great deal of information on labour, material, and transport costs that to the modern eye looks out of place. So, for example, we learn from William Johns that the Cornish miners employed on the Mushkaf Tunnels were paid Rs300 per month, that dynamite costs were around Rs13 per foot of tunnel excavated,[138] and from Charles Cole that 175,500 (14,625 dozen) candles were used during the construction, at a cost of Rs7925.[139] All these figures allowed James Ramsay to give a detailed tabulation of the cost per foot of all twenty individual tunnels on the Mushkaf-Bolan Railway; most cost Rs250–300 per foot, with the cheapest being the cut-and-cover tunnels at around Rs200 per foot and the costliest being Panir at Rs425 per foot. The overall cost of the line was a little over Rs20million, or Rs340,000 per mile.

Stephen Martin-Leake was interrogated by the audience of his paper about the costs of using caissons and islands for sinking the foundations at the Rupnarayan Bridge. In his paper, he was quick to point out, "During the greater part of the year, this side of India is subject to severe cyclones; and many cyclonic storms occurred during the progress of the work. No reliance could therefore be placed upon a defined working-season, and arrangements had to be made to guard against all conditions of weather and tide. Everything in connection with the work had to be of a very substantial nature. This added to the outlay on the bridge, and renders it difficult to compare the cost with that of other Indian bridges."[140] Ultimately the bridge cost Rs3.8million, compared with Rs5.9 million for all the large bridges on the Mushkaf-Bolan Railway.

Canals

Engineering Modelling

Nowhere was the issue of accounting higher in the engineers' minds than in canal irrigation works where the costs and revenues were carefully factored into the engineering design. India's irrigation canals were arguably the greatest engineering works in the country, requiring huge investments and affecting millions of lives. A 1911 article in the *Coopers Hill Magazine* described the lofty goals of the irrigation engineers:

> Away down south lies a desolate dreary tract of country supporting with difficulty half a dozen souls to the square mile. By means of the canal which will pass through these headworks the whole of this tract, covering nearly 1,650,000 acres,

will be converted into a fertile land supporting a vast and prosperous population. If the irrigation engineer has to live a life of isolation, if he sometimes feels that his labours are inadequately recompensed and his achievements passed unrecognised, he has at least the satisfaction of knowing that he is a man who does things, that his work contributes to the sum of human happiness.[141]

At Coopers Hill, all students first learned about canals from Descriptive Engineering and from the subsequent Hydraulics course. Such was the importance of irrigation that the *Roorkee Treatise* devoted twelve chapters to the subject, although the Descriptive Engineering gave it less prominence. The Course of Hydraulics and Mechanism largely dealt with the theoretical aspects of liquid flow in pipes, but also had sections on water flow in rivers and canals.

The greatest network of canals was in the Punjab, a name loosely translated as "five rivers" – the Sutlej, Beas, Ravi, Chenab, and Jhelum – flowing south-west to the Indus like spread fingers. The land between the rivers, called *doabs*, is excellent for farming except for its low rainfall. Irrigation canals running from the higher ground to draw water from the rivers and transport it into the doabs had been in use for 1500 years before they were massively expanded by the British. By 1903 almost 16 million acres in British India as a whole were irrigated by government canals, out of a total irrigated area of 44 million acres (for comparison, the total area of cultivated land was 226 million acres). The total expenditure on these canal works over eight decades of public works was nearly Rs430 million.

Set against this massive outlay was the revenue generated by the canal system, for the enterprise was intended to be profitable. And such it was – total revenue for all provinces in 1900–1 exceeded expenses (interest on the capital and operation costs) by Rs20 million.[142] These revenues were collected from the individual cultivators of the land on the basis that they were literally reaping the benefits of the irrigation. The amount paid depended not on the actual volume of water used, which was hard to determine, but on the land area and the type (and hence value) of the crop. The amount charged was typically 10–12 per cent of the value of the crop, perhaps Rs3–5 per acre for wheat in the Punjab and up to Rs25 per acre for sugar cane on the Deccan Plateau.[143]

These direct "productive" benefits of the irrigation canal system – increased crop production, population growth, and revenue – were attractive enough for the government to downplay the known disadvantages, such as swampy, malarial land caused by poor drainage and salt deposits from the evaporating water.[144] There were also the indirect benefits resulting from more reliable harvests that irrigation allowed, such as a general increase of prosperity and the supposed mitigation of famine (the so-called protective function).

In some areas, such as on the Afghan border, irrigation was used as a deliberate strategy "of winning the lawless tribes of those parts to more peaceful pursuits, and of providing an interest in agriculture, thereby diverting their minds from their more usual practices of war and murder."[145] Such was at least part of the intent of the Upper Swat River Canal on which William Sangster worked. The project was commenced only ten years after the siege of Malakand, a conflict in July 1897 in which local Pashtuns, angry that their lands had been divided by the new Afghan-Indian border, assaulted British encampments in the Malakand Pass.

While the *Times of India* proclaimed that "the engineer brought peace and plenty to a barren land,"[146] the views of Sir John Benton, architect of the canal, were more pragmatic: "The whole of the works for about the first nine miles of the canal lie in tribal territory beyond the British frontier. The tribesmen will benefit by receiving irrigation to such portions of their land as are commanded,[147] and they have been favourably disposed towards the canal project throughout; at the same time there have been ceaseless efforts to make money out of the works while under execution."[148]

John Benton and the Punjab Triple Canal System

Amongst the last major canal works in the Punjab was the so-called Punjab Triple Canal System, known for the "boldness of the conception and of its execution."[149] It was carried out by one of Coopers Hill's most illustrious alumni, Sir John Benton, whose knowledge, courage, driving force, and "freedom from paralysing perplexities" set him apart.[150] The following outline is taken from Benton's epic 1915 paper to the Institution of Civil Engineers that won him the Telford Gold Medal, the institution's highest award for a technical paper.[151]

Because the more straightforward and remunerative canals had already been built, it became increasingly difficult to construct new canals that fitted into the vast web of waterways and yet would be financially viable. The ultimate goal of the Triple Canal was the irrigation of 1.5 million acres in the Lower Bari Doab Region,[152] a scheme that had been contemplated even before the 1903 *Report of the Commission on Irrigation*.[153] But the only available source of surplus water was the Jhelum River, which meant building a sequence of three canals to transfer the water from west to east across several of the large rivers, but "unless this were done, the surplus water of the Jhelum must be allowed to run to waste forever in the ocean, whilst the arid tract of the Lower Bari Doab must be deprived of irrigation for all time."[154] This transfer was needed mainly in the colder, drier months from October to March, when the second, *rabi*, crops were grown. During the earlier, *kharif*, crop the rains and the meltwater from the Himalayas were sufficient.

Because irrigation canals were intended to transport water, they were constructed with a gradient (unlike navigation canals). Generally, the headworks of the canal where water was taken from a river was at the highest elevation feasible, and the route of the canal maintained its height for as long as possible – giving "command" of the lower areas – so water could be distributed to the fields.

The combined considerations of terrain and water supply therefore necessitated building three canals to transfer water from where it was plentiful to where it was needed:

The *Upper Jhelum Canal* (opened 1915) took water from the Jhelum River and transferred it into the Chenab River so there was enough water to supply irrigation needs downstream. It also irrigated the Upper Jech Doab that it traversed.

Because the downstream needs of the Chenab were now met by the supply of the new Upper Jhelum Canal, water could now be taken from the upper reaches of the Chenab by the *Upper Chenab Canal* (opened 1912) and carried through the Rechna Doab to the Ravi River. To generate revenue, the Upper Rechna Doab was irrigated along the way.

The *Lower Bari Doab Canal* (opened 1913) then drew water from the junction of the Ravi and the Upper Chenab Canal and used it for the final goal of irrigating the Lower Bari Boab.

Irrigation was needed in these areas because the "lift" required to bring water up from wells in many places was too large. Benton estimated that the area of cultivatable land that the three canals would irrigate was about two million acres, approximately half of which was in the Lower Bari Doab. From the estimate of area and the types of crops to be grown, Benton was able to determine the maximum capacities needed for the canals. Water was measured in "cusecs" (cubic feet per second), and the "duty" was the area irrigated per cusec; Benton's example was 200 acres per cusec, but the figure depended on the needs of a particular crop. Lastly, the number of days of irrigation required for the crop to mature was called the "base." However, the canals lost water by evaporation and by absorption through the unsealed bed. Both depended on the surface area, and a typical total loss was eight cusecs per million square feet of canal. For the Triple Canal System as a whole, the loss was approximately 25 per cent.

From estimates of the required capacities, the canal could be sized and the route for the main line and the distributaries that carried the water out towards the fields could be calculated. The width and depth varied between the three canals and along their length, but typical values were 200 feet and 10 feet, respectively. The bed slope of 1 in 6666 was carefully determined to achieve the necessary flow rate to prevent silting.[155] At the village level, the land was divided into twenty-five-acre parcels, each with a service road and culverts for the water courses. Finally, each parcel was further subdivided into one-acre fields to which water was supplied by channels along the edges. The total length of the three main canals was 339 miles, with an additional 184 miles of branch canals, 2,633 miles of distributaries, and 19,489 miles of water courses.

The construction of these canals was a massive, complex undertaking, as il-lustrated by a description of the site of the headworks on the Lower Bari Doab Canal where the canal crossed the Ravi River:

> To the uninitiated observer an engineering work of this kind seems at first an ut-terly confusing muddle. The noise and the dust of the *kunkar* [limestone] mills, the clang of the steam hammers in the workshops are as nothing to the shrieking of engines, the hoarse cries of the workers, the hurried rushing hither and thither of thousands of men and women on the works themselves. Looking down from the high bank on either side the works present the sort of purposeless disorder, the frantic confusion, seen in a disturbed ant-heap. Heavily laden trains slide squeak-ing down the inclines, or laboriously puff upwards with loads of sand and earth; donkey engines grunt and snort at their tasks; steam pumps drain off a ceaseless flow of water....
>
> Hundreds of masons toil steadily at their allotted task of brick-laying; and everywhere thick as bees in a hive, are the ubiquitous coolies – men and women – clad in the gayest of colours, hurrying to and fro with their tiny basket-loads of earth or bricks or lime. It is an astonishing scene of ceaseless activity, a moving picture of colour to delight the eye of an artist.[156]

At the peak of construction in 1911–12 the Triple Canal works employed 76,000 workers and 13,000 donkeys.[157] Collecting this many people was a great challenge, particularly because the region was then suffering from a terrible outbreak of plague that killed 10 per cent of the 22 million population. This de-mand also increased the cost of labour by 50 per cent and so the entire system came in almost £2 million pounds over budget, at almost £7 million. Neverthe-less, a return on investment of at least 7 per cent was still expected.

At the time of the official opening of the Upper Jhelum Canal, the *Coopers Hill Magazine* listed thirty-four college graduates who had worked on the pro-ject, including Benton himself, and William Sangster.

In a community that relied on the canal for its very existence, all the power rested with the British authorities, the different branches of which were not always in alignment. Moreover, the juxtaposition of irrigation officials moti-vated by efficiency and the traditional practices of the Indian communities led to unforeseen social and political implications within the communities themselves.[158] For example, the rapid growth of these towns drew craftspeo-ple from established communities, while the extraction of water higher up the rivers led to shortages downstream, both to the detriment of those older villages.[159]

Reading Benton's paper, it is remarkable that he seems able to quantify everything: the rainfalls, river flows, land areas, crop needs, and canal ca-pacities. On one hand it is a masterful engineering analysis – breaking the larger problem down into manageable pieces, developing a mathematical model, collecting the necessary data, refining the solution, and building the highly complex system. This is how engineering deals with large problems and why the shift from a purely practical education to theory was necessary and inevitable.

On the other hand, it stands out for the high degree of abstraction and the complexities it ignored. One is still left with an impression of hubris. Nature was there to submit to human wishes. Torrential rivers could be reduced to a single number of cusecs, and water transferred between them at will. The behaviours of people and crop were subjugated to the "engineering vision of environmental domination."[160] Remarkably, though not without its problems, the scheme worked broadly as planned.[161]

Some of this apparent contradiction lies in Benton's need to present the material to a non-specialized audience. Also, being a scientific paper, it would have been de rigueur for Benton to emphasize the theoretical aspects of his work. Even so, the thirty-four Coopers Hill engineers, and many others, would have had detailed knowledge of every section of the work and resolved myriad implementational details that could never have appeared in Benton's account. Moreover, the ideals of the engineering theoreticians notwithstanding, the practical design and management of the canal system relied on a local interpretation of the situation. Knowledge of the actual water flows and the detailed requirements of the farmers was essential to canal operations in much the same way as were the interpretation and administration of the forest regulations (see chapter 6).[162]

Also absent from the report was any analysis of potential negative effects of the project. The focus was strictly on achieving the assigned task rather than on assessing and remediating environmental or social consequences. Many modern engineers now argue that it is incumbent on members of the profession to have sufficient breadth of understanding to consider such issues during the design process.[163]

Towards Diversity

The projects described in this chapter illustrate why George Chesney designed Coopers Hill to be at the forefront of a more theoretical approach to engineering education. In the midst of this transition, William Cole wrote, "The permanent-way inspector of to-day – so far as my knowledge of him goes – is not merely that vague product, a 'practical man'; he does not necessarily prefer guesswork to certainty; he has not a supreme contempt for a simple and intelligible mathematical method; he has not a serene confidence in the fallible process of 'putting it in by the eye'; and he is not one jot less practical than the plate-layer of the old school, although he has had the advantage of a better education."[164]

The wide range of knowledge and skill sets required by engineers in India is apparent even from the brief descriptions of selected projects by Coopers Hill engineers given here. While some issues were universal – accounting, labour, climate – there was a high degree of specialization in each branch of the PWD. As theoretical understanding deepened, practical experience grew, and projects became increasingly ambitious, each field itself became more complex, and the body of knowledge needed to be a competent engineer expanded commensurately. Electrification brought an entirely new dimension, along with the need for power plants and transmission systems. Dams started including hydroelectric generating stations, so they were no longer purely civil engineering projects. Despite this reality, all the Coopers Hill engineers took essentially the same set of courses at the college.

Engineering was on the cusp of a major expansion of subjects and fields of application. Cars, aircraft, and wireless communications were just around the corner. Coopers Hill was beginning to feel the strain of having to incorporate these newer disciplines into a single, fairly short course of study. It was the beginning of the end for the engineering generalist. Although there are still general engineering science programs, sub-disciplines have proliferated – Engineers Canada lists more than ninety English-language engineering program types that have been accredited in Canada since 1965. No longer could engineers only be categorized as "wet," "dry," or "rapid";[165] the career paths for the holder of an engineering diploma were diversifying rapidly.

There were also signs that the student body itself was diversifying. Increasing numbers of non-European students were coming to study at Coopers Hill, drawn by its reputation as the "premier engineering College of the Empire."[166] Some were Indian and returned home as engineers, while others travelled from elsewhere to learn skills of value to their countries.

The next chapter examines how Coopers Hill's response to teaching new engineering disciplines and its increasing role in offering a general engineering education contributed to the college's eventual closure.

8

Crisis, Diversification, and Closure: 1896–1906

New President, John Pennycuick

At Coopers Hill, the retirement of President Taylor in 1896 marked the start of a period of crisis that lasted for the next decade and ultimately led to the college's abolition. A number of factors contributed to this state of affairs, including technological advances and the increasing number of university engineering programs, but the immediate issue was financial sustainability. As described earlier, the New Scheme designed by George Chesney and implemented by Alexander Taylor attempted to offset the Indian PWD's lower demand for engineers with students seeking a general engineering education and telegraph and forestry courses. These measures brought in additional students but also incurred greater expenses. The challenges of dealing with these issues defined the terms in office of the college's third and fourth presidents, John Pennycuick (pronounced "penny-quick") and John Ottley.

In appointing Alexander Taylor's successor, the government took the advice offered by Sir Charles Bernard in a personal addendum to the 1895 Committee of Enquiry's report. Bernard believed that the large cost of a "separate salaried president" was unwarranted for so few students and recommended that in future the position should be combined with that of a senior teaching professor. So when Colonel John Pennycuick was appointed president in September 1896, it was on the condition that he would also serve as professor of construction upon Callcott Reilly's imminent retirement, with no additional salary.[1]

Pennycuick was apparently an excellent choice for college president. He had served with distinction in India, having been educated at Addiscombe just before its closure. He was transferred from the Madras Engineers to the PWD and was promoted to executive engineer by 1866. Later he rose to deputy chief engineer and undersecretary to government in the 1880s, and head of the Madras Public Works Department for five years before his retirement in January 1896.

Pennycuick had been a member of the Faculty of Engineering at the University of Madras since 1891 and served as its president (or dean) for several years.[2]

As an engineer, Pennycuick was most known for the "great and beneficent" Periyar Diversion project. Flowing west from the mountainous spine of southern India, the Periyar River was fed by heavy annual rainfalls. In contrast the eastern side of the mountains, drained by the Vaigai River, was arid. The diversion engineering works were supervised by Pennycuick and carried out mainly by Coopers Hill engineers between 1888 and 1895.[3] They saw the construction of a large masonry dam – "one of the most extraordinary feats of engineering ever performed by man"[4] – across the Periyar River. Water was then diverted into the Vaigai River valley by means of a tunnel through the hills: "Then did the headwaters of the Periyar River appear in full flow on the eastern slope of the Southern Ghats."[5]

The distribution of this water irrigated approximately 160,000 acres and brought such prosperity that farmers of the region still revere Pennycuick's name.[6] The work was hampered by floods, wild elephants, malaria, and cholera. On medical grounds, the labourers were offered *arrack*, a strong local alcohol, and Pennycuick recorded that "had it not been for the medicinal virtues of *arrack* … it is difficult to see how the Periyar Dam would ever have been built." Apparently, the cost of the workers' drinks was reported in the accounts as "pegs for the workmen" in order to escape the eye of the examiner of accounts while still remaining truthful.[7]

At Coopers Hill, the plan to amalgamate the roles of president and professor of construction was not popular with everyone. For example, a former student wrote to *Indian Engineering* saying (unfairly), "One would have thought that the duties of administrating such a large College would have been sufficient to tax the abilities and time of any ordinary man, but apparently Colonel Pennycuick longs for more worlds to conquer." In particular, the correspondent asked, how could Pennycuick have the time to assist students as Reilly had, "under the warming influence of a decanter of sherry and a box of cigars?"[8]

As it turned out, Pennycuick could not do so. Two years later he was relieved of his teaching duties because the two roles were "constantly clashing."[9] So, in September 1899, Dr. Arthur Brightmore was appointed to the position of professor of engineering and surveying. Brightmore's expertise was in the construction of aqueducts for the waterworks of Liverpool and Birmingham.[10]

Brightmore's appointment was made by the secretary of state over the objections of Pennycuick, who promptly resigned. In a letter to the editor of *Engineering*, Pennycuick expressed his dissatisfaction at the interference of the secretary of state who had, against his wishes, mandated both the appointment and the amalgamation of the professorships of engineering and surveying into a single position.[11] Furthermore, he believed that Brightmore "did not possess the practical experience necessary for the post," by which he probably meant that Brightmore had no experience in India. Pennycuick continued, "I declined to be a party to a measure which I consider inimical to the interests of the college, and of the department for whose benefit it exists, … and have therefore resigned."[12]

Pennycuick's letter was reprinted in the *Coopers Hill Magazine* for October 1899, an action that the Board of Visitors later considered to be inappropriate because the letter was "disparaging the qualifications of a professor newly appointed to the College, and accusing the Governing body of injustice."[13] Arthur Hicks and Francis Matthews, the magazine's co-editors, were questioned by the visitors about how this had happened. They each stated that they had requested a piece from Pennycuick about his reasons for resigning, and that the former president had replied, "No, I would rather not do that, but I am sending all the facts to '*Engineering*,' and you can reproduce that if you like."[14] This they had done, with some reservations, but their intention had not been to criticize either Professor Brightmore or the secretary of state.

Brightmore himself was of the opinion that the reproduction of Pennycuick's criticisms in the magazine undermined his authority in class and led to an organized campaign by the students to disrupt his lectures.[15] Brightmore and his Coopers Hill colleagues appear in the 1903 photograph reproduced in figure 8.1.[16]

Enrolments

Much was made by detractors of Coopers Hill of the low numbers of assistant engineers appointed to India relative to the expense of maintaining the college. Figure 8.2 shows the enrolment statistics for the college between the introduction of the New Scheme in 1883 and 1900.[17] It is clear that the overall number of students in college (a number that includes forest students not on tour) never reached the intended figure of 150, instead averaging 125 over the period. By contrast, the average number appointed to the PWD annually was fewer than 18. The combined effects of the New Scheme and the introduction of forestry produced the desired increase in the number of resident students between 1884 and 1887. But in the following years there was considerable volatility in the number of applicants, causing the overall enrolments to be almost periodic. The causes of this are not clear. In 1895, the Board of Visitors attributed declining admissions to the reduced number of positions in India.[18] They also pointed to a lack of clarity on the career prospects for graduates who did not qualify for the PWD, at a time when the demand for engineers in England was also weak.[19] However, the number of available appointments did not change greatly during this period. The varying engineering admission figures may simply have been because the college responded to each dip by temporarily increasing its recruitment efforts.

Figure 8.1. Coopers Hill staff, probably late 1903: (*left to right, back row*) Edgar, Harbord, Matthews, Woods, McKenzie, Browne, Hicks, Hopps, Campion, (*middle row*) Collett, Templeton, Giffard, Aitchison, Hay, McElwee, Fisher, Tickell, Lodge, (*front row*) Perret, Burrough, Groom, Schlich, Ottley, Minchin, Brightmore, Boyes, Seeley. (© British Library Board, photo 297/2)

For the Engineering Branch in particular, a comparison of the numbers of students applying for admission and those entering the college shows that the acceptance rate was very high, averaging 84 per cent. As a direct consequence, 30 per cent of these students on average failed to complete their diplomas. Data from the late 1890s suggest that most of these students left the college before the end of their course, while only two to four students per year were unsuccessful in the final examinations.[20] Admitting poorly prepared students to boost revenue is a classic temptation for struggling institutions whose financial viability depends on student tuition. A few of those marginal students ultimately may have been successful but, as Coopers Hill data show, most would not.

A second temptation might have been to lower academic standards so fewer students dropped out. However, this does not appear to have been the practice at Coopers Hill, and standards were maintained by the use of external examiners and by removing unsuitable men before they completed their studies. That said, the quality of the graduates was judged mainly from those top performers

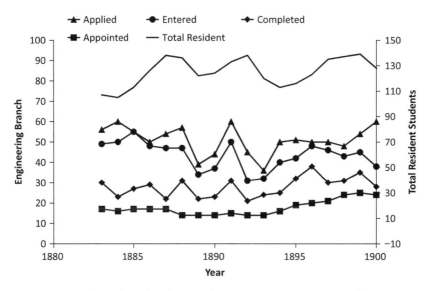

Figure 8.2. Yearly number of students in the Engineering Branch who applied to the college (triangle), entered the first year (circle), successfully completed their diplomas (diamond), and were offered appointments in India (square). The total number of students living in college (including Forest students but excluding those on the Continental Tour) is also shown.

awarded Indian appointments. The disparity between this top group and some of the other graduates seems to have been quite large at some points in the college's history, leading George Minchin to comment, "Below the [top] twelve, they are very often, a good many of them, highly qualified; but you cannot be certain in the case of the extreme men."[21]

John Ottley, Last President of Coopers Hill

The man appointed to address these issues and to "restore discipline"[22] at the college was Colonel John Walter Ottley, formerly inspector general of the Indian Irrigation Department. He had reportedly been in the running for college president in 1896, which may account for the speed of his selection. Ottley was appointed in October 1899, just three weeks after Pennycuick resigned.[23] He is shown in figure 8.3 presiding over the 1902 Coopers Hill prizegiving. Many years later *Indian Engineering* opined that Ottley was the "best President Coopers Hill ever had," supposedly knowing more about the engineering needs of India than Chesney, being a better administrator than Pennycuick, and both a

Figure 8.3. Lord George Hamilton (*on podium, left*) and John Ottley (*podium right*) present a student prize, Coopers Hill Prize Day, 1902 (*Daily Graphic*, 3 July 1902, 4)

better engineer and administrator than Taylor.[24] This view is highly debatable since the fallout from Ottley's reform initiatives hastened closure of the college. In hindsight, reform was needed at Coopers Hill and it was high time for a reinvigoration of the academic side of the college. The data above show that the entrance examination provided, at best, only a rudimentary screening of applicants. Pennycuick thought that 10 per cent of the first-year students were "unable to follow the course of instruction with advantage to themselves, while they obstructed the progress of their better-instructed comrades."[25] Ottley believed that this estimate was too low and was in favour of continuing Pennycuick's strengthening of the entrance examination.

However, Ottley also believed that the Coopers Hill course of instruction was a significant part of the problem and set about a comprehensive reform of the academic curriculum. He swiftly expressed his initial impressions in February 1900: "The present system is antiquated, illogical, and quite out of date. What is required is a thorough overhauling of the whole system. If this were done, I feel quite certain that we could devise a much better curriculum and could provide not only for instruction in Railway and Canal and Electrical Engineering, but also for many other important matters, such, for instance, as waterworks, drainage, &c., … The first thing is to settle what we want the men to learn, and then to decide how best to afford the necessary instruction."[26]

Ottley's *Memoranda on the Educational Course* was submitted to the Board of Visitors on 13 June 1900. He proposed sweeping changes to the curriculum, the construction of additional laboratories, and the reorganization of the academic program, which would require dismissal of several faculty members. The analysis and proposed solutions made a great deal of academic sense. However, they proved to be politically disastrous because Ottey developed and implemented them in high-handed fashion, with little input from his staff.

Working Time: Ottley calculated that Coopers Hill students received formal instruction for an average of 26.3 hours per week (ranging from 33.9 to 14.4 hours per week, including drill, during the first and third years, respectively). He contrasted this to Chesney's curriculum, under which students worked 42.25 hours per week. By developing a new timetable, Ottley was able to increase the average to 32.4 hours per week, thereby gaining 549 hours of instructional time over the three years. This was achieved partly at the expense of private study time, which Ottley regarded as providing "most objectionable facilities for copying of notebooks and the fudging of plans and plates" (a view that not all shared).

Teaching Workload: Ottley implied strongly that professors were not spending enough time teaching. A table compiled by Dr. Matthews provided data for each instructor, which ranged from 7.1 to 17.8 hours per week. These were somewhat misleading because almost every person had additional responsibilities, usually related to laboratories, surveying, or electrical engineering. To save money, Ottley advocated a system for engineering whereby the professor would teach large lectures and an assistant would give tutorial instruction. This arrangement is adopted by many modern universities for the same reason.

Engineering Curriculum and Workshops: While he advocated many minor adjustments to the curriculum, Ottley's main change was to make electrical engineering compulsory. He also believed that there should be significantly more practical instruction. For this, he proposed to give the old mechanical laboratory to the Telegraph program and to build a new one attached to the existing workshop.

Mathematics: Ottley believed that there was a consensus that the knowledge of mathematics by Coopers Hill students left "much to be desired." Accordingly, he proposed "stiffening up" the entrance examination, teaching mathematics in all three years, and hiring instructors to assist the mathematics professors.

Sciences: While physics instruction was left unchanged, Ottley thought that too much emphasis and expense was placed on chemistry when all an engineer needed to know about the subject was how to "interpret the results given to them by professional chemists." This could be taught by "one man on much less pay" than the costly arrangement of Prof. McLeod, Dr. Matthews, and their staff.

Evaluation: Lastly, Ottley attempted to simplify the marking scheme and realign it with the hours of instruction and priorities of the subject matter. He also gave marks for physical exercise and student discipline (in place of the fine system then used).

On 16 July, a month after the submission of Ottley's document, the public works secretary of the India Office called for a Board of Visitors meeting for the following week; despite the short notice, eight of the nine members were in attendance to consider the proposals. The visitors added some suggestions of their own but approved the great majority of Ottley's reforms, particularly the new timetable and curriculum, the amalgamation of the workshops, and the new grading system. Moreover, they expressed their "appreciation of the thoroughness with which [Ottley] has dealt with the subject." Ironically, in view of future events, they also anticipated that "the Royal Indian Engineering College is about to enter on a career of increased usefulness."[27]

In due course, the secretary of state for India, now Lord George Hamilton, gave approval of the extra costs for the new mechanical laboratory (£1056) and a carpenter's shop proposed by the Visitors (£882). On 2 November 1900 he also approved the board's academic recommendations.

Beginning of the End

Professors Dismissed

John Ottley's proposals marked the beginning of the end for the college.[28] It was fully understood by Ottley, the Board of Visitors, and the India Office that implementing these reforms would necessitate the "retirement" of seven long-standing members of the college's teaching staff:

- Herbert McLeod, professor of chemistry (thirty years)
- Arthur Heath, assistant professor of engineering (twenty-five years)
- William Stocker, professor of physics (seventeen years)
- Thomas Hearson, professor of hydraulic engineering (sixteen years at Coopers Hill)
- Philip Reilly, demonstrator, Mechanical Laboratory (ten years, son of Prof. Callcott Reilly)
- Thomas Shields, demonstrator in physics (eight years)
- James Hurst, lecturer in accounts (twenty years on a fee basis)

In a letter dated 14 December 1900, Horace Walpole at the India Office[29] instructed Ottley to "inform these gentlemen that the Secretary of State, while regretting the necessity for reducing the Staff, and fully recognizing the value of their past services, is compelled to give them notice that they will be required to vacate their appointments at the end of the Easter term next," which would be in early April 1901.[30]

Ottley did so in brutal fashion by simply forwarding Walpole's correspondence, together with a cover letter that read:

Coopers Hill, Englefield-green, Surrey, Dec. 17, 1900.

Sir,– I have the honour to forward for your information a copy of a letter, P.W. 2,531, dated 14th inst., from which you will see that I am instructed to convey to you the decision of the Secretary of State for India in Council that you will be required to vacate your appointment at this College at the end of the next Easter term.
I have the honour to be, Sir, your obedient servant,
John W. Ottley, President R.I.E.C.[31]

The gross insensitivity of the timing and manner of this notification is astonishing from someone of Ottley's reputed administrative ability. Moreover, by forwarding Walpole's letter in its entirety, he drew attention to the hitherto secret proposed remodelling, the Board of Visitors' report, and the subsequent correspondence with the India Office. Ottley later claimed that he had no intention of "doing anything abrupt or harsh, or to in any way hurt anybody's feelings." Because everything was done in a great hurry, he claimed, he did not have time to write individual letters, for which he was "exceedingly sorry."[32]

After a miserable Christmas, the seven professors jointly prepared a response to the secretary of state. They requested the appointment of an "independent committee of experts in scientific engineering education and college management" to review the working of the college. Furthermore, they desired that "the whole of the teaching staff may be allowed to state their experience to that committee." In conclusion, the memorialists noted that "no member of staff has been consulted respecting the changes proposed in the curriculum."[33] Until they received the copy of Walpole's letter, in fact, they were "in ignorance" of the nature of those proposals and the board's recommendations.

Hearson, on behalf of his colleagues, sent the contents of Walpole's and Ottley's letters, together with their own memorial, to the *Times*, which printed them on 3 January 1901. These documents were mailed on 31 December 1900, very shortly after the memorial was sent to the India Office, so they clearly had no intention of awaiting an official response before going public.[34] In the event, Walpole did not reply until 15 January 1901, and by then he could only respond that the secretary of state considered the Board of Visitors to be "fully qualified" to deal with the management of the college and that he saw no need for an enquiry by another committee.

The publication of these documents triggered a flood of outraged letters in the press and a flurry of official activity over the next six months.[35] Writers repeatedly noted the quality and importance of Coopers Hill and overwhelmingly supported the dismissed instructors. The abrupt dismissal of nearly half of the engineering teaching staff of the college,[36] long-standing educators whose performance had not been criticized, was seen as "heartless brutality."[37] J.A. Ewing from Cambridge University wrote, "To suggest that the dismissal of men like Prof. McLeod and Prof. Hearson can make for efficiency is preposterous."[38]

Editorials questioned Ottley's qualifications as an "educationalist and man of science."[39] A letter from "Three Students" went so far as to call Ottley "a costly incubus."[40] Other students distanced themselves from the insult but endorsed their general dissatisfaction.[41] Writers also criticized the secretary of state for conceiving how "a system in which a college president has absolute control over all educational details could possibly be successful."[42]

But the complaint that gained most traction was that the dismissals were an affront to British scientific education and to science itself. The eminent physicist Lord Kelvin (William Thompson) wrote to the *Times* on 12 January 1901 in support of the professors and coordinated a petition to the secretary of state reiterating the request for an enquiry. A total of 374 people signed the petition, including many of the country's highest-profile educators and scientists.[43]

Somewhat sourly, Ottley wrote, "I understand a good many men refused to sign the memorial for an enquiry because it so clearly prejudiced the case. Many others appear to have been led away by the glamour of Lord Kelvin's name."[44] Nevertheless, faced with such a public and high-profile show of support, Lord Hamilton had little choice but to meet with a deputation of the memorialists. A well-prepared Hamilton met with Kelvin and thirteen other distinguished scientists on 12 February to address their concerns, which essentially reflected those already expressed in the media. Hamilton responded with a summary of the reasoning set out by Ottley and the board. The *Times* printed an account of the meeting the following day, which was the first time that any of the facts had been made public.

Faced with this deluge of new information, the deputation had no further comments and withdrew. The next day, however, Kelvin expressed dissatisfaction with Hamilton's response, particularly that the impossibility of implementing the new curriculum without dismissing the professors had not been adequately demonstrated. Ten days later, in anticipation of questions in Parliament, the government issued a "blue book," entitled *Correspondence Relating to the Remodelling of the Studies and the Retirement of Certain of the Professors and Lecturers,* that set out the official sequence of events.[45]

On 8 March 1901, the *Engineer* reacted to the publication of the blue book with a highly negative leading article. Taking Ottley's numbers at face value, the article was venomous in its condemnation of the college's mismanagement, the slackness of the professors, and the low student workload. In part the first paragraph read, "The story told is the story of a blunder, and of something worse. It is difficult, indeed, to understand how the system, or want of system, under which the work of the College was carried on could have been permitted to come into being at all; or to exist for a single year.... In one word, the existence of the college at all cannot be justified; it should be reformed off the face of the earth."[46] The *Engineer* did not get its wish immediately, but such public criticism effectively started the countdown to the eventual abolition of Coopers Hill.

Not a University

The aftermath of the deputation revealed much about the Government's attitude towards Coopers Hill. As reported by the *Times*, Hamilton said that he did not want to leave the deputation with the idea that the India Office had been "discourteous or harsh" to the dismissed instructors. As far as he was concerned, "they had entered into a contract which the Office had observed, but beyond that contract had no vested rights."[47] That is to say, they were ordinary employees not university academics, despite the title "Professor."

One of the deputation's complaints was that the curricular changes should not have been sanctioned until the Coopers Hill teaching staff had been consulted. Hamilton responded, "This practice may be the custom in collegiate management ... but it certainly does not prevail in the administration of the great public departments of the country."[48] In private, the views of the matter were even stronger. In preparation for meeting the deputation, the undersecretary of state for India, Sir Arthur Godley, advised Hamilton, "The College is not a University or an institution for the advancement of science. It is maintained at the cost of Indian revenues for an exclusively Indian purpose, of a strictly practical kind; viz. the training of candidates for the Public Works and Forest Services."[49]

Ottley believed the same and accused some of his professors of "playing at being a university, whilst Cooper's Hill is nothing of the sort; it is a technical school, and not a university at all."[50] Godley further commented that the members of the Council of India looked "at the matter practically and in the interests of India; not in the interests of science, and still less in the interests of scientific individuals."[51] In the view of the government, therefore, Coopers Hill was not a university but a government training establishment with the sole purpose of providing competent engineers for the Indian service.

These examples highlight an inherent contradiction in the structure of Coopers Hill. To the professors, the public, and the technical community at large, it had all the appearances of an academic institution. It offered academic courses and credentials taught by distinguished professors who published textbooks and research papers. Successful students received awards, fellowships, and their academic diplomas. Students and professors were even required to wear academic gowns.

To the government, however, it was one of several public departments funded to provide a specific service. Institutions such as Chatham (Royal Engineers) and Sandhurst (army) had operated in this way for many years without confusion. Unlike those establishments, however, the majority of engineering students at Coopers Hill were there for a general education.

Even if we accept Ottley's explanation that his heavy-handed treatment of the dismissed instructors was a lapse of judgment brought upon by time pressure (although more than six weeks elapsed between the secretary of state's approval of the dismissals and Ottley's letter of notice), there remain his previous twelve months of non-consultative work on the curriculum. It seems more probable that his military background had conditioned him to a hierarchical mode of command that aligned better with the government's conception of Coopers Hill than the public's.

Interviewing the Professors

A tangible outcome of the deputation was that Lord Hamilton asked the Board of Visitors to meet with the college's academic staff to hear their opinions of the changes. He also proposed adding university representatives (from Oxford, Cambridge, and London) to the visitors to help reduce the gap between the perspectives of the academics and the authorities. Lastly, Hamilton asked the new visitors to "enquire into and report upon the working, discipline, and constitution of the College, and the relations of the Visitors, President, and Teaching Staff."[52]

Accordingly, over three days in March 1901, the Board of Visitors questioned the seven dismissed instructors and some of their colleagues about the proposed reforms. That evidence, apparently recorded almost verbatim, and the visitors' report (dated 25 March 1901) were again published in a parliamentary paper.[53]

The goals of the Board of Visitors were clearly to diffuse the public reaction, reaffirm their decisions, and allow the college to move ahead without delay. Therefore, the visitors' report changed little, with the exception of retaining William Stocker as professor of physics until the needs of the new curriculum could be fully assessed. Even so, Ottley did not "very much care" whether Stocker stayed or not.[54] The evidence did, however, reveal the deep divisions between the president and the staff, and the visitors were forced to admit that "the very unsatisfactory state in which matters stand, and have stood, we fear, for some time past, has been forced upon our attention."[55]

Ottley noted that the third-year students (who were disrupting Brightmore's classes) "*may* still give trouble" but that they would be "rid of them" at the end of the following term.[56] Of the professors, Ottley believed that, after the dismissals, only George Minchin (professor of applied mathematics) and Percy Groom (lecturer in botany and entomology) were actively opposed to him.[57] Behind the scenes, Minchin had written to the Royal Society to express his concerns about its role in the affair and to suggest that the president of the society, Sir Michael Foster, should invite Pennycuick to discuss the matter. This they declined, stating that to do so would be contrary to the society's position of non-involvement in the Coopers Hill dispute.[58] This position in itself speaks to the complex political considerations at play. In addition, the number and triviality of notes that Minchin sent Ottley during this period suggest that Minchin expressed his opposition by an annoying adherence to the letter of Ottley's rules.[59]

Thomas Hearson was singled out by the visitors for particular "condemnation" on account of a letter he had sent to the *Engineer* to rebut its scathing editorial of 8 March.[60] In his letter Hearson asked the publication's readers "not to be deceived by the perverted outside view which has been shown in the official report." He noted the high success rates for students taking the ICE exams,[61] took issue with the financial calculations, and gave his estimates of the actual student and professor workloads. Lastly, Hearson suggested that the "motive for presenting such a misleading account" was merely to justify the authorities' unpopular actions.

With regard to Ottley's conduct, the visitors expressed doubt about whether communications between him and the teaching staff had been as "free and cordial" as they could have been. Nevertheless, they were "not prepared to blame him" because of the difficult task he had faced.[62] Indeed, it is plausible that this task had in fact been given to the new president by the India Office.

Working, Discipline, and Constitution of the College

In response to the third part of Lord Hamilton's instructions, the Board of Visitors established a committee to investigate the operations of the college comprising six visitors, as well as a representative from the University of London. This committee reported back to the visitors on 12 July 1901, in time for implementation in the coming academic year.[63]

To remove all doubt about where the ultimate control of the college lay, the committee's report started with a statement from the secretary of state for India that "the authority of the President and of everyone else at the College is derived directly or indirectly from him." Rather belatedly, it was also to be understood that "any measure which for one reason or another is likely to be publicly challenged, is, if possible, to be referred to him [the secretary of state] before being adopted."[64]

Unsurprisingly, the committee recommended that the ultimate authority should indeed continue to lie with the secretary of state. Likewise, the committee reinforced the status quo that the purpose of the Board of Visitors was advisory to the secretary of state. To do otherwise, they argued, would be to turn the visitors into a distinct governing body, inevitably leading the college away from serving the need of India, which would be inappropriate for as long as the funding came from that country. They proposed that in future the "character of the questions to be referred to the Board of Visitors be defined by a rule passed by the Secretary of State in Council."[65] Here, the committee had in mind the inconsistency between the recent remodelling, which had been referred to them, and the abolition of the Practical Course in 1898, which had not.

With regard to the appointment of the president, the committee advocated a long-overdue reform: "We are inclined to doubt the expediency of limiting the appointment to [officers of the Corps of Royal Engineers], and we consider that it would be desirable, in case of future appointments, to attach weight to educational experience."[66] Along with this, the College Regulations needed to be revised to remove standing orders of "too military a character." Five years previously, at the time of President Taylor's retirement, Professor Minchin had privately expressed the same view: "I suppose that we shall get some wretched superannuated General as a President, although some of us will endeavour to get a representation made to the Secretary of State for India in favour of a scientific man.... I wish that you could do something to take this College out of the hands of the Old Generals. It is disgraceful."[67]

In light of the evidently poor communications between the president and the teaching staff, the committee proposed the reinvigoration of the so-called College Board, a group consisting of the senior professors in the college that had all but lapsed. In future it should meet monthly to discuss all matters "connected with the studies of the College."[68]

The last of the major recommendations was that the new equipment and facilities needed to run the revised curriculum, particularly in electrical engineering, should be procured forthwith so as to be ready for the coming session. In keeping with the established governance structure, the Board of Visitors endorsed the findings of the committee and forwarded their report to the secretary of state, who in turn approved and communicated their instructions to President Ottley.

McLeod's Diary

Herbert McLeod's diary for December 1900 and the early months of 1901 conveys a strong sense of the anxiety and confusion in the college. McLeod received Ottley's letter of dismissal at around midday on Monday, 17 December 1901. Throughout the day, word spread amongst the seven who were dismissed and the colleagues in whom they confided. He had intended to wait until the following day before telling his young wife Amelia (Min), but she overheard Heath talking about it, and McLeod had to explain – she bore it better than expected but was still "very much distressed." The next day in the Professors' Room, other members of staff "were all in consternation" about the news.

The affected men lost no time in taking action. McLeod in particular was very well connected in London's scientific community,[69] and over the next few days he wrote to his mentor, Lord Salisbury, and influential scientists such as Lord Rayleigh, Norman Lockyer, William Preece from the Board of Visitors, former Coopers Hill colleague Sir Geoffrey Clarke, and Arthur Rücker, secretary of the Royal Society. At a regular committee meeting, he also consulted his colleagues William Tilden and Henry Armstrong at the Chemical Society of London, who said the dismissals were "unprecedented" and advised the men to "fight it out."[70] Other scientific colleagues started writing to express sympathy about the "high handed action" at the college.

McLeod also communicated with the college president to find out whether Ottley agreed with the decision to fire him and to ask if he had done his work to Ottley's satisfaction. Ottley's frustratingly disingenuous responses were that he was not permitted to say whether or not he agreed with the decision, although it would soon become clear that he did, and that he "was not competent to give a certificate of [McLeod's] capabilities."[71]

The professors started drafting their memorial to the India Office on 19 December at a gathering in Hearson's quarters. On Boxing Day the rest of the Coopers Hill staff declined to sign the memorial but gave permission for it to say that they approved of it. In parallel, Shields met with the editor of the *Times*, and the seven memorialists started to seek a prominent supporter to write to that newspaper in support of their case. They initially asked Minchin if he would approach his friend George Fitzgerald, but Minchin suggested William Unwin instead. Unwin declined, so McLeod next asked Henry Armstrong. Despite another conversation with McLeod in which he expressed sympathy for the memorialists, Armstrong replied that he had been "strongly advised not to write a letter,"[72] hinting at a systematic opposition to the case of the dismissed staff by the authorities. This supposition is reinforced by an entry in McLeod's diary a few days earlier reporting that the president had told Mrs. Edgcome that "he had heard we were going to make a stand and it would be the worse for us."[73]

On 3 January, McLeod noted the publication of their memorial. He spent the day sharing copies around and waiting in Sergeant Lanning's lodge for letters to arrive. Over the next two weeks, McLeod, Minchin, and Hearson scanned the newspapers for fresh letters and articles. As early as 7 January 1901, McLeod went to London to visit Sir Norman Lockyer, prominent scientist and astronomer, which may have ultimately resulted in the letter of support from Lord Kelvin in the *Times* a few days later. Thinking that William Preece seemed "angry with the memorial," McLeod also visited him on 7 January. Preece told McLeod that when the Board of Visitors approved Ottley's curriculum changes it had been under the impression that the internal College Board had discussed the matter beforehand.

McLeod received from Philip Reilly a copy of the parliamentary blue book of correspondence related to the curriculum remodelling on 2 March. He and Minchin spent the next few days making notes and drafting criticisms. He also visited the House of Commons to meet his local member of Parliament and recorded that Hicks and Matthews were to go before the Board of Visitors about the publication of Pennycuick's letter in the *Coopers Hill Magazine*.

McLeod was examined by the Board of Visitors on its first day of interviews, 14 March 1901. Afterwards he felt that he had been well treated by the Visitors, and even though Preece had attacked him about the memorial, the chairman stopped it. He was left with an understanding that they wanted to "diminish the chemistry and possibly close the College."[74] Those who gave evidence were allowed to correct proofs of their interviews, and although the transcript states that McLeod handed in a statement that day, it appears he was given time to finalize it, because he recorded in his diary working on the draft on 23 March, two days before the visitors' final report.[75] He devoted the major portion of his statement to the proposed changes to the course of chemistry instruction and the evolution of his own duties. He closed by noting the "sudden and unexpected" manner of his dismissal and that such treatment must fall "very heavily on oneself and one's family." He had four children at the time, the eldest being eleven.

Ultimately all this activity changed little. McLeod received a letter from Ottley on 1 April to confirm that his appointment would terminate on 3 April, the day after the *Times* reported the publication of the parliamentary blue book containing the visitors' report. Perhaps McLeod suspected all along that his efforts would be futile, because he had started enquiring about other positions early in the New Year. But as he would write later to Ottley, his long service left him "too old to obtain another appointment of a permanent character."[76] Instead, a part-time appointment was created for him by his friends at the Royal Society to direct the preparation of a catalogue of the society's papers for a salary of £250 per year. The project was expected to take five years, but it ultimately lasted more than twenty.[77]

On Saturday, 4 May, McLeod described his final day at Coopers Hill, almost exactly thirty years since his appointment. After breakfast with Mrs. Groom, he and his wife met the movers who had packed up their possessions and went around the empty house finding things the men had forgotten to take.

They collected mail from the post office, paid bills, and arranged a cab to the station. They also went to say goodbye to the Edgcomes, saw Mrs. Harbord and Mrs. Shields, and later in the day met Hicks and the Lodges.

After lunch with the Minchins, McLeod "took the lab keys and left them at Matthews' house." This must have been a sad task for McLeod, who had designed, built, and supervised these laboratories since before the college had opened. On their final visit to the college, McLeod gave his new address to the Porter's Lodge while his wife left calling cards at the Ottleys. Finally, they caught the train to London, and the last personal connection to Coopers Hill's earliest days was severed.

Closure Avoided

Between its March and July reports, the Board of Visitors undertook one other item of business. On 1 April 1901, they reported to the secretary of state that, before the new university representatives were appointed, the visitors had met to discuss the delicate question of whether Coopers Hill College should be continued.[78] In a succinct but powerful document, the visitors gave five reasons for closing the college that were "at least entitled to examination":

a) That although the college may have been established to train a sufficient number of qualified engineers for India, "many excellent engineering schools and colleges" now existed that might be able to satisfy the current needs.[79]

b) "That much inconvenience results from Government undertaking the administration of an educational college."

c) "That the limitation of Indian appointments to Cooper's Hill College men prejudicially restricts the choice of Government in filling various posts."

d) That the expensive education at Coopers Hill prevented "the sons of Indian officials from following in their fathers' steps."

e) That Coopers Hill did not "pay its way."

Although the Board of Visitors presumably considered itself to be acting proactively on the matter, one member, Hugh Leonard, declined to sign the report. He argued that it was not his place, as a visitor, to be "party to an attack of this kind on the College" without having been asked for advice on the matter by the secretary of state.

Fortunately for Coopers Hill, the secretary of state was "not disposed at the present moment to raise the question of the abolition of the College."[80] Lord Hamilton believed "with some doubt" that because both he and the visitors had approved Ottley's reforms, the new curriculum should be given a fair trial. However, the letter sent by the undersecretary of state to the visitors on

Hamilton's behalf concluded ominously, "He will, however, not lose sight of the arguments which the majority of the Board of Visitors have placed before him, and will regard the maintenance of the College in the light of an experiment, the continuance of which can only be justified by its success."[81]

Diversity in Engineering Careers

The events of 1901 took place in the context of an increasing variety of engineering employment demonstrated by Coopers Hill Engineers and the changing nature of the college's student body. Ironically, it was indirectly President John Ottley's attempts to modernize the curriculum and facilities to meet this demand that led to the government enquiry that ultimately recommended closing the college.

The great majority of Coopers Hill engineers worked in the traditional areas of civil and mechanical engineering – roads, railways, bridges, water works, foundries, and docks. Elements of mechanical engineering had been in the curriculum since its beginning in 1871, and the subject had gained importance with the introduction of mechanical design and laboratories. Ottley's biggest curricular change in 1901 was the introduction of mandatory courses in electrical engineering. By doing so, he was responding to a trend towards an increasing diversity of engineering disciplines that has lasted to the present.

Electrical Engineering

Before Ottley's changes, Thomas Shields, the lecturer in electrical engineering, estimated that all Coopers Hill students received about 100 hours of tuition in electricity and magnetism. The dozen or so students who opted for the third-year electrical engineering course gained an additional 225 hours. After 1901 the "electrotechnology" curriculum at Coopers Hill essentially consisted of three courses in each of the second and third years, totalling 230 hours.[82] In the second year, students learned about electricity and magnetism, basic circuits, and measurement instruments and their calibration, with a focus on electricity. In the third year there was an advanced theory course, a second electrotechnology course focusing on "the applications of alternating currents," and another Practical Course on the testing of instruments and machines. Topics in these courses are summarized in table 8.1.

At the same time, several Coopers Hill professors were researching new electrical phenomena. The most energetic was George Minchin, the popular, ebullient professor of applied mathematics. He conducted early experiments

Table 8.1. Electrical Engineering Courses after 1901 Revisions (1902–3 RIEC Calendar)

Year	Subject	Autumn term	Easter term	Summer term
2	Physics	Electricity & magnetism; electrical potential & energy; Gauss's theorem; dielectric constants	Magnetic theory; magnetic forces & potential; magnetic induction; susceptibility & permeability; magnetic properties of iron; permanent magnets	Resistance; Wheatstone bridge; basic circuits; inductance; time constants; reactance & impedance; thermo-electricity; primary cells
	Electro-technology	CGS units; practical units of measurement; EMF; measurement instruments; magnetic force & induction; hysteresis loops	Magnetic circuit; conductors & insulators – properties of various materials; temperature effects; practical measurement apparatus	Dynamos – components and construction; Electromotors – mechanical and electrical aspects; testing dynamos and motors; efficiency
	Physical Laboratory	Basic material properties	Lenses; practical use of measurement equipment; resistivity; cells	Measurements of capacity, inductance, hysteresis; thermoelectric circuits
3	Electricity & Magnetism	Dielectrics & capacitors; AC circuits; long submarine cables	EM waves; Hertz's experiments; EM theory of light; radiation detectors; cathode & Röntgen rays[83]	Electrolysis; ion migration; primary power cells; absolute measurements of resistance & current
	Electro-technology	Electric lighting; electric arcs; searchlights; incandescent lighting; house wiring; power generation & distribution; secondary power cells	Single-phase alternators; motors; transformers; safety & testing; design of alternators & transformers; AC power distribution; polyphase alternators & transformers; induction motors; polyphase power distribution	Electric railways; brakes; controllers; lightning arresters; electric fans, lifts, cranes; electric welding; electrochemistry
	Electro-technical Laboratory	Instrument calibration; incandescent lamps; magnetic measurements on dynamos	Testing motors, alternators & dynamos; properties of secondary cells	Transformers; polyphase systems; induction motors

on the electrical measurement of starlight[84] and proposed the concepts of solar cells[85] and electronic cameras.[86] He was also investigating radio transmission at the time of Marconi's famous demonstrations.

Ottley assumed the need for students to learn more electrical engineering, but Charles Reynolds, former director general of the Indian Telegraph Department, explained further:

> Our knowledge of electricity is admitted to be in its infancy; at the same time there is no branch of science which is attracting more attention now-a-days and in which greater progress is being made. Again the immense continent of India offers an unlimited scope to the economic application of the powers of electricity, not alone in telegraphy and telephony, but in electric lighting, power, and traction, and possibly in many other fields which will open to future generations. It is therefore a matter of vital importance to the Government of this country that in its Telegraph Department it should possess some officers who have been trained in the most advanced school in this science.[87]

It wasn't so much that the civil engineers needed to know about electricity immediately but that the electrical age was clearly coming, and a knowledge of it would be required by all engineers soon. Minchin stated that he had advocated for better electrical engineering instruction as far back as 1885.[88] The late 1890s had already seen the construction of India's first hydroelectric generating station at Sidrapong, near Darjeeling, in order to generate electricity for lighting. In 1902, power was generated at Sivasamundrum from the Cauvery Falls, and in 1905–6 Richard Thorp (CH 1887–9) constructed a small hydroelectric plant to supply several tea-processing factories in North Travancore.[89] The scale of the projects increased swiftly, and in 1911 Lord Sydenham of Coombe (who as George Sydenham Clarke had taught at Coopers Hill) opened the ambitious Tata Hydro-Electric works that supplied power to Bombay's mills and factories.[90]

From that time onward, many water supply projects also included power generation. At William Sangster's Malakand Canal Tunnel, not only were the rock drills powered by hydroelectricity during construction, but the 300-foot drop from the tail of the tunnel to the valley below was designed for power generation (built in 1938). As chief engineer for India's Hydro-Electric Survey, George Barlow (CH 1883–6) produced the first comprehensive report on the water power resources of India.[91] And in 1937, when Vincent Hart (CH 1900–3) was engineer-in-chief of the gigantic Cauvery-Mettur Dam, hydroelectricity generation was a key feature of the project.[92]

However, at the turn of the twentieth century, it was the Telegraph students who became the first graduates from Coopers Hill effectively to practise electrical engineering.

Ernest Hudson's technical instructions from the Indian Telegraph Department consisted of a series of short, practical briefings for officers in the field. The earliest is dated 1887 and concerns "duplex" telegraphy capable of sending messages simultaneously in both directions down a single wire using a cleverly constructed signalling relay.[93] Others in the late 1880s deal with fault-finding, conductivity of wires, and other basic techniques. A decade later is a similar document on "quadruplex" signalling,[94] for which the circuitry, operation, and testing were correspondingly more complex.

At the end of the of the 1890s, when Reynolds was making the remarks above, many of the department's instructions concerned telephones. Its 1897 observations on the convenience of users indicate the novelty of the device: "Experience has shown that the Telephone is often much more appreciated by subscribers if it can be used without having to rise from the chair or move from the office table.... In offices in Europe it is not unusual for a subscriber to have two telephones – one in his clerk's office and one on his own office table.... [T]here are some subscribers ... who look on a set of apparatus reserved for their sole use as essential on sanitary grounds."[95]

The nearly seventy pages of telephone instructions for Telegraph officers reveal a wide array of manufacturers and patent designs. Each contained delicate proprietary components that were difficult to maintain and adjust.[96] Many of these pages have revisions pasted over the original version or slips of paper with additions glued into the margins. Not only were the instruments more complex, but they also needed better lines and specialized exchanges to direct calls. Moreover, these instruments were in the hands of the public instead of trained operators, so servicing them became more problematic. So, in calling for the creation of an "electrical engineering" branch of the Telegraph Department, the Telegraph Committee's 1907 report noted, "The enormous increase of traffic must be met in part by the use of modern instruments of intricate mechanism.... The working of these instruments needs more than ordinary electrical knowledge, and great loss and dislocation of traffic are the consequence if their mechanical efficiency is not effectively maintained. The rapid advance of electrical science also imposes upon the Department the obligation of keeping its electrical knowledge up to date, and for this purpose a body of specialists is required."[97]

One of these "modern instruments" was wireless telegraphy. In the early 1900s, Maurice G. Simpson (CH 1885–7)[98] studied wireless telegraphy at the British Admiralty while on leave from the Telegraph Department. He then started his own experiments when he returned to India. The first was the establishment of a successful transmission over the 45 miles between Saugor (Sagar) Island at the mouth of the Hooghly River and a ship in the Bay of Bengal.[99] Between August 1904 and February 1905, Simpson and John Parker (CH 1899–1902) used this experience to construct an operational 305-mile wireless link from

Port Blair on the Andaman Islands to Diamond Island in the mouths of the Ir-
rawaddy River, south of Rangoon. Their apparatus was of the Lodge-Muirhead
design with the antennas at each station mounted on four 150-foot supports
constructed like ships' masts. These were ultimately driven by "3 H.P. engines
with alternators." Initially the challenge was in getting the equipment delivered,
shipped to the various stations, and installed. Later, the weather proved to be
the main problem, with transmissions interrupted by electrical storms, equipment damaged by lightning, and the impossibility of working in the temporary
station "on account of the noise and rain blowing through the tent." Simpson
claimed that the reliability of the wireless system was similar to that of a land
line of the same length passing through jungle, and that "as ... experience has
shown us how to so improve the methods of construction and maintenance of
the land lines as to minimise the interruptions in number and duration so it
is to be expected that experience will show us how to improve our methods of
wireless telegraphy to attain a similar result."[100]

Concerned by the competing interests of the Admiralty, Lloyd's of London
(because of shipping safety), and the Marconi Wireless Telegraphy Company,
the Government of India decided to retain control of wireless communications
under the auspices of the Telegraph Department. The first applications of wireless telegraphy were mainly for communications with ships, in "isolated and
unattractive spots" that were difficult to connect by land lines, and for strategic
military applications.[101] By 1918 there were thirty wireless stations operating in
India, but message transmission was still too slow to complete with the well-es-
tablished ordinary telegraph system.[102]

Although Simpson's official title was chief electrician, his 1954 obituary in
the *Times*[103] described his position as "electrical engineer-in-chief," which gives
a more accurate view of what his work entailed. The trend of telegraph work
away from the line construction of Hudson's early career towards electrical instrumentation was clear; it was to this evolution that Coopers Hill needed to
respond.

Mechanical Engineering

President Ottley's 1901 curriculum changes effectively strengthened the college's coverage of electrical engineering, railway engineering, and irrigation
at the expense of mechanical engineering, which was largely ignored. Indeed,
Ottley's only real mention of the subject in his proposal was that he did "not
favour the 'turbine design' [course] from any point of view."[104]

In 1871 it was avant garde to cover mechanical engineering in a civil engineering college. Descriptive Engineering devoted substantial time to
prime movers (boilers, steam engines, and locomotives). Mechanisms (e.g.,
gears, cranks, drive-belts that connected the prime movers to the tools) were

addressed in both Descriptive Engineering and in Applied Mechanics. In his opening address to the college, George Chesney listed mechanics as one topic about whose inclusion in the curriculum "there can be no doubt." His argument was that, unlike in England where specialist contractors were available, the Indian engineer had "often to run alone" and therefore needed a broad knowledge that included the basics of the mechanical tools he used.[105]

Little changed over the following two decades, although a turbine design project was added. By 1902–3 the "prime movers" sections had even been taken out of the first-year Engineering course. Steam engines were still extensively covered in the second-year workshops, and there was a small section on "Gas and Oil Engines: Common Types; Method of Action and Management." The third-year Applied Mechanics course did include material on oil and gas engines but continued to focus heavily on steam.[106]

By the end of the nineteenth century, however, other types of engines were rapidly becoming mainstream. For example, the Electro-Technical Laboratory installed in 1902 at Coopers Hill[107] as a result of Ottley's remodelling was powered from twenty-four-horsepower Crossley internal combustion engines that ran on coal gas. These drove two dynamos that each generated more than 120 amps at 110–140V. In addition to operating the laboratory, these were designed to supply the 100V electric lighting system that was installed the following year in the Lawrence Corridor by the fourth-year students, each room having two lamps.[108]

Motorized transport was familiar at Coopers Hill at an even earlier date. Neville Minchin, son of professor George Minchin, recalled taking a ride on an early motorbike:

> It was about that time [1897], however, that 1 made my first real contact with a motor vehicle; a motor tricycle (De Dion, 1 expect) which was kept in the college bicycle shed by one of the students. His name was Frank McClean.... He was very kind-hearted (and still is), and he promised the small boy that when he could get the machine working he would give him a ride.... And so it came to pass that I stood up behind him on this vibrating machine and we sped across Englefield Green at 15 or perhaps even 20 miles an hour.[109]

Internal combustion engines had been available in cars for some years already and would soon become popular for another branch of transportation: aviation. In December 1903, as the secretary of state for India was contemplating the future of Coopers Hill (see chapter 8), a historic event in human technology took place at a windy field in North Carolina. The Wright Brothers' first powered flight, and their European tour to demonstrate their flying machine, captured the imaginations of several Coopers Hill men and led them into a completely new field of engineering. Foremost amongst them was the above-mentioned (Sir) Francis Kennedy McClean (CH 1894–7),[110] one of the earliest

Figure 8.4. Frank McClean flying through Tower Bridge (*Illustrated London News*, 17 August 1912, 246)

Britons to fly with Wilbur Wright and later became a pioneer of British naval aviation. McClean achieved some notoriety in 1912 for being the first person to fly up the River Thames and through Tower Bridge (figure 8.4).[111]

However, these mechanical engineering advances were not reflected in the Coopers Hill curriculum. Ottley's remodelling led to the dismissal of long-time professor of hydraulic engineering and mechanism, Thomas Hearson, in order to unite the instruction of engineering under a single professor, Arthur Brightmore. Relations were strained between Hearson and Brightmore because, according to the former, Brightmore was "very reticent" and "detached" from the rest of the staff. The root cause was more likely that Hearson felt himself to be better qualified for the position of senior engineering professor, saying "that a man with no experience at all in teaching should be considered to be my superior on absolutely no evidence is, I think, somewhat monstrous."[112]

Brightmore's assistant would be Richard Woods, whose expertise was in structures, not mechanics. The alignment of Brightmore's and Woods's expertise with the new curriculum content speaks to a strategy that favoured other subjects over mechanical engineering. Given the constraints on instructional time, there was little alternative. In the event, Coopers Hill would not last long enough for this decision to matter. If it had survived, however, the need to supply engineers for the twentieth century's mechanization would have soon required Coopers Hill to further diversify its academic content and perhaps to offer different engineering specialties.

Student Diversity

Indians in the Public Works Department

Appointments of Indians into the senior establishments of the Public Works, Telegraph, and Forest Departments were affected by a complex mixture of prejudices, vested interests, discrimination, political dogma, and economic considerations. Some practices may have had some legitimacy, but Indians were for the most part systematically deprived of choice in the matter. In the 1870s, one of the fathers of Indian nationalism, Dadabhai Naoroji, took aim at the poor record of the PWD in recruiting Indian engineers, presenting examples where qualified Indian graduates from Roorkee and other colleges had been excluded from the PWD. When Coopers Hill was established, M.E. Grant Duff spoke in Parliament: "It is said, too, that we are excluding the Natives from competing. So far from this being the case, young Englishmen are obliged to pay for being educated for the Public Works Department, while young Natives of India are actually paid for allowing themselves to be educated for that service, and the scholarships available for that purpose are not taken up."[113]

Naoroji was scornful of what he saw as an attempt to hide the true situation, that because qualified Indian graduates from Roorkee and other colleges had been repeatedly and spuriously excluded from the PWD they saw no point in taking up the scholarships: "The Engineering graduates have absolutely no future to look forward to, and it cannot be expected that candidates will be found to go up for the University degree if there be absolutely no likelihood of subsequent employment."[114]

Not until after the First World War did the ranks of these departments see a significant upturn in the number of Indian officers.[115] It was not that Indians were considered incapable of the technical aspects of engineering and forestry but that they were thought to lack the supposed "finer qualities" of the Englishman. Even then, some European engineers found the cultural differences a challenge, even if they respected their Indian colleagues' professional competence.[116] During the 1901 College Board of Visitors investigation, Charles Reynolds, formerly director general of the Telegraph Department, summed up this prevailing British perspective:

It need hardly be said that it is absolutely necessary for the Government of India to recruit a portion of the Superior Officers of the Telegraph Department from England; not alone is this necessary to obtain men scientifically trained in a better school than is available in India, but in the Telegraph Department, as in most branches of the public service, India requires a proportion of officers of good physique and of English home and public school training to impart tone and morale to the service generally and to set a good example in pluck, endurance and conscientious work to those in the department who have not had similar advantage in bringing up.[117]

It was also widely believed that it was unwise for Indians to supervise their countrymen. In 1936 Charles Lane (CH 1900–3) expressed typical and hypocritical concerns about increasing Indian autonomy: "Can it yet be said that reliance may with any confidence be placed on Indians in high positions of administrative responsibility to completely relegate all caste and racial intri[g]ue and prejudice to the background and to accord a square and unbiassed deal in the recognition and due reward of meritorious service, professional ability and administrative qualification for the higher appointments within their gift?"[118]

Foresters, on the other hand, were of the opinion that Indians lacked interest in and appetite for the rigours of life on jungle inspection.[119]

For all this discrimination, Indians were supposedly entitled to compete for the senior ranks of the PWD on an equal basis with Europeans, the three sources being:

1 From Officers of the Corps of Royal Engineers on the Indian establishment;
2 From passed students of the Royal Indian Engineering College;
3 From passed students of the Indian civil engineering colleges.[120]

After Coopers Hill was closed, category 2 was replaced by two new categories: "Selected candidates appointed in England by the Secretary of State" and "Deserving Upper Subordinates."[121] However, this simple statement hides a wealth of complexity. The first was the allocation of available PWD appointments to the three sets of establishments. While the numbers and proportions varied over the years, more than half were from Coopers Hill, and the Indian colleges' allocation went mainly to Roorkee, with one or two each to the colleges at Sibpur, Poona, and Madras.[122]

A second complexity arose because the engineering classes at Indian institutions themselves contained students from a variety of backgrounds. Of the 211 engineers who passed out of Roorkee between 1876 and 1895, 66 per cent (139) were of English background, whereas 34 per cent (72) were Indian.[123] By representing one-third of an already small allocation, the true opportunity for Indians to join the PWD was quite limited. When the Government of India had proposed closing Coopers Hill in 1879, one of their reasons was a desire to promote the appointment of Indian engineers. In November 1882, the viceroy of India, Lord Ripon, responded not by agreeing to the closure but by mandating that appointments from the Indian college should be awarded preferentially to "pure Natives." With such small numbers of allocations, he argued this was "almost the only way" to achieve that goal. Unsurprisingly, the "Roorkee Resolution" was not popular with Anglo-Indians and was withdrawn two years later. To compensate for the greater number of candidates, the allocation to the Indian colleges was increased to 9 from the planned 5 or 6.[124]

Any hope that the Roorkee Resolution may have given for equal opportunities for Indian engineers was dispelled by the report of the Public Service Commission 1886–7, presided over by Sir Charles Aitchison. The commission observed that certain types of public works, such as the maintenance of buildings and roads, did not require "highly trained Engineers" but that all engineers were currently paid on a common scale. It therefore believed that economies could be made by dividing the Engineering Establishment of the PWD into two branches. The Imperial Branch would be for works requiring the "highest Engineering skill and training obtainable in England," while the Provincial Branch would be for "construction and maintenance … of works not ordinarily calling for high Engineering skill" and would be staffed from the Indian colleges.

The commission stated its expectation that this system would further "the larger admission of qualified Natives of India to employment in this important Department." However, it went on to recommend that, while the terms of service for the Imperial Branch should be attractive enough to recruit officers from England, the "pay, furlough, and pension should be fixed for the Provincial Branches without reference to those of the Imperial Branch."[125] In other words, the terms of service needed to be sufficient only to attract engineers in India. When the recommendation was implemented, the salaries in the Provincial Branch were fixed at two-thirds of those in the Imperial Branch.

A similar demarcation had been made in other Indian services, but the implementation for Public Works presented challenges; the work could not be divided on technical grounds because large engineering projects would call for both Imperial and Provincial engineers. Dividing the work on a geographical basis was also clearly unworkable. As a result, the Imperial and Provincial branches co-existed equally on a single list with the only difference being the salary. Despite this supposedly equal status and the greater number of available positions, this "constituted a serious set-back in the position and prospects thus far enjoyed by Indian engineers in the Public Works Department."[126]

It has been said that "Cooper's Hill exerted an important, limiting and adverse influence over colonial engineering colleges," starving them of resources and preventing their growth.[127] While this is undoubtedly true, recruitment to the Imperial Branch continued to be primarily from England, even after the closure of the college. The true culprit was therefore the attitude, exemplified by

Charles Reynolds, that prioritized British pluck and education. Like the Indian colleges,[128] Coopers Hill was itself shaped by the policies of the Public Works Department, whose demands they were all designed to serve.

The Imperial and Provincial services remained distinct though various reorganizations (including separating the lists in 1908 and reintegrating them again in 1912), despite the recommendation of the 1915 Public Service Commission that the two branches should be fully merged. This was not changed by the India Act of 1919 or the creation of the Indian Service of Engineers, and it was not until 1935 that the modern all-India service took shape. The history of the Indian Forest Service followed a similar path.[129]

Admission to Coopers Hill

Rules for the admission of Indians to Coopers Hill also changed over time. In the early years when admission guaranteed an appointment to the PWD, the college was open only to "British-born subjects."[130] From 1880 to 1899, the college was "open, to the extent of the accommodation available, to all persons desirous of following the course of study pursued there."[131] To enter the PWD, however, the candidate had to be a British subject and successful in the competitive examination.

The description changed considerably under the presidency of John Ottley, and the prospectus stated specifically that the college had been established "for the purpose of training British subjects of European race for the service of India." The entry of Indians, it continued, was exclusively through the Indian colleges but, although they had no right of admission to Coopers Hill, they could be admitted if there was space and were able to compete for a maximum of two PWD appointments per year. If more than two Indian students qualified in a given year, only the top two would actually be given positions. However, as Ottley observed, "that is a thing that, practically, would never happen."[132] The reason is apparent from the historical numbers of non-European students in the college.

Non-European Students at Coopers Hill

At least fifty-five students of non-European heritage studied at Coopers Hill, some 3 per cent of the total.[133] About half were from India and the remainder were predominantly from Egypt, Siam, and South America. Figure 8.5 shows that the numbers of non-European students at the college increased steadily over time, with more than 40 per cent of the total number completing their studies in the final six years.

Eighteen of these graduates joined the Indian public services, again mostly in the period 1895–1906. Even then, they represented an average of less than

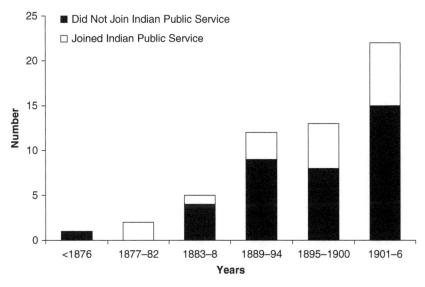

Figure 8.5. Estimated numbers of non-European students leaving Coopers Hill, 1871–1906

one per year, half of Ottley's maximum allocation. The rest of the overseas enrolments were "extra" students, there to learn engineering or forestry skills for use in their own countries. One barrier faced by Indian and other overseas students was the prohibitive cost and inconvenience of studying in England. Typically, the Siamese and Egyptian students were therefore supported by their governments, as were those from the native states of India, such as Hyderabad and Mysore.[134] Applicants from the native states were not considered to be "British born" and so could attend the college only if space permitted and were not eligible for the two appointments allocated to Indian students.[135] Such candidates had to pass the entrance examination and be recommended by the Government of India.

These men often rose to positions of authority, so, for example, Fazul Mooraj's (CH 1892–5) appointment as PWD secretary to the nizam's government was celebrated at a grand party attended by Hyderabad nobility and European officials. "As dusk approached the gardens were brilliantly illuminated, conjurors and musicians amusing the guests."[136] As well as serving on the City Improvement Board, Mooraj was a justice of the peace, a member of the local Masonic Lodge, and trustee of Muslim educational charities. Nawab Ali Nawaz Jung Bahadur (known as Mir Ahmed Ali as a student, 1896–9) served in the same capacity.[137]

A letter dated 1947 from Bangkok to the *Coopers Hill Magazine* provided information about Siamese students who attended the college.[138] The correspondent first described how his personal name by which he had been known at Coopers Hill (1903–6), Sanra Sirn, had been replaced by his official name-cum-title, Phya Sarasastra Sirilakshana.[139] Sirn was one of three Siamese attending Coopers Hill between 1903 and 1906. Mom Chao (Prince) Chat Cho (C.C.) Sak joined the Railway Department in 1906 but died of cholera five years later aged twenty-nine. Graduating from the last class before the closure, the third was Mom Chao Chalart, who served in irrigation and railways before being appointed to the Ministry of Commerce and Communication. President Ottley maintained a correspondence with the Siamese Legation in London about the progress of these three students. In January 1906, Sirn seemed to be progressing the best in his studies, while C.C. Sak struggled because of a poorer knowledge of English and a strongly suspected addiction to opium.[140]

Sirn himself went to the United States after Coopers Hill, obtaining a master's degree from the University of California before returning to Siam via Japan. After serving in various departments, he retired in 1932 from the post of commissioner general of State Railways.

PWD Appointments

The first two non-European students appointed to the PWD were Boli Narayan Borrah (1877) and Ali Akbar (1882). Borrah, a non-resident student, was the first Indian to pass out of Coopers Hill. He was posted back to his native Assam in 1878 as an assistant engineer and made executive engineer in 1890. He became a prominent figure in the Assamese community and in 1900 was awarded the Kaisar-i-Hind medal for Public Service in India.[141]

Ali Akbar crammed for the Coopers Hill entrance examination at Ashton's with twelve other aspiring engineers. A contemporary recalled that he was "a really good sort" and learned his English from reading "a whole edition of Scott's novels."[142] After graduation he joined the PWD in Bombay, became a member of the Institution of Civil Engineers, and became superintending engineer in 1900. He died in 1923.

Rustum Kai Kushro Nariman was one of the earliest Parsis to attend Coopers Hill (CH 1896–9). He was appointed to the Punjab Irrigation Department and became known internationally as an expert on rivers and canals. He took the unusual step of dating and authenticating his travels by opening Post Office savings accounts wherever he went during his career, a collection that ultimately spanned destinations such as Batavia, Singapore, Venice, Taiping, Vienna, Venice, Paris, Moscow, Stockholm, and others.[143]

Prominent among the Indian graduates of Coopers Hill was Sir Ganendra Prosad Roy (CH 1891–4), one of only two non-European graduates to receive a British honour.[144] Roy was educated in England before entering Coopers Hill, where he had only one or two Indian contemporaries. He was the first Indian to enter the Telegraph Department and made his way to the top by a combination of hard work, ability, and, one imagines, persistence against the prejudices of the time.

Roy's feisty attitude is apparent in the event dramatically dubbed by the *Amrita Bazar Patrika* newspaper "The Feni Dak Bungalow Incident." In reality this was a minor contretemps over the occupation of said dâk bungalow at Feni.[145] Roy, then twenty-nine years old, and his subdivisional officer were staying at the bungalow. According to the report, the local overseer had asked Roy repeatedly to make at least one room available for the coming visit of the district engineer and civil surgeon. This was a reasonable request since the bungalow belonged to the PWD, not to the Telegraphs. Although Roy was technically in the wrong, he refused to vacate either room when the visitors arrived and "lost his temper and is reported to have used unparliamentary language."[146]

The article's critical tone towards Roy was perhaps simply because the author believed him to have been in the wrong. But the fact that the district engineer and the civil surgeon were also Indian suggests another potential difficulty faced by Indians who succeeded in the British system – being perceived to have "sold out." A decade later, the *Amrita Bazar Patrika* (Indian owned and increasingly nationalist) accused Roy of being an "anti-Indian Indian" over his testimony to the Royal Commission on Public Services in India. The newspaper was angered by Roy's statement, as officiating director of Telegraphs for Eastern Bengal and Assam, that 75 per cent of future telegraph engineers should still be recruited from England. The paper quoted Roy as saying it was difficult to get men of the "right sort" from India because there was "no place where they could be properly trained."[147] Roy's admission under questioning that he had not visited the schools at Roorkee or Sibpur and had no idea of their equipment further undermined his credibility in the eyes of the *Amrita Bazar Patrika*.

We have no way of knowing the accuracy of the quotes attributed to Roy, which were apparently recorded by someone present at the meeting. The official minutes of evidence published the following year[148] state that Roy did indeed propose the 75 per cent number and that he "had seen nothing of the engineering colleges in India." But the minutes also show that Roy's opinions were fully shared by his colleagues. There was widespread agreement, including by those with knowledge of the country's colleges, that Indian training in the rapidly advancing field of electrical engineering was inadequate.[149] It was generally believed that Indian engineers should have a chance to study in England, and that those who did should be treated the same as the English engineers. Roy was therefore advocating the establishment view. Even if he had privately thought otherwise, it would have been very hard for Roy to have publicly stated differently. Having spent his career fighting British prejudices, he was now seen by some of his countrymen as too British.

At the time, Roy was one of only three Indians in senior positions at the Telegraph Department, the others being Mukand Lal Pasricha (CH 1899–1901) and Baba Sundar Singh (CH 1896–8).[150] As chief engineer in 1922, Roy organized the successful telephone conversation over the 703 miles between Calcutta and Nagpur, representing the longest transmission in the world at that time without repeaters.[151] Roy was also involved in John Newlands's 1907–8 remodelling of the Telegraphs as well as the amalgamation that formed the Department of Posts and Telegraphs in 1912. Roy served for two years as director general of that department, during which time he was knighted.[152] He ultimately retired to Pinner in north-west London with his wife, the daughter of Surgeon-Major S.C. Goodeve Chuckerbutty, one of the first Indians to complete medical studies in England and pass into the Indian Medical Service.[153]

The 1903 Enquiry

Context

In the early years of the twentieth century, then, Coopers Hill was starting to attract an increasingly diverse mix of students from around the world, as well as British students seeking a general engineering education without taking an appointment in India. John Ottley's curriculum remodelling and organizational changes had strengthened the academic course of study. The public fuss over his implementation was dying down and there were signs of a renewed sense of purpose in the college. Moreover, Ottley was confident of his ability to find good positions for other students not destined for India:

> We get appointments in all sorts of places for the others. Egypt takes a good many; the Uganda Railway has taken several; Hong Kong took two or three; and so on in different places. They apply to me from all over the world for men. Those that I cannot get an appointment for at once I place with an engineer in this country; their parents pay a fee of 60*l*., and they remain there for a year at any rate, and at the end of that time I invariably get them appointments – generally before the year is out.[154]

Academically, the future for Coopers Hill was therefore looking good. However, in order to continue offering the new curriculum developed by Ottley, further expenditures on buildings and equipment were required. In 1901 members of the Board of Visitors drew attention to the need for renewed facilities, saying that "a considerable expenditure must be sanctioned without loss of

Table 8.2. Members of Committee of Enquiry into Expediency of Maintaining Coopers Hill

Sir Charles Crosthwaite	Chairman, Public Works Committee of the Council of India
Sir James Mackay	Member, Council of India
Sir William Arrol	Engineer and member of Parliament
Professor Arthur Rücker	Physicist and principal, University of London
Sir Thomas Higham	Former inspector general of irrigation, Government of India

time for the provision of more lecture rooms, of models, of instruments, and other appliances, without which engineering cannot be properly taught."[155] As a result, over £31,000 was spent on the new mechanical, electro-technical, and carpentry workshops, as well as repairs and improvements to the buildings and sanitary system.[156]

By 1903, plans were being prepared for further upgrades, including a drill hall, additional lecture rooms, geological museum, laundry, and baths, with an estimated further cost of £40,000. Moreover, they could expect the need for capital investments to be ongoing: "There is, as is well known, no finality in the equipment of laboratories and workshops, and the maintenance of the high standard of efficiency which is essential to the success of a technical institution like the Royal Indian Engineering College will necessitate recurring calls for expenditure on tools and apparatus."[157]

Faced with the unappealing prospect of having to justify such continuing large capital expenditures to Parliament, the secretary of state, Lord George Hamilton, on 12 May 1903 finally took the step of establishing a formal Committee of Enquiry to determine whether it was "expedient" to maintain Coopers Hill. If the answer was affirmative, the committee was to report on how the quality of recruits to the service could be improved. If negative, it would recommend means by which the system should be terminated and how future PWD and Forest recruits should be obtained. The committee consisted of the individuals listed in table 8.2, all with Indian and/or scientific backgrounds. Over the following three months the committee examined the training provided by Coopers Hill, its relevance to the needs of India, and the possibility of recruiting qualified engineers and foresters from other sources.

Coopers Hill Education

Coopers Hill was established in 1871 because proponents argued that it was the most effective way of supplying enough comprehensively educated engineers for India's needs.[158] By 1906, however, opponents claimed that other British institutions had developed sufficient capacity and program quality to meet the demand. In the intervening thirty-five years, did the existence of Coopers Hill spur the growth of those other engineering schools, or did the college become a backwater that the others surpassed? The minutes of evidence given to the 1903 Committee of Enquiry reveal much about perceptions of the college's success.

Over five days, the committee questioned witnesses from the Coopers Hill academic staff (who were also members of the internal College Board), former students, the past presidents of the Institutions of Civil and Electrical Engineers, prominent non–Coopers Hill engineers from India, and representatives from the University of Edinburgh and the Central Technical School in London. As the Government of India would later point out, relatively few of the witnesses and submissions were entirely disinterested in the matter at hand. It is also questionable how familiar these senior men would have been with the actual abilities of recent Coopers Hill graduates.

The committee also collected fifteen written responses to a set of questions circulated to other British engineering schools about their ability to meet India's projected needs. These institutions had little interest in supporting the college's case and all believed there would be no difficulty supplying the requisite number of engineers from other sources.

At the time Coopers Hill was established, the engineering programs of other British universities and colleges were small, poorly resourced, and suffering from competition with the entrenched pupilage system.[159] After the closure, William Cawthorne Unwin, who had left Coopers Hill for the Central Technical College in London in 1885, went so far as to say that in 1871 there was "no engineering school in Great Britain of reputation or directed by anyone recognised as of standing in the Profession."[160] By contrast, the Coopers Hill's curriculum was based on the latest ideas of the theory and practice a young engineer needed, relatively well resourced, and with a strong sense of identity and purpose.

For example, in an early report on the progress of the college, George Chesney wrote, "Professor Reilly's course in applied mechanics promises to be of a very thorough and comprehensive kind, and very different to anything that has been systematically taught in this country before."[161] By this, Chesney probably meant that Reilly's treatment was more theoretical, because he later wrote, "While in such a subject as mathematics there are abundant text-books …, in Engineering almost everything has to be created in the way of systematic theoretical instruction; in Hydraulic Engineering, for example, there is nothing in the English language fit to be employed as a text book."[162] As the Board of Visitors reported in 1882, the result was that Coopers Hill "enjoys some unique advantages as a place of education, and supplies a kind of training which exists nowhere, at least in the United Kingdom."[163]

By 1903, however, several witnesses to the Committee of Enquiry testified to the substantial progress that engineering education had made in the previous thirty years. "There has been very great progress [in engineering education]. It is an entirely different thing now," said Sir John Wolfe-Barry, past president of the ICE.[164] Unwin testified that in his opinion, good though Coopers Hill's education had been at the beginning, the course he was currently offering in London was better. This seemed to be the consensus of the witnesses. Sir Benjamin Baker, for example, thought, "At one time, of course, Coopers Hill was, in my opinion, in advance [of other technical colleges], but I should hesitate to say whether it is now." Professor Hudson-Beare, from Edinburgh University, thought that the technical training at Coopers Hill was "exactly identical" to that at other colleges. Only Col. Sir William Bisset considered the college to be "much better than ever it was."[165]

So the general opinion was that the training at Coopers Hill had indeed been superior at the beginning, but that other institutions had now caught up. There was no suggestion in 1903 that the education at Coopers Hill was in any way sub-standard, a sentiment that the eminent Sir William Preece summed up: "Coopers Hill training in pure engineering has been very good, and I have never heard any complaint of the engineering."[166] Charles Reynolds and Charles Pitman both considered the revised telegraph instruction entirely satisfactory, and prominent botanist Sir William Thiselton-Dyer was happy with the teaching of the Forestry curriculum (although he was dissatisfied with how students were selected).

Opinions varied on how much Indian specialization the curriculum at Coopers Hill really contained. Sir Benjamin Baker was the most enthusiastic, claiming that one could still meet engineers of great experience in India who would make blunders "for want of technical training" that no Coopers Hill graduate would ever make.[167] Naturally, all three former students supported this view, but it was Frederick Hebbert, one of the first batch of graduates, who was questioned most extensively about this issue. He spoke of the special requirements imposed by the monsoons, for example, or the deep sand foundations needed for railway bridges, to illustrate the specific focus of Coopers Hill.[168] It should be noted that his answers reflected the curriculum as it had been in the past, not the most recent revision. Several witnesses commented that, because the whole thrust of Coopers Hill was towards service in India, the graduates knew what to expect in that country, had invested financially and emotionally in a career there, and were less likely to quit the service from "being disgusted with the life" than engineers from elsewhere.[169]

However, the witnesses left the overall impression that while the top Coopers Hill men were as good as, if not better than, those from elsewhere, the average quality was not demonstrably superior. Several witnesses thought that the problem was not intrinsically with the Coopers Hill education but with the narrow pool from which students were selected. George Minchin commented that Coopers Hill was a "rich man's college."[170] Sir Alexander Rendel (consulting engineer to the secretary of state) agreed, saying that the pool of applicants was severely limited to those who could afford not only the annual £180 fees (which included the mandatory accommodation), but all the private school fees beforehand. There were also incidental costs to consider; Eustace Kenyon had received an allowance of £40 per year while attending the college but thought that £50 would have been "fairer."[171] In contrast, fees at other engineering schools were typically only £20–30 per year, ranging up to £50–60 for the two London colleges, King's College and University College (without accommodation).[172]

Supply of Engineers

Because of the significant rise in the numbers of students and the perceived increase in the social status of engineering, opponents to Coopers Hill argued that better engineers could now be recruited by open competition. It would be at lower cost to the students and zero cost to the government. Possibly they could be given what specialized training they needed afterwards. The committee's report included responses from fifteen other engineering schools that together represented an annual supply of 400–500 students.[173] The consensus was that there was plenty of capacity to supply the needs of the Indian PWD if Coopers Hill were to be abolished.

How had this change in engineering education come about? Frederick Upcott was of the opinion that "the establishment of Coopers Hill College has been one of the great factors in raising the standard of all the other engineering colleges."[174] The catalyzing effect of Coopers Hill was part of the story, but other factors were also responsible. These included an increasing industrial demand for formally educated engineers,[175] an expectation of more clearly defined qualifications, and the resultant increase in engineers' social status. The adoption by the ICE in 1896 of a "well thought out scheme of examinations"[176] for membership in the institution also helped formalize the technical qualifications. By 1903 Wolfe-Barry estimated that 90–95 per cent of new associate members had either taken the ICE's exams or achieved the equivalent qualifications at a college or university, including from Coopers Hill. Coopers Hill men apparently performed well in these examinations. In 1904, all eighteen who took it passed, with L.E. Becher placing first in the country.[177] Ultimately, the final twenty-five engineering students to graduate from Coopers Hill in 1906 sat the ICE examinations in order to qualify for positions in India; at least twenty-three of the twenty-five passed, because they received their PWD appointments.[178]

One driving force for the advances in engineering education had been the growth of engineering colleges in the larger industrial British cities, where support from local businessmen and opportunities for practical training was plentiful.[179] Unwin drew attention to this in his interview with the Committee of Enquiry, when he commented that the students and professors at Coopers Hill "lost a good deal by being isolated."[180] In this, Unwin echoed Reynolds's evidence from the 1901 telegraph instruction review, in which he commented that residence at Coopers Hill did not "bring the telegraph students in contact with the electrical world generally."[181]

What Chesney had originally seen as an advantage – a self-contained residential college "rais'd above the tumult and the crowd"[182] – had since become a hindrance to its progress.

Esprit de Corp Reprise

Each C.H. man is proud of all C.H. men

F.E. Dempster (CH January–July 1878)[183]

While it had been important since the beginning, the value of esprit de corps had since grown into one of the principal merits of the college and a strong reason for maintaining it. Noted alumnus C.W. Hodson (CH 1871–3) said that one reason the college was founded was to "redeem" the service from "the selfish and chaotic state in which it was before Coopers Hill was formed."[184] He was playing to his audience at the Coopers Hill Dinner, but this was a common concept that bound Coopers Hill men together, of "keeping up the reputation of their old college."[185] A prominent PWD engineer before Coopers Hill agreed, reportedly writing, "I always enjoyed meeting Coopers Hill Men. Their advent into the P.W. Department raised it all round from the depressed plane to which it was sinking."[186]

The matter of esprit de corps was therefore high on the committee's agenda, and many of the witnesses were asked about it. There was general agreement that esprit de corps was a good thing for the Indian service, although those in the state railway companies were more inclined to be dismissive. E.H. Stone, chief engineer of the East Indian Railway but with Public Works experience, made the point that upholding a "honour of the service" was important because of the responsibilities placed on the engineers: "An engineer has great powers, especially in up-country places out of the way; he deals with large amounts of money, much of his work is with native contractors, and it is very important that for such conditions you should have a high standard of qualifications not only as an engineer, but also as a man of honour and a gentleman."[187]

However, according to Upcott and Glass,[188] these high principles did not lead to Coopers Hill men being especially efficient or zealous in their duties compared with other officers. Set against all this was the original concern about "*esprit de clique*" expressed during the House of Commons debate in 1871, or as Sir Alexander Rendel put it at the enquiry, "Its worser part is a kind of trades unionism."[189]

On balance, then, the general feeling of the witnesses was that, valuable though esprit de corps might be, it did not on its own warrant the expense of keeping up Coopers Hill. Quite reasonably, considering the evidence they heard, the Committee of Enquiry summed the matter up thus: "We admit that [considerations of esprit de corps] are not without force. To our minds, however, the disadvantage of keeping together a few young men destined for one special service, in an isolated place like Coopers Hill, outweighs them."[190]

The Committee Verdict

Buried in paragraph 30 of the committee's final report, dated 17 August 1903, was the verdict that Cooper's Hill men dreaded:

> The arguments for and against the maintenance of the College have been now stated. After giving very careful consideration to them, we are forced to the conclusion that the maintenance of the College is no longer necessary; that it cannot be efficiently and usefully maintained without further capital and current expenditure of considerable amounts; and that this expenditure would not be justified by the results. For we are of opinion that not only can men as good be obtained by other means, but that the narrow field of selection to which Coopers Hill is restricted must exclude many of the best men from the Indian service. We unanimously recommend, therefore, that the Royal Indian Engineering College should be closed.

Many of the arguments against the college were financial. The committee did not comment on the overall quality of the education provided by Coopers Hill. However, it was of the opinion that the curriculum was only modestly specialized for Indian needs, mainly in surveying and public works accounts, and these could be provided equally well by other institutions.[191] Even though it effectively owed its existence to the 1901 remodelling, the committee's report made no mention of Ottley's new curriculum or to Lord Hamilton's assertion at the time that the revised curriculum should be "fairly tried."

The new curriculum had been designed, by implication at least, to be more relevant to India, but the first students had not yet even graduated from it. The committee's judgment about the college's training for India was therefore based entirely on the past and took no account of the significant political and financial capital recently invested in its improvement.

There was a strong sense that the government had lost its appetite for the "inconvenience" of "maintaining an educational college" with all its associated complexities, as exemplified by the 1901 controversy. It therefore wanted to rid itself of the responsibility as soon as possible, regardless of the actual merits of the case.

It is noteworthy that, so far, the Government of India, on whose behalf this process was ostensibly being conducted, had not had any input whatsoever into the discussion.[192] But now the secretary of state for India, George Hamilton, wasted little time in forwarding 100 copies of the report to Lord Curzon, the governor general of India.

By the time the Government of India responded on 22 October 1903, Hamilton had resigned from his office, to be replaced by St. John Brodrick (later Viscount Middleton). The memorandum from Lord Curzon and the other five members of the Government of India was lengthy, detailed, and a striking contrast to its previous negative or lukewarm opinions of Coopers Hill. They first commented that when the testimony of potentially prejudiced witnesses was eliminated, "the amount of disinterested evidence is not very great." What there was appeared "to be rather in favour of maintaining the College on account of the *esprit de corps* engendered, to which they attach considerable importance."[193]

Curzon's response pointed out that both the Indian and home governments supported many colleges, so the "inconvenience" argument was irrelevant. While the Government of India did not disagree that the number of engineers they needed could be supplied by the open market, they were "confident that they would not be so well suited to our requirements as those we receive from Coopers Hill." The response went on to speak of the multiple roles engineers played in the administration of public works and in liaising with the Indian population, duties laying "great stress on their being gentlemen in training, integrity, and tact, besides being competent Engineers to design and build the works." The Government of India did not believe that graduates from other institutions, without the "Indian atmosphere" of Coopers Hill, could perform these duties.

With regard to finances, their response was clear: "Efficient Engineers are the essential requisite, and for these the Government of India is prepared to meet the necessary cost." It concluded that the Government of India desired to record its "unanimous protest against the closing of the College at the present time."[194] This was a reversal of the position they had held for the previous thirty years.

Although Brodrick was in the difficult position of having to decide between two opposing perspectives, the exclusion of the Government of India until after the enquiry was completed had effectively sealed Coopers Hill's fate. The matter was not put before the Council of India until 1 March 1904 when Brodrick and ten of the twelve members of council met to consider a motion to close the Royal Indian Engineering College. It was carried by six votes to four.[195] Two of the six in favour were authors of the Committee of Enquiry report and two members of the council were absent. Recalling that the original proposal to establish the college had only squeaked through by six votes to five, with many absences, the irony of similarly dubious tactics being used to abolish the college is inescapable.

The *Times* announced the details of the closure on 23 June 1904, saying that the upcoming entrance examination would be the last and that the college would close permanently at the end of the 1905–6 session. On the very next day, Herbert McLeod wrote in his diary, "Ottley has got his knighthood!"

The Hour of Farewell

Through years to be the Thames will flow,
    With winter flood and summer stream,
But faces that it used to know
    Will pass away as in a dream.

Unbroken silence now will cling
Where joy and laughter used to ring.

As each succeeding year passed by,
    Fresh sons were nourished by your hand
Till now you are condemned to die
    By those who fail to understand.

The Nation scorns your school and meads;
What Nation knows its Empire's needs?

They founded you to undertake
    A task that others could not do,
And gladly for the Empire's sake
    You cultivated pastures new.

Now rivals whom you've always led
Are called to do your work instead.

So we must bid a fond farewell
    With gratitude that is your due,
This saddened thought on which to dwell
    To-morrow has no place for you.

But all the sons o'er whom you sway
Look back with thanks on yesterday.

By H.I.B, June 1904[196]

Sale of Coopers Hill

Once the final students left in 1906, the disposal of the college's assets proceeded swiftly. The myriad details included such disparate issues as the transfer of 100,000 young trees from the college nursery to Oxford,[197] disposal of the farm's livestock, selling the floor coverings of the president's quarters, disposal of the bursar's records, and compensation for the domestic staff. Laboratory equipment was transferred to Indian colleges and the National Physical Laboratory, student club assets were sold and the proceeds of nearly £600[198] transferred to the new Coopers Hill Society (see chapter 9), and of course the estate was put on the market.

Despite several enquiries, the property was not sold until 1911, when it was bought by Lord and Lady Cheylesmore as a private residence. The sale removed "one of the most important landmarks in the history of British Engineering," and the planned demolition of lecture rooms, laboratories, and workshops led The *Times* to admit that "the last hope of resuscitation must be abandoned."[199]

The college's official annual expenses had cumulatively shown a slight surplus (or modest deficit if the Forest students continental tour was included)[200] in the years since the restructuring in 1884, so this was not the major issue. The financial argument for closing the college was therefore based principally on its inability to cover from revenues the interest costs arising from this investment. But this does not survive scrutiny. The Coopers Hill estate ultimately realized a price of just £20,000.[201] This sum was substantially less than the £55,000 originally paid to purchase the estate, and less even than the £31,142 spent in the final few years (1901–3) on upgrading the facilities. Without interest, the total capital investment in Coopers Hill College over its lifetime had been £167,578, including the initial purchase and alterations, and subsequent investments. By recovering so little of the total investment at the final sale, almost all of the capital debt (and the interest on it) remained outstanding. Closing Coopers Hill therefore saved the government only the interest on the £20,000, a mere £600 per year at 3 per cent.

So in 1906, the government effectively gave away a fully operational engineering college with an international reputation, bought at considerable expense. One has to wonder at the haste with which this tremendous asset was dismantled and why, even if it did not want to run an engineering college itself, the government did not find a better use for its facilities. Indeed, as far back as 1888 the Special Committee on Home Charges had suggested an arrangement "to relieve the Secretary of State from the responsibility involved in maintaining the College at Cooper's Hill as a Government institution, ... by transferring the establishment to some public body or private corporation, subject to a return on the capital outlay of 128,000*l*. in the form of rent."[202]

Was the closure of Coopers Hill inevitable? In the long run, of course, political events in India would have meant a major shift in focus for the college at the very least. But in the shorter term, it is hard to escape the idea that a cleverer politician than Sir John Ottley would have avoided the pitfalls of the early 1900s. The public scrutiny arising from Ottley's mishandling of his reforms, exacerbated by the haste of the dismissed professors to involve the press, undermined the already shaky political will to defend the college's financial record. It is difficult to imagine George Chesney, for example, either making these mistakes or not defending the college with the utmost vigour and imagination. Rollo Appleyard, once Minchin's assistant and later instructor in telegraphy, certainly thought so: "After General Chesney, military officers of less knowledge and tact have failed as Presidents. They exercised everything but discretion. They grumbled to the India Office, they cherished fads, they wasted official time, they wasted money, they did not limit their ideas to the teaching of civil engineering. The India Office might have tried the experiment of appointing as president a properly qualified civil engineer."[203]

A different Board of Visitors, one less steeped in the more traditional branches of engineering in India, might have recognized the value of Coopers Hill for training the country's engineers for the new, diversified world. Even then, it would have required the secretary of state to think of the issue beyond his focused Indian mandate. This narrowness of perspective can be traced back to Sir Arthur Godley's advice to Lord Hamilton in preparation for his 1901 meeting with Lord Kelvin's delegation: "If we are not allowed ... to manage our own College for our own purposes in our own way, we prefer to shut it up."[204]

Under different leadership Coopers Hill might have continued for another twenty years, during which time, it would have needed to adapt to the new political climate, perhaps becoming independent from the government and offering engineering courses on its own or in partnership with another institution. Alternatively, it might have evolved into a specialized training establishment for the armed forces. But at least the college would have had an opportunity to adapt that was denied it by the abolition in 1904.

9

The Coopers Hill Society

The Saddest and Most Dreary Work

Disposal of Assets

The Royal Indian Engineering College was closed, as far as the students were concerned, on Saturday, 13th October [1906].

So began the lead article in the *Coopers Hill Magazine* from December of that year. The closure was marked by the publication in the *Times* and the *Engineer* of an open letter from Horace Walpole to John Ottley summarizing the accomplishments of the college and the reasons for its closure. About the former it said, "It is clear … how successfully the College has fulfilled the purpose with which it was founded, besides serving the general interests of the Empire at large." The letter closed by thanking Ottley and the staff for the "highly efficient working of the College in past years" and for "the zeal and loyalty" with which they brought their labours to "a satisfactory and successful conclusion."[1] It is hard to believe that anyone involved with the college regarded the outcome as either satisfactory or successful.

In an environment where esprit de corps was by intention a core tenet, the closure was seen by the Coopers Hill men very much as a betrayal by their leadership. "Our dear old College has been abolished by a cruel edict"[2] and was "sacrificed for a few hundred pounds per annum."[3] During his speech in the college theatre on that final prizegiving, William Brodrick, the new secretary of state for India, said that "when Coopers Hill was wanted, Coopers Hill did its duty, and they might all be proud of it." This was presumably intended positively, but like Walpole's letter, it must have rung hollow to the Coopers Hill men in the audience.[4]

Following the prizegiving, the engineering students had an abbreviated summer session, taught by a skeleton staff, during which they took the ICE exams to qualify them to leave for India as usual in October. President Ottley remained in residence in the college for a few months after the official closing to dispose of the college's assets (also see chapter 8).

One of the most challenging tasks was to deal appropriately with items that had been donated to the college.[5] As soon as the closure was known, General James McLeod-Innes wrote to the president to ask what would become of some books he had donated to the college library a few years before, with the implication that he would like them returned.[6] Soon afterwards, Herbert McLeod asked for the return of the chapel altar cloth he had donated when he thought "the College would last for all time."[7] Thereafter, Ottley kept up a steady correspondence with the India Office about these requests and others, including the fate of the Forest Museum's specimens and the chapel fittings. These matters became especially important as prospective purchasers started inspecting the Coopers Hill estate.

Ottley was left with the unenviable task of contacting the surviving relatives of those who had made donations – Lady Chesney with regards to the prayer kneelers that had been given by the ladies of the staff of the college when the chapel opened in 1874, Lady Clarke about the altar vases, and Lady Taylor about the brass altar cross. To Lady Chesney, Ottley wrote, "I know you will sympathise with me in my present task of winding up the estate. It is without exception the saddest and most dreary work I have ever been called upon to perform."[8] She replied cordially: "I am indeed full of sympathy for you – it is to my mind a horrid piece of destruction."[9] Privately she may have felt less charitable about Ottley's part in the undoing of her late husband's work.[10]

Another difficulty concerned the memorials to some thirty individuals that had been erected in the college chapel over the years, as it was considered inappropriate to sell them with the estate. These were mostly dedicated to former students, although one was to a group of five staff who had died between late 1891 and early 1892 (including Wolstenholme, Whiffin, and "Tommy" Eagles, commander of the Volunteer Corps). Many of the memorial plaques were returned to the families of the people concerned. The remaining six windows and five brasses were removed from Coopers Hill and re-erected in Egham Parish Church at the expense of the India Office (see figure 9.1). The nearby St. Jude's in Englefield Green was considered too small and dark.[11] This process involved considerable work tracking down and corresponding with the relatives on one hand and arranging the transferral of the windows and remaining plaques on the other. One of these was the memorial to E.A. Lewis, the student who had died in 1884 in a rugby football accident.

Figure 9.1. Coopers Hill memorial windows and brasses re-installed in St. John's Church, Egham (with kind permission)

A related issue concerned the silver chalice, cups, and patens used during communion in the college chapel. These had been made in 1816 for the East India Company's Haileybury College and used there until it closed in 1858. They were given to Coopers Hill by the India Office in January 1874 when the chapel was built. Upon the closure of Coopers Hill, they were returned to the India Office and Ottley wondered "where and how it will next be used."[12] In communicating this to the Coopers Hill Society, Ottley commented how curious it was that this silverware had "witnessed the rise and extinction of two such celebrated centres of education." In fact, less than a week after Ottley wrote his letter to the *Coopers Hill Magazine*, the master of Haileybury School (which occupied the buildings of the old Haileybury College) was approached by the son of a former chaplain at Coopers Hill with the suggestion that the school should request the return of the communion plate to its former home. That request was approved, and the vessels were "carried in and used for the first time at the Choral Celebration on Easter Day,"[13] less than six months after their last use at Coopers Hill.

Engineering at Oxford

By the time of the 1906 prizegiving, the forestry students, along with Dr. Schlich and Mr. Fisher, had already left Coopers Hill to continue their studies at the University of Oxford. Because of its status as essentially the only British forestry school, the original 1903 committee on the future of Coopers Hill had recommended that the Forestry program should be relocated to one of the universities at Edinburgh, Cambridge, or Oxford. After due consideration Schlich recommended the last, based on the availability of appropriate courses and of nearby woodlands suitable for practical training. Edinburgh was never really in the running and Cambridge had initially said they could not provide the required woodland. However, Cambridge belatedly mounted a spirited campaign once they realized the probability of the program going to Oxford.[14] Their argument was essentially that there should be equal treatment of the two ancient English institutions and that each should establish its own forestry school.

However, the authorities did not want to split the funding, preferring instead to continue the Coopers Hill system as unaltered as possible. It helped that, with encouragement from Oxford's Hebdomadal Council, St. John's College was willing to redirect endowment funding of £200 per year from the Sibthorpian Professorship in Rural Economy to a new professorship of forestry. Negotiations continued into early 1905 when the decision was finally taken to adopt Oxford as the new home for the Forestry School (to Cambridge's "profound dissatisfaction"[15]). The issue of favouring a single university with the training of Indian foresters was raised in the House of Lords. Proponents observed that Coopers Hill had already exercised a monopoly for the previous twenty years and the objection was narrowly defeated.[16] When the Forest men transferred in July 1905, the *Coopers Hill Magazine* regretted the necessity of bidding them farewell. It complimented them on their academic success and opined that the average forester had been "athletically, at least, equal if not superior to his Engineer *confrère*."[17] High praise indeed.

Oxford did not have an engineering school at the time. Proponents of engineering within the university had been in discussion with St. John's College about using the same Sibthorpian funds to endow the first professorship in engineering. This was not the first time that engineering's interests had been overlooked by Hebdomadal Council, and the forestry decision spurred men such as John Townsend, professor of experimental physics, and Leveson Vernon-Harcourt of University College London (an Oxford graduate) to renew their calls for an engineering professorship. Oxford, they said, was now almost the only "first-rate" university without engineering. The breakthrough came two years later with the appointment to the Oxford chancellorship of Lord Curzon (the former viceroy of India who had opposed the closure of Coopers Hill). Serious fundraising commenced and by October 1907 the university had approved the professorship of engineering.[18] In this way, the abolition of Coopers Hill was a catalyst for the establishment of engineering at Oxford.

In October 1908 Nichol Mackenzie (CH 1875-8) and newly appointed instructor in surveying at Oxford, captured the sentiment of the time: "The University [of Oxford] is starting its Engineering School on sound lines, and ... it may before long supply both India and the Colonies with Engineers up to Coopers Hill standard."[19]

The Coopers Hill Society

The closing of Coopers Hill precipitated the creation of an extraordinary fellowship of men lasting nearly six decades, through two world wars and the independence of their beloved India.

Through 1904 and late into 1905 the college waited uncertainly first for news about its fate and, when it was known, about arrangements being made for the students and the Coopers Hill estate. After the announcement of the closure in *Coopers Hill Magazine* for July 1904, rumours circulated that the college would be used for a hotel, a school run by the Imperial Services Trust, or a college for training female teachers.[20] For a while, the most credible and popular rumour was that the School of Military Engineering would be transferred from Chatham to Coopers Hill.[21] In July 1905, the *Coopers Hill Magazine* reported that it had heard that the Royal Engineers had "entire sympathy" with the sentiments of the CH men, and that the chapel memorials would be "considered as a sacred charge." There were supposedly even plans for where the barracks and commandant's house would be constructed.[22] The lack of definitive news stretched into 1906, initially with a strong sense that the RE move was all but decided, but then with decreasing certainty. Eventually the June 1906 issue of the *Coopers Hill Magazine* reprinted the "sad announcement" in the *Times* that the property was for sale.

In parallel with the news, or lack thereof, regarding the college's future, the thoughts of Coopers Hill men started turning towards the future – how would they maintain their connections and celebrated esprit de corps in the absence of the college that linked them? At a Coopers Hill dinner on 12 July 1905 it was first mooted[23] that all OCH men should join a common London club so they could meet and dine together. While this idea initially had some appeal, it was soon realized that this would be unworkable; people were already members of a wide variety of clubs, membership was costly, and it was hard to find a single club that catered to the disparate professional and social interests of all Coopers Hill men. So very quickly, the alternative notion emerged of forming a new Coopers Hill Society.

The first mention of a society came in the November 1905 issue of the *Coopers Hill Magazine*, its editorial announcing that the matter would be discussed at the upcoming dinner and trusting that "the endeavours to form a C.H. Society, on a sound and firm basis acceptable to all, will meet with the success that they deserve."[24] Ernest Hudson was amongst the first to express his interest in joining the society; while on combined leave in England, he recorded in his diary writing to Arthur Hicks on 15 November 1905, which must have been immediately he received the magazine.

At the CH dinner on 6 December, the matter of the society was duly discussed.[25] Some draft rules were generally approved but there was no agreement on the appropriate balance between the annual subscription amount and the scale of its physical headquarters. Arthur Hicks and William Cole were deputed to prepare a circular asking CH men for their opinion.

While work was being carried out behind the scenes on the formalities of establishing the society, letters and reminders in the *Coopers Hill Magazine* kept the issue in front of its readers. The magazine at the time (edited again by Arthur Hicks, who was back at the college as an instructor in drawing, estimating, and building construction) contained a curious juxtaposition of themes – uncertainty about the future, plans for the Coopers Hill Society, and the everyday business of current and former students. By May 1906, some 550 men had joined Hudson in expressing interest in the new society, with replies still arriving from overseas. Most responses favoured a lower-cost option.

As it turned out, so many men were interested in joining that it proved possible to achieve the best of both worlds. At a meeting on 11 July 1906 at St. Ermin's Hotel, 105 CH men approved the draft constitution and brought the Coopers Hill Society into formal existence. The first entry in the minute book of the society was dated 19 September 1906 and in Arthur Hicks's slanted hand recorded the society's mandate: "The scope of the Society is to promote good fellowship, to keep C.H. men in touch with one another by means of social gatherings & by the publication of the Magazine, to keep sundry mementos of the R.I.E. College, to administer any funds entrusted to its care, to keep a register of those of its members who seek employment, and generally to further the interests of C.H. men at home & abroad."[26]

The annual subscription was to be a reasonable 10s.6d., with options for compounded life memberships of between seven and five guineas, depending on age. The society would rent a room in St. Ermin's Hotel. A large central committee was elected to coordinate its affairs, including a network of local committees, and with General Sir Alexander Taylor as its first president. Arthur Hicks was elected secretary and treasurer. Over the next few months, while the magazine reported the final closure of the college, the eight sporting challenge cups, rugby, cricket, and boating photographs and Bal Masqué programs (formerly hanging in the canteen) were transferred to the Coopers Hill Room. Pictures of the college presidents were hung, books were donated from the college library, and photographs were taken of the large boards that honoured student achievements. A register of members' addresses and a visitors' book were also added, along with Indian and engineering periodicals and official gazettes.

Coopers Hill Still Lives Vigorously

The spirit of the moment was captured by Ernest Shadbolt, one of the first cohort of students, at the 5 December 1906 dinner:

It is not for me to pronounce a funeral oration, for Coopers Hill is not dead. It has not passed away into nothing but a mere memory. Oh no, it is something very much more than that. For as the years have gone by, the name, or phrase, has come to mean to us not merely the place itself but the sum total of all the thoughts and ideas, and activities, that have emanated from it, and which have influenced the minds and moulded the characters of its sons. In this sense Coopers Hill still lives vigorously though the rooms may be deserted and the corridors may be silent.[27]

It was this spirit that carried the society through its first two decades, while memories of the college remained fresh and many of the members were still on active service in India and elsewhere. For these men, life went on very much as it had before because the semi-annual dinners and other social gatherings continued as usual. As in the other services, there was a strong sense of hierarchy, with the most senior men presiding over gatherings and being elected officers of the society. For obvious reasons, the most active members tended to be retired men based near London. The majority had also been engineers who had served in India. That they were engineers was a function of both the relative numbers and seniority, because the other disciplines had not been added to the college until later. That they had served in India was also by virtue of their numbers, but also because their services in India provided a common identity not present amongst graduates employed elsewhere.

Figures printed in 1910 (reproduced in table 9.1) showed that 70 per cent of all graduates served overseas and nearly 50 per cent joined the Indian PWD.[28]

Table 9.1. Employment of Coopers Hill Graduates

Graduates		1623
Whereabouts known (incl. staff)		1568
Members	820	
Non-members	533	
Deceased	215	
Graduates appointed to imperial services		1126
Indian Public Works Department	751	
Indian Telegraph Department	84	
Indian Forests	172	
Army in India	66	
Other Indian services	15	
Egypt and Sudan	38	
Graduates employed elsewhere		379

Of the remainder, many would have worked for Indian state railways and other enterprises not strictly considered to be government service.

At its peak, membership in the Coopers Hill Society reached half of all graduates, so its appeal was broad, presumably because the members enjoyed receiving the magazine to keep up with their fellows. Although Ernest Hudson participated very little in the society's affairs, he diligently filed his copies of the magazine and annual report. Attendance at the half-yearly dinners, open to members only, started off at over 100 while garden parties and "At Homes," to which ladies and guests were welcome, were considerably larger. Equally often, however, the pages of the magazine lamented the disappointing turnout at various events and was adept at suggesting plausible reasons. Annual general meetings of the society were less popular and a turnout of twenty was typical.

For many years, dinners were held on Piccadilly Circus at The Criterion or Café Monico (figure 9.2). They typically included former college servants to

Figure 9.2. Annual Coopers Hill Society Dinners followed the format of the Fourth Year Dinner shown here. This dinner was held on 17 October 1903 at the Criterion Restaurant, Piccadilly Circus, London, with Arthur Hicks in the chair. "Auld Lang Syne" was sung after a "capital dinner" and "enthusiastically received" toasts. (Canning Collection, CSAS. Described in the *Coopers Hill Magazine*. *CHM* 5, no. 2, November 1903, 22)

announce names and maintain order. Prominent amongst these was Sergeant George Lanning (1846–1919) who joined Coopers Hill in 1885 as sergeant-instructor for the "O" Company Royal Berkshire Volunteers. Then, after nearly twenty years of army service, he become the college's night porter in the winter of 1890–1.[29] Photographed standing next to Hudson in 1888 (figure 4.2), Lanning appears the archetypal weather-beaten, no-nonsense military sergeant of indeterminate age. Lanning's impact on the college was such that Mrs. Pennycuick was led to remark that Coopers Hill was "a collection of young gentlemen run by one, Sergeant Lanning, with the help of a President ... and a few Professors."[30] After the closure, Lanning stayed on in the entrance lodge at Coopers Hill for a few years to keep an eye on the place before moving in with his daughter in Egham.

Coopers Hill garden parties were much lighter affairs than the dinners, attracting members and their wives from further afield. The parties in the late 1920s were particularly popular because Lady Cheylesmore, the new owner of Coopers Hill, permitted the gatherings to take place in the grounds. More than 200 guests attended on the first such occasion in 1928, 80 travelling in chartered coaches from Bailey's Hotel.

The Coopers Hill Society activities carried on essentially as usual during the First World War. But because a significant number of members were still of an age to fight, and still more had children of enlisting age, the magazine was filled with war records, stories, honours received, and of course fatalities. The society coordinated the collection of India Office Forms of Service by which men could volunteer their technical services (such as constructing defence works, repairing and superintending railways, and working on the telegraphs), some 160 of 200 eligible men returning the information sheet.[31] It also collected donations for the Prince of Wales National Relief Fund, raising £750.[32] For the first time since the 1890s, CH dinners were not held.

After the armistice, the society considered how best to remember the men who had given their lives. Several proposals were considered, including a prize of some kind, a marble memorial, and a sponsored hospital bed. Over £1000 was raised and the members overwhelmingly voted to establish an engineering prize. In addition, £28 of accumulated interest was used to make and install a small brass plaque bearing the names of the twenty-four members of the RIEC who were killed in the war.[33] This plaque can be seen on the balcony wall of Egham Parish Church.

Arrangements for the Coopers Hill War Memorial Prize (consisting of a medal, parchment, and monetary award) were a little involved; one prize was to be awarded annually by the Institution for Civil Engineers for the best essay submitted by one of its members, while a second was to be offered in rotation by the Institution for Electrical Engineers, the School of Military Engineering at Chatham, and the University of Oxford Forestry School. Over the next nine years, however, the ICE managed to award the prize only twice, for lack of applications. In response to the society's expressions of dissatisfaction, the ICE made a proposal to award the medal for the best paper published each year in its journal. The society's 1933 Annual General Meeting agreed to this and the prize is still awarded in this way.[34]

Shortly after the establishment of the prizes, a member suggested that the recipients should be made honorary members of the Coopers Hill Society. The president at the time, Carleton Tufnell (CH 1875–8) responded firmly against the proposal, saying, "One can imagine the feelings of the last ten or twenty survivors of the Society, when they find themselves, years hence, sandwiched in with a lot of honorary members, unknown to them and to each other, with no connection whatever with Coopers Hill traditions."[35] Ironically, the decision during the First World War to offer members' widows honorary memberships had just this effect, and by 1960, 64 of the Society's 154 members were honorary.

Arthur Hicks

No history of Coopers Hill would be complete without mentioning the "fifty-three years of unbroken friendship" and service provided by the extraordinarily dedicated Arthur Hicks: "He founded the Society, he continued the magazine, he arranged the annual dinners, the garden parties, the winter meetings, and kept himself in touch with Coopers Hill men all over the world."[36]

Hicks's association with Coopers Hill began in 1875 when he entered the college, squeaking in at forty-fourth place in the entrance examination. However, he graduated fourth in his year, with prize for design and a scholarship in applied mathematics. He was also stroke of the college rowing eight, a member of the band and the President's Orchestra, and a founder of the first college magazine, the *Oracle*. After serving his professional year working on the extension to the Chatham Dockyard he was posted to the Buildings and Roads Department in the Punjab. He spent seven years in the Indian service before resigning for "private reasons." In 1892, when the death of Tommy Eagles left the position of instructor in geometrical drawing and lecturer on estimating and architecture vacant, Hicks competed successfully with twenty-seven other applicants for the opportunity to rejoin Coopers Hill.

As instructor, president of Coopers Hill Boat Club, tennis player, and founder and editor of the *Coopers Hill Magazine*, Hicks was in his element. He was in charge of building many new facilities, including laboratories, a billiards room, workshops, and residences for college staff. For a man so steeped in the lifeblood of the college, its abolition must have been devastating. Hicks's anxiety showed in his magazine editorials as he bemoaned the absence of hard news on the college's fate, clung to promising rumours, and eagerly embraced and promoted the foundation of the society.

As the society became a reality, its annual reports indicated how many individuals had visited the room in St. Ermin's Hotel. For example, 1912–13 saw 495 visits by 192 individuals, and Hicks was present on no fewer than 200 days. During that time, Hicks worked for a firm of patent agents, eventually becoming a partner. He and his wife had four sons and two daughters, and he had the exceptionally painful duty of reporting in the magazine the deaths of two of the sons. A third son was crippled, while one of the daughters served as an ambulance driver in France and Italy for the duration of the war. Of course, with more than eighty CH men having sons serving in the armed forces, the Hicks family was not alone in their loss, and the pages of the magazine listed these tributes along with the accolades and honours of CH men and their sons.

By any measure, the Coopers Hill Society was a success, and much of that achievement was due to the persistence of Arthur Hicks. He resigned from his post as secretary and treasurer for medical reasons in 1927, thirty years after the *Coopers Hill Magazine* first appeared. At the society's annual dinner, Carleton Tufnell presented a watch and a cheque to Hicks, saying, "He has done more for Coopers Hill and Coopers Hill Society than any man living."[37]

Despite his resignation, Hicks continued to drop by the Coopers Hill Room, as he put it "to see how my baby is getting on." His last visit was just a few weeks before his death on 28 January 1928. His memorial in the magazine continued the analogy, likening Hicks to a nanny looking after the Coopers Hill men "with sad eyes when they were bad, and with glad eyes when they were good, regarding their antics with kindly tolerance, and when they fell asleep for the last time tucking them away and writing nice obituary articles about them, telling the world what fine fellows they were."[38]

Career's End

Retirement

The closure of the college and the establishment of the Coopers Hill Society coincided with the time that the earliest graduates were starting to retire. In 1905, Newcombe, who had retired three years previously, wrote, "Some think that retirement to England with nothing to do, or rather with the opportunity of doing just what one likes best, is an ideal existence. It is not found so by those who have led for many years an active life, and are still vigorous and in the enjoyment of good health."[39]

Such was the intention of the great majority of Coopers Hill men and other British officers working in India when they approached the required superannuation age of fifty-five. A few, like Hugo Wood, decided to stay in India. Ooty, where he settled, became something of a retirement centre for British people after Indian independence. Many, like Ernest Hudson and Lionel Osmaston, retired at the earliest opportunity. Both Hudson and Osmaston returned to England four times during their twenty or so years in India, managing to stay in touch with friends, family, and society with a frequency that was impossible a generation earlier. Others served much longer; Eustace Kenyon and William Sangster each served more than thirty years, and John Benton was in India for nearly forty years.

> One rarely serves now more than six or eight years before taking one's first furlough. That, however, is quite long enough to awaken one to the reality of the exile; for the visit to England, pleasant as it may be in many ways, is generally to some extent a disappointment. The absence has been long enough for changes which have come gradually to those who have stayed at home, but are startling to the one who has been away while they were going on.... This may be partly due also to the fact that there is a change in oneself which is not quite satisfactory to them.[40]

From a life of some importance in India, perhaps with authority over huge areas of land, complex engineering projects, or large numbers of people, retirees could easily lose their sense of purpose. They could no longer support the numbers of servants they were accustomed to. Britain was crowded and had fewer opportunities for hunting anything, let alone tigers. And the winter was "so cold, so cheerless, so unsettled."[41] Moreover, their experiences set them apart from others. They might unconsciously speak with the "chi-chi" accent of the East.[42] Moreover, British people tended to be uninterested and uninformed about India. James Best's experience of being ignored and misunderstood was commonplace:

> Those who have served India, some rising to very high positions, retire to comparative obscurity when they settle down again at home. They have given of their best to the country they have served, and prefer peace in their retirement. For this reason Anglo-Indians, though men of ability and administrative experience, seldom make their voices heard in their native land. So they are ignored in the more weighty matters of state policy. They are often considered too narrow in their outlook even when the policy of India is under consideration. Men who have spent their lives in the villages and cities of India are told that their experiences are too narrow to be of use in framing a policy suitable for the whole country.[43]

The entire experience of a family retiring to London after many years in India was captured in all its anticipation, uncertainty, small victories, and tragedy by Alice Perrin (1867–1934) in her novel *The Anglo-Indians*. First published in 1912, the book was a "story of a distinguished administrator, who after the splendour of a great Indian career, retired to die of cold and boredom in a London suburb." There's more to the story than this, of course, especially concerning affairs of the heart, but a key theme is the longing for past days in India from which the Fleetwood family was severed by the "guillotine of completed service."[44]

With Alice's heavy involvement in London's literary scene[45] and her husband's responsible job, the story does not appear to have been autobiographical. Indeed, Perrin poked fun at herself when Mr. Fleetwood considered occupying himself by writing a book, and at the system in which an "engineer, a doctor, a parson, or even a soldier has more chance of employment [than a civil servant] when he comes home, but then he hasn't been treated so well financially out there."[46] However, *The Anglo-Indians* would have been based on her observations of many similar real-life stories amongst the ex-British-Indians resident in London.

The range of occupations for Coopers Hill engineers after retirement from India was quite large. Some, like Charles Perrin and Lionel Osmaston, accepted other positions in England. Others assumed the lifestyle of "gentlemen," taking up farming or living on the family's means; although we know little about Ernest Hudson's forty years after retirement, this seems to apply to him. Many undertook work in the community, such as voluntary work for the church and political parties or serving as justices of the peace. And yet others pursued artistic careers as authors, painters, or photographers.

Where to Retire?

Most retirees, of course, settled in England. For example, the March 1939 membership list of the Coopers Hill Society shows that 90 per cent (293) of the 325 retired members lived in the United Kingdom and Ireland. Of these, 64 per cent (187) lived in London and southern England (66 and 121 individuals, respectively).[47] A handful of members listed addresses in each of Australia and New Zealand, Africa, and Asia.[48]

Alice Perrin's Fleetwood family had to decide whether to "settle at Ealing, or one of those places where Anglo-Indians most do congregate – for a common reason that they are cheap to live in and close to London."[49] For Coopers Hill men, Cornwall, Devon, Hampshire, and Kent were all popular retirement destinations (with a few in Ealing). The shared experiences and vocabulary tended to bring retirees together, and some enclaves of British-Indians were well recognized – Cheltenham (with its private school attended by, for example, Bertram and Lionel Osmaston), Tonbridge, and Eastbourne.[50] All these places were well represented in the address list of the Coopers Hill Society's membership. Coopers Hill drew many students from Scotland and Ireland, so it is not surprising to find many men returning to those countries on their return from India.

The *Coopers Hill Magazine* occasionally published advice from members (particularly between the wars) on the pros and cons of retiring to various parts of the world. The main concerns of retirees were "(a) language difficulty, (b) servants, (c) purchasing power of the sovereign, (d) climate, (e) work or employment, (f) education of children, (g) starting children in life."[51] For example, £2000 could buy a 100-acre farm in the Annapolis Valley of Nova Scotia, but the village schools had only one nineteen-year-old teacher in charge of seventy children. Moreover, "ladies who smoke and wear one-piece bathing suits are still apt to be frowned upon."[52]

In speaking about how to overcome the difficulties of starting a fruit-growing business in Kelowna, British Columbia, one such article summed up the problem: "The obstacles that must arise in the mind of a man of middle age who is contemplating a plunge into a business strange to him in a new country are only too obvious; no wonder that they keep many fretting in inactivity at Cheltenham or Ealing who would be glad enough to embark on a larger life if only they could see their way over the beginnings."[53] The men's primary concern was "finding ways and means of making both ends meet should they retire on an inadequate pension,"[54] so cost of living was an important element of these articles.

A settler in New Zealand wrote a long piece about that country in 1922. Amongst its many virtues, it noted, "Lack of servants will probably deter Anglo-Indians from coming to New Zealand more than anything else" because "few Anglo-Indian ladies would care for sweeping and scrubbing their floors, cooking their breakfast and dinner, and lighting their fires."[55]

For low taxes and cheap labour, Rhodesia was recommended, and unlike New Zealand it had the advantage of big game hunting. Sensibly the author suggested that a long visit to the country before deciding to settle would help to avoid "land sharks, dud farms and dud neighbours."[56]

Pensions

After a long and demanding service in India, an engineer's much anticipated retirement would be funded by a government pension. The salary of an engineer in India was not noted for its generosity, so the pension was a significant part of most men's post-retirement income. Its amount and security were therefore of widespread concern. Unfortunately, the Coopers Hill engineers had issues with their terms of service, starting almost on day one and lasting until the ends of their careers.[57] Essentially, they believed that to attract applicants to the college in 1871 the government had made attractive promises in speeches before opening day and in the early prospectuses, but had not kept them. Calling it "the glittering bait of the highly mendacious first prospectus,"[58] the Coopers Hill men identified three issues:

Exchange rate: While all forms of compensation were expressed in rupees, the college prospectus had specified an exchange rate: "10 rupees are nearly equivalent to 1l. sterling." This applied to salaries, allowances during leaves, and pensions. This last was particularly important because the majority of retirees would live in England with costs in pounds sterling. The difference was significant: in 1871, £1 was equivalent to about Rs10 as the prospectus stated, but by 1900 the exchange rate had fallen to Rs15.[59]

Promotion: A table had been given to prospective Coopers Hill applicants illustrating the lengths of service historically required to reach the various engineering ranks. However, the PWD had continued recruiting engineers from elsewhere while waiting for the first graduates from Coopers Hill. So when the first batch joined the service, the higher ranks were already full and promotion was significantly slower than the table suggested (and, perhaps worse, slower than other Indian services). The difference between the reality and the prospectus was estimated to be almost four years at the level of superintending engineer first class.

Leave rules: The Coopers Hill men had been led to expect leaves and allowances "on a level with the two great Indian services, the Civil and the Military." However, it rapidly became clear that the rules for them did not permit them to count time on furlough towards their years of service, whereas men in other services could count up to four years.

On this last point, a deputation of the first students approached President Chesney almost immediately after their arrival in college to request that the Civil Service leave rules should apply to them. He agreed and approached the secretary of state about the matter, and the prospectus was changed (although the section on furloughs did not appear until the 1877–8 calendar). Eighty-five early graduates from Coopers Hill wrote petitions, mostly to the viceroy, about their circumstances. Improved rules were put in place in 1883, and the issue of exchange rates for furlough pay was addressed, at least partially, in 1889. But these did not always act retroactively, and it was not always clear which officers were included and why, leading to further discontent.

The government responses would prove to be a long and complex sequence of omissions, partial solutions, delays, and denials. The unhappiness of the first few groups of Coopers Hill men apparently affected their relations with colleagues, although their persistence ultimately led to improvements for all. The matter was still dragging on at the turn of the century when these men were starting to retire. In correspondence between the secretary of state and the Government of India, the latter reviewed the various versions of the prospectus, assessed strengths of the engineers' case,[60] and estimated the cost of addressing their concerns. The shortfalls in overall salaries calculated for the most senior men by the Government of India were tens of thousands of rupees, with the average being more than Rs30,000 (around £2000).

Far from thanking the Government of India for taking the trouble to lay out and analyse this complex situation for him, the secretary of state rebuked them for doing so without asking and refused to reopen the matter.[61] Notwithstanding that hope, the pensions issue came up again in front of the 1912–17 Royal Commission on Public Services in India and a rearguard action was still being waged as late as 1923, as evidenced by several outraged leading articles in the *Coopers Hill Magazine* in the early 1920s.[62] In due course, time and old age naturally rendered the issue moot.

From afar, the engineers' grievances over the leave rules and promotion seem justified, although it was difficult to address the latter. Their case regarding the exchange rate was perhaps technically correct, but the wording in the prospectus reads less like a guarantee and more like a statement of current fact. Nevertheless, the Government of India's conclusion was that "the earlier prospectuses up to that of 1872 contained clear statements equating rupees and pounds." It is highly unlikely that the government had any intention of fixing the exchange rate, which would have advantaged Coopers Hill men over all other services. This, as the secretary of state was quick to point out, made it unacceptable. Whether the ambiguous wording was designed to mislead applicants to Coopers Hill, as the college graduates believed, is impossible to ascertain.

Life Expectancy

There is a temptation – particularly when contemplating the early deaths of Sangster's friends Benwell and Ridell, Hudson's friend Cather, and many others – to think that going to India was bad for an engineer's health. In December 1877 the *Oracle* clearly felt similarly: "Looking back on the past year, we cannot fail to see that illness has been rife amongst our members in India.... Some few have returned through failing health, and others have had to find shelter in the hills of India to avoid the ravages of fever."[63]

At the start of an 1880 lecture to students at Coopers Hill entitled "On Preservation of Health in India," Sir Joseph Fayrer quoted George Chesney as saying, "I have been very much struck with the amount of sickness to which our young engineers appear to be subject. A certain excess over the English average is of course to be expected in India, and engineers in particular are subject to exposure and malarious influence in an exceptional degree, but the number of cases of ill health among these young men seems to be altogether beyond what might be set down reasonably to these causes."[64]

Fraser's Magazine agreed and painted a very depressing view of the engineering profession in India. Its concluding paragraph stated, "Ruined health is almost certain; death – an early death – is very probable."[65]

This stung the *Oracle* into changing its mind: "The ruined health argument has too often been refuted to need much notice, and we only have to look around us to discover men who have passed their lives in India, and whose health compares favourably with that of the denizens of the 'noble rooms of the India Office.'"[66] Indeed the great longevity of Percy Tottenham, the Osmastons, and Anthony Wimbush lend support to that argument.

Alfred Newcombe introduced his chapter on furlough and retirement rather depressingly: "Of the fifty young engineers appointed in 1874 from Coopers Hill College, ten left the service for various reasons after a few years in India, eleven died (and among these were some of the most zealous and efficient), and some reached their time of retirement after having suffered much in health. On the whole, it may be said that only about half satisfactorily reached the end of their Indian career."[67]

This was a little too pessimistic. In 1962, one of the final acts of the Coopers Hill Society was to update the record of its members' years of death. While this record just lists the dates of death of thirty-eight of the first fifty men from Coopers Hill, only seven occurred before Newcombe wrote his book.

The society's record contains data on 948 of the 1623 men who passed through the college, arranged by the date they entered. Assuming they were all aged 18 on entering, the average and median ages at death were 61 and 66 years, respectively. These are quite consistent with the life expectancy of a twenty-year-old Victorian man.[68]

Figure 9.3 shows the ages at death of Coopers Hill graduates. It is important to note that approximately ninety men were still alive when these data were collected, so the numbers for ages seventy-seven and older are significantly underrepresented in this chart.

This figure also plots the distributions for the ages at death of men in the general population of England and Wales in 1875 and 1925.[69] The shape of the distribution for the first decade of Coopers Hill entrants (1871–80) is similar to that of the 1875 general population, showing a larger tail towards the younger age range. For the last decade of college entrants (1896–1905), the distribution resembles more closely that of the 1925 population.[70] This evidence does not, therefore, indicate strongly that engineers in India were more likely to die young than men in the general population.

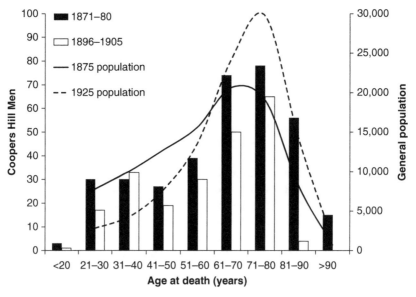

Figure 9.3. Distribution of ages of death of Coopers Hill graduates who entered the college during the indicated decades (columns). Distributions of ages of death for men in the general population of England and Wales in the years 1875 and 1925 (lines) (Data from "Mortality in England and Wales: Average Life Span, 2010," Office for National Statistics, 17 December 2012)

Attitudes towards India

Disagreements

It is risky to ascribe identical values to all individuals at all times, but Coopers Hill engineers, telegraph men, and foresters in India generally shared a common patriotism and belief in the value of their labours. Sir Alex Taylor reminded students of this in his farewell address: "The great incentive will always remain the same – your sense of duty and honour, and your loyalty to your Government."[71] This was, after all, the purpose of the college, so students were by inclination, upbringing, and design the committed implementers of British colonialist policies, with all that entails. Their duty, as they saw it, was to support the objectives of the British Government of India by accomplishing their tasks to the best of their professional abilities, which as a group they did competently and diligently. Hindsight leads us to question those objectives and to reject colonialism's injustices and atrocities, but we can appreciate their personal abilities and technical achievements without agreeing with their guiding purpose.

The *Coopers Hill Magazine* was officially apolitical,[72] and so it did not directly address the transformative events in India leading up to, and following, Indian independence in 1947. However, some idea of the attitudes of Coopers Hill men towards increasing self-governance can be obtained from incidental information, particularly their dinner speeches. Again, these are the views of individuals, but they provide a general sense of the topics of interest being discussed.

With the exception of mentioning periodic reorganizations of the services, there was in fact little commentary on affairs in India until after the First World War, presumably because the situation remained familiar to the magazine's readers. The first suspicion that there might be old and new guards within the society arises in the wake of the slaughter of Indian civilians at Jallianwalla Bagh, Amritsar, on 13 April 1919.

Ostensibly fearing a major rebellion,[73] but also with the intention of "creating a moral effect" in the Indian population, troops under the command of Brigadier-General Reginald Dyer had opened fire without warning on a large crowd of civilians pinned by his soldiers inside the Jallianwalla Bagh walled garden. The egregious firing continued for ten minutes and expended 1650 rounds of ammunition, resulting in at least 379 deaths and probably many more. Dyer's actions were investigated by the Hunter Commission, and he was heavily censured for his "grave error" and a "mistaken conception of his duty." Indian members of the commission denounced his conduct as "inhuman and un-British."[74] The British House of Commons also condemned Dyer, and the event has since come to epitomize the very worst aspects of British rule in India.

Nevertheless, there was considerable public support in Britain at the time for Dyer's actions in, as they saw it, the defence of India. The *Morning Post* raised an enormous fund of £26,317 to support him.[75] The issue was raised at the 22 June 1920 annual general meeting of the Coopers Hill Society (after the publication of the Hunter report but before the House of Commons vote). A member proposed that those present who had retired from government service "should convey by means of a letter, to Brigadier-General Dyer ..., their appreciation of the prompt and resolute measures which he took to quell the disturbances and to express their sympathy with him in regard to the orders passed on his actions."[76] However, the meeting was split, so it was eventually decided that attendees of that evening's society dinner would be given the opportunity to sign a draft of such a letter. Frank Fowler (CH 1875–8)[77] in the chair mentioned the "deplorable event in the Punjab a year ago" and drew members' attention to the letter. Of the 107 diners, about a third (38) signed. It was duly forwarded to Dyer, who responded that he would "greatly cherish the document."[78]

Far from furthering British interests, Dyer's callous actions hastened the end of British rule by undermining trust in the government and boosting the Indian independence movement. Reflecting on changes that had taken place in India since the war, F. Austen Hadow (CH 1892–5) raised the issue of trust in his speech to the society's dinner in 1922. He first remarked that if senior members could return to India, they would notice the large number of Indian officers in all services, and that "Indians of good social standing" had adapted well to the civil engineering profession. Hadow then commented on the "growth of racial hatred for which the political agitator has been mainly responsible, and which has led to so much distrust of British rule amongst many of the ignorant classes." After Dyer's actions at Jallianwalla Bagh, could he really have been surprised at this? Hadow's conclusion from this hypocritical statement was that men of "the right stamp" were still needed from Britain to undertake the work of the empire, saying, "Men and Empires are made by facing difficulties, not by avoiding them."[79]

Joseph Coates (CH 1890–3) gave a rather more realistic assessment of the situation in his 1925 speech.[80] After expressing his understanding for "Indians in their desire to take a greater part in the government of the country," he quite reasonably urged the careful selection and thorough training of Indian recruits for the technical services. Coates naturally agreed with Hadow that the right kind of British men were needed in India but believed that the characteristics of "tact and courtesy" were also necessary to guide the changes in India sustainably.[81]

By 1929 a new concern was being raised by Nathaniel Pearce (CH 1898–1901): that "sinister influences" outside India, meaning Russia, were using the "disgruntled youths of India" to disseminate Communism to their "misguided fellow countrymen." Again, he believed that British officers needed tact and sympathy more than technical qualifications to "lead, guide, and control this spirit of progress in India." Maybe, he suggested, a new RIEC was needed to provide the necessary training.[82]

According to their own experiences and prejudices, Coates, Pearce, and even Hadow acknowledged that change was underway in India, and that the technical services needed to adapt to these times. This view was at odds with that of the retired members who stood "aghast at the prospect of the goal towards which things are tending" in India, namely that the country was not ready for democracy.[83] So when the Indian Empire Society was founded in London in 1931 to resist constitutional reform in India, the Coopers Hill Society's committee sent a letter of support to the press on behalf of retired members. Hadow was appointed to the new society's executive, where he joined men of like mind, such as Lord Sydenham of Combe and Michael O'Dwyer (who as lieutenant governor of the Punjab had been a strong supporter of

Brig.-Gen. Dyer's actions at Amritsar). While "not unmindful of the ability and sterling qualities of the many thousands of Indians" who served with the Coopers Hill Society members, the letter contained what today would be considered overtly racist generalizations. Perhaps revealing their true concern, the letter was also clear about the need to protect "the vast sums of British money" at stake in India.[84]

It is not clear how many retired members actively supported the letter signed by Carleton Tufnell and Frank Dempster on their behalf, but ten "definitely opposed" it.[85] We hear no more about the Indian Empire Society, although letters from its executive appeared in the national newspapers over the next few years. After the deaths of several of its leading members in the early 1930s, including Hadow (d. 1932) and Sydenham (d. 1933), its mantle was taken up by the India Defence League, which counted Rudyard Kipling, Winston Churchill, and many MPs and peers amongst its membership.

With the passing of the Government of India Act in 1935, public works, telegraphs, and forests fell under the jurisdiction of elected provincial councils. While this caused some uncertainty at the time,[86] Col. Charles Colbeck (CH 1900–3) was able by 1937 to take a more philosophical view: "It is hardly to be expected that that those who were brought up in and knew the best of the old regime should feel happy about the new." He further remarked on the "curious coincidence that the passing of the last of the Coopers Hill men from the service of India should coincide with the passing of the old regime."[87] Insofar as the retiring Coopers Hill men embodied a generation brought up with a specific mindset, it was probably no coincidence at all.

Referring to the difficulty of implementing the democratic government "with that mixture of peoples of which India is composed," Colbeck stated, "One must sympathise with Indians who want to get free from tutelage and to stand on their own feet, even when it means treading on our toes. It would perhaps be a poor tribute to our tutelage if they did not."[88] Colbeck meant well, but this type of paternalistic view of the British "tutelage" crept into the discourse when it was clear that self-governance was there to stay.

Remarkably, there was no mention of Indian politics in the pages of the *Coopers Hill Magazine* for the following two decades, not even of the great changes brought about by independence and partition.[89] There were probably several reasons for this. In the first place, the primary vehicle for reporting these opinions had been the chair's speech at the annual dinner, but dinners were discontinued during the Second World War and not reinstated in their previous formal style when the war was over. Second, the leaders of the society were now of a generation different from those who had established the Indian Empire Society. They had different experiences and expectations, although the inscription on a statue of Sir Alex Taylor describing him as a "hero of the Indian

Mutiny" suggests that they still clung to aspects of the old world view. Lastly, after the war, the magazine really did cater only to "things of interest only amongst ourselves"[90] – social functions, announcements and reminiscences, membership lists, and of course, obituaries.

As the tenth anniversary of independence approached, Sir Percival Griffiths spoke to the 1956 Garden Party at Coopers Hill of India's excellent progress since 1947. He described the country's "new self-respect" and "dynamic spirit" and its "great and ambitious plans" for infrastructure development. This, he said, had been made possible by the "firm technical foundations laid by the Imperial Services," and that "Coopers Hill men could justly feel that they had played a great part in building a new and vital India."[91]

Self-Reflection

While much of the rhetoric of engineering in British India emphasized duty and loyalty, a few colonial engineers displayed concern for the broader social and technical context of their work. A prominent example was Sir William Willcocks, an irrigation engineer in India, Egypt, and Mesopotamia, who was an advocate for environmentalism and the local people affected by his projects (as well as a critic of Coopers Hill).[92]

In contrast, there is little written evidence that Coopers Hill graduates reflected on the bigger picture of their work in India or elsewhere. This is the more surprising, considering their close collaboration in the field with Indian staff, facility with the language, and familiarity with the country. In more than sixty years of publication, the *Coopers Hill Magazine* printed not a single article that reflected on, or examined thoughtfully, the impact on Indians themselves of the work of its members. Even in later years, long after the people concerned were retired, even quite lengthy reminiscences were wholly jingoistic. For example, the ninety-three-year-old Walter de Winton concluded his accounts of a career in India with his firm belief that "no-one can deny that [the English] are a great race, and British rule in India is to my mind the best illustration of it."[93] If anything, these retrospectives became progressively focused inwards on the past doings of their comrades rather than reflective on the larger context or legacies of their work.

As described above, there are indications from the speeches at Coopers Hill dinners that broader Indian affairs were indeed discussed privately between members who were not always in agreement. It may therefore have been that the successive secretaries of the society (Arthur Hicks, Frank Dempster, Stephen Babington, John Cameron) excluded articles on such contentious subjects. In published books, the closing description of the "efficiency" of the

British administrative system in India, written by the otherwise generally perceptive Alfred Newcombe, is typical.

> Imperial and provincial roads, railways, irrigation, and other public works, the post-offices, telegraphs, forests, sanitation, education, hospitals, police, and the currency, have been placed on a sound basis and are in vigorous working order. The Agricultural Department systematically fosters and improves Indian agriculture by collecting and distributing information, introducing new processes, new staples, instruments, manures, rotation of crops, methods of storing fodder, improved sugar-mills, and better breeds of cattle and horses. The introduction of the indigo, tea, and mining industries, and of railways, the revival of the cotton industries, and of irrigation on a large scale, are all due to British enterprise. A hundred years ago the North-West Provinces were a desert and infested with robbers. They are now a garden, well irrigated, in tranquillity, with numerous schools, hospitals, post-offices, &c., and covered with a network of railways. [94]

Unusually, Hogarth Todd made some effort to describe the impact of his irrigation work through the eyes of "Nathu, the Indian Cultivator," but only to the extent of demonstrating that Nathu benefitted greatly from Todd's intervention.[95] There is a strong sense that it was not the "done thing" for Coopers Hill men to question publicly the official legacies of their work.

Reviewing the diaries and unpublished reminiscences on which sections of the present book are based, it is striking how little self-reflection even they contain. Ernest Hudson was curious about India and its people, but in the somewhat distant manner of a tourist. He also loved associating with the engineers he encountered and was very interested in the people he met and the European social dynamics. Hudson was diligent at his work but seems to have taken a rather narrow view of what it entailed, focusing on accomplishing the specific task at hand and satisfying his superior officers. What little information the diaries contain suggests that he saw himself primarily as serving European telegraph needs rather than believing that he was doing the right thing for the Indian people.

An overall impression from the available evidence is that Hudson's attitude was very typical of Coopers Hill engineers. Where they did consider the broader impact of their work, they did so with statistics – millions of acres irrigated, forest productivity, freight carried by railways – rather than the experiences of the people. This supports, or was perhaps one of the origins for, the stereotype of engineers as being "in the back room," working on the technical solutions while others were concerned with the purpose and motivation. Except for unusual examples like Willcocks, engineers were not ones to make a fuss.

Hudson's diaries (and to a lesser extent those of Lionel and Selina Osmaston) were still written somewhat formally, as if they expected others to read them. Generally brief and factual, they may not have been seen as the place to confide anti-establishment thoughts on the nature of their work in India. The personal letters of Eustace Kenyon were markedly less formal and do contain some commentary on the affairs of the day, so letters may have been where such thoughts were expressed. Unfortunately, too few personal letters from Coopers Hill men survive in the archive to make a proper assessment.

The great love of the foresters for India's forests and wildlife is apparent from the published works of Stebbing and Bertram Osmaston, the Osmaston diaries, and Wimbush's unpublished recollections. As discussed in chapter 6, forest officers were aware of the impact of their regulations on local villagers and to some extent tried to balance the needs of each. Nevertheless, there were strong overtones of "tough love," in the sense that firm forest regulations required difficult adjustments now but would bring the country long-term benefits later. Stebbing in particular was thoughtful on the value of training Indian forest officers and on the need for conservation. Even he, though, demonstrated the success of British policies in terms of the economic productivity of the forests.[96]

One author who stands out from the majority for his thoughtful comments was forester James Best, although he was interested in the broader picture for India rather than the impact of his own work. Despite his thoughtful arguments, the modern reader would not subscribe to all he wrote in his final two chapters of *Forest Life in India*, particularly concerning race and Indian membership in social clubs. But for his time, his ideas on bridging the gap between Indian and British cultures, respect for Anglo-Indians, and Indian self-rule were pluralist and progressive.[97] He also speculated interestingly on the impact of motor vehicles on Indian society and criticized the decisions of policymakers with no in-depth knowledge of the country.

Best believed that senior officials did not necessarily give their true opinions of official policies. Referring to the KCIE honour, he wrote of the officials that "the time when they are expecting the coveted 'K' is generally the time when they are asked for their opinions."[98] This would not have been applicable for the majority of Coopers Hill men, but fear of repercussions from superiors might well have led officers to keep quiet while they were in active service.

In all likelihood, a changing mixture of these factors and others may have accounted for the absence of apparent self-reflection for different men at different career stages. Towards the end, in the changing context of the mid-twentieth century, it may have simply been that elderly men had no wish to re-examine the raison d'être for their life's work, being content instead to sustain old beliefs and relive past glories.

The Society's Later Years

Membership and Existential Crises

From the beginning it was of course recognized that there was only one fate for the Coopers Hill Society – extinction. It was also understood that the society's revenues would peak early on and then decline, so care was taken to invest the subscription fees in the early years in order to keep activities running for the younger men in later years. The wives of the Coopers Hill presidents were admitted as honorary members in 1908, and in 1916 as a result of the war, widows or close family members of any deceased member could also be invited to join the society. The outcome was that the ratio of members to honorary members decreased with time. From the peak of 816 in 1912, general membership declined at a steady rate of 15 per year to 101 in 1960. Attendance of members at the annual gathering fell commensurately from 100–140 in the early years to fewer than 20 when the society was disbanded.

Given the circumstances, it is not surprising that the Coopers Hill Society experienced occasional existential crises. The first came in the late 1920s and was prompted by a string of years in which expenses exceeded revenues. Annual deficits were in the £50 to £140 range, compared with total annual expenses of close to £400.[99] These deficits were covered by transferring funds from the roughly £3000 of investments, although these had also depreciated. At first, the situation was considered to be acceptable, but as the deficits continued the financial position was explained to the membership in the March 1926 magazine and discussed at that year's annual general meeting. The largest budget items were Coopers Hill Room (£100, plus £20 for fires and servants); secretary's salary (£100); printing the magazine and annual report (£65); and the tea parties (£29). While small economies could be made here and there, the rent for the room (moved to Bailey's Hotel after the war) was a clear target, especially since only 111 visits were recorded during 1925.

To make his point during discussions about whether to continue paying for the CHS Room, one author told the "pathetically humorous" fictional tale of the "Father of Coopers Hill":

The writer described himself as the only survivor of the once-numerous band of C.H. brothers, and stated that, in spite of the continuous diminution of the members of the Society, the room in which they used to meet had still been maintained. When the time for the Annual Meeting came round, he attended that as sole representative of the Society, voted himself into the chair, addressed the meeting (himself), and finally passed a vote of thanks to the Chairman (himself) for presiding. Then, taking a last fond look at the various trophies displayed in the room and the pictures on its walls, he tottered out – a bent, withered, silver-headed figure – and down the exit passage of the hotel. As he passed along, he was greeted with the admiring glances and respectful salutations of those he encountered. Noticing this a strange lady asked the hall-porter who was the venerable old gentleman and was told he was the "Father of Coopers Hill." "But who was Coopers Hill?" she further enquired.[100]

Most people at the AGM in 1926 supported keeping the room, thinking that without a home base the society would start to disintegrate. They further observed that with £3000 capital and a deficit of £100 per year, they could continue for many years to come. Ultimately the decision was for the committee to poll members asking whether the society should operate strictly within its means or carry on as usual. If the latter, would they prefer increasing subscriptions, paying out of the capital, or a one-time levy of £2. By January 1927, only 200 votes had been received and the magazine's editorial urged responses by noting, "Let us keep before us the fact that we are a large body, with strong mutual interests; and, although our youngest Members may be over 40 years of age, we never have been delicate blossoms; and that our interest in the Society is not on the wane."[101]

The results of the referendum were presented at the June 1927 AGM and showed that about 70 per cent of the membership voted, with a majority of five to one preferring to continue the current expenditures. A strong majority also opted for the voluntary levy. More than half the membership contributed their £2 and the accumulated amount was invested. This measure kept the Coopers Hill Room open until the end of March 1937, the year that also saw the discontinuation after thirty years of the separate annual report and its incorporation into the magazine. With the closure of the room, the challenge cups were lodged with the East India and United Service Club, photos not wanted by members were (regrettably) destroyed, and books were donated to the Red Cross.[102]

A four-year interruption in the society's functions caused by the Second World War prompted the next reconsideration of the Coopers Hill Society's future in 1945. A "cheery gathering" for lunch on 30 October was followed by the AGM, which considered proposals developed by the committee the previous summer:

1 That having "served the purpose for which it was intended," the society should be wound up in 1946, or that consideration of this proposal should be postponed until 1950;
2 The Society should continue for as long as funds lasted and there were enough members "who desire to and can conveniently meet"; or
3 That membership be opened to officers currently serving in India even if they had not attended Coopers Hill.

A motion "That the Society should carry on" was passed unanimously. However, cost-cutting measures were also approved, including reducing the magazine to one issue per year (although it went back to two after a couple of years). In the years before the college's closure, eight issues of the magazine were produced per year. The frequency fell to three to five for the following two decades, to two per year for the 1930s and 1940s, and just once per year in the 1950s.

Later Years

By the 1950s, the *Coopers Hill Magazine* consisted mainly of a brief annual report, a list of members, a relentless flow of obituaries, and occasional reminiscences. In 1951, the future of the society was raised again, as agreed in 1945. Again the eighteen members attending the AGM decided to carry on "for some time at least."[103] The society still had investments valuing more than £2100, but the membership urged "all possible economy." But when that economy extended to eliminating obituaries from the magazine (to reduce printing costs), the membership protested, and they were reinstated.[104] After all, for the elderly and less mobile readership, this must have been one of the main reasons for continuing to read the rather slender and depressing publication.

However, the Coopers Hill Society was given a boost in 1954. The principal of Shoreditch College, a teacher training institution that had recently been relocated to the Coopers Hill property, invited members to hold a reunion at Coopers Hill for the first time since 1928. For the first time also, "lady guests" were invited, and 116 sat down to lunch. Members "had the opportunity of looking over their familiar haunts in the old buildings, which had not been much altered, even the names of the corridors being still the same as in the days of the R.I.E.C."[105] In his speech of appreciation, society president, Sir Alexander Rouse (CH 1897–1900), rejoiced that "Coopers Hill, as a College, has been reborn." He also noted with regret the demolition of the New Wing of student accommodation, whose window bars had "provided an illicit entry for roisterers who desired to avoid the porter's lodge after hours" (as president Taylor had suspected).

Over the next few years a strong relationship was established between the principal of Shoreditch and the society, with annual reunions at the college until 1960 and the installation of a plaque to commemorate the Royal Indian Engineering College. There was talk of a small museum to the RIEC, and members were asked to contribute photographs and other memorabilia. The society also retrieved one of its challenge cups from the now-renamed East India & Sports Club and donated it to Shoreditch College to be awarded once again to a student for athletic achievement. This was followed by the remaining cups.

A major project of the society's latter years was the repatriation of a statue of General Sir Alex Taylor. It had originally been erected near the Mori Gate in Old Delhi in 1916 to honour Taylor's part in recapturing the city following the Indian uprisings. Arthur Hicks had coordinated the Coopers Hill Society's involvement in the original installation. Rouse and Hugh Keeling (CH 1884–7) had supervised in its erection in India.[106] In 1957, the ten-year anniversary of Indian independence, socialist parties in India had been agitating for the removal of foreigners' statues deemed "offensive or provocative," with threats of "direct action" if this was not done.[107] The socialist parties were jubilant when the Indian government removed Taylor's statue during the night.[108]

In 1959 Taylor's granddaughter established a group to retrieve the statue. She contributed £100 and P&O was persuaded to provide free shipping. The Coopers Hill Society agreed to contribute funds, provided the statue was to be re-erected at Coopers Hill. London County Council agreed to accept the statue as a gift as long as they would incur no costs. Although the cost to the society was greater than expected, the statue was duly re-erected next to the main driveway of Coopers Hill. It was unveiled on 23 July 1960 by Alexander Taylor, General Taylor's great-grandson, before nearly eighty guests.

There it remained as Shoreditch College gave way to the Runnymede Campus of Brunel University, which in turn was sold for housing development in 2008.[109] Taylor's statue was then removed to the Brompton Barracks of the Royal Engineers, where it currently resides.

Final Meeting

It feels from the records as if the Coopers Hill Society closed abruptly, but the process actually took a year. Because it provided a sense of completion, perhaps of coming full circle, the unveiling of Taylor's statue precipitated the end. At the Fiftieth Annual General Meeting, held later on the day of the unveiling, Arthur Astbury made a proposal: "There is a time when a Society such as ours should decide to hold its final meeting, rather than dwindle to inevitable extinction. Accordingly I propose that this notable occasion should be regarded as our last organized social function and that we should decide to settle our affairs."[110]

Views were divided. Rouse spoke for many when he expressed the view that "the society should carry on as long as there were members left to do so." As their predecessors had done before them, they decided to put the matter to the members and then decide at the next annual meeting.[111]

The 1961 issue of the *Coopers Hill Magazine* that reported the unveiling would be the last, after more than sixty years of continuous publication. A typed report dated 1 September 1961 described how, after the 1961 reunion at Coopers Hill, the AGM had considered the results of the referendum: twenty-one votes for carrying on, forty-two votes for closing. "It was unanimously decided by the Meeting to comply with the wishes thus expressed." They also decided that the society's remaining funds, nearly £1000, would be used to endow an award at Shoreditch College. A final memo by J.G.P. Cameron on 25 August 1962 detailed the winding up of the society finances.

Longevity

Many years earlier, R.M. MacGregor (CH 1900–3) talked about the society created when the East India Company's college at Haileybury was closed: "The College was closed in 1860 and the Society lasted till 1900. I remember quite well just before I went to Coopers Hill, my father returned from the annual dinner and said it had been decided to close down, because at each dinner one looked around to see who was no longer there."[112] The Coopers Hill Society lasted half as long again. What was responsible for this extraordinary longevity?

It should be understood that the Coopers Hill Society was not representative of the entirety of the college's 1623 graduates. At its peak, the membership comprised only just over half the total alumni. The half it did represent were primarily those whom the college was established to train – of the 439 members in March 1939 (the first year in which relatively complete professional affiliation was identified), 89 per cent indicated service in India, other parts of the empire, or the armed forces (compared with 69 per cent of all graduates).[113]

Like any alma mater, Coopers Hill as a place was held in great affection by its students, but the sentiment was particularly powerful at Coopers Hill. Joseph Coates remarked in 1925, "I doubt if the former students of any college in the world feel the same affection towards their College as we do for Coopers Hill. I suppose this affection is partly due to the fact that the College was small and to its isolation. This made us feel we were a select family party, to which we were proud to belong."[114]

This comes across clearly in student recollections – the interactions the president and professors and their wives, who lived on site, very cordially supported. Hudson reports in his college diaries taking tea with the Eagles and Pomphs, and he sailed with Taylor. George and Emma Minchin were very popular, and the magazine's tribute to Lady Taylor described her as "the kindest, most sympathetic and indulgent of friends."[115] There was football, rowing, golf, and tennis, and garden parties, just as Coates had said.

As discussed earlier, much of the credit for maintaining the Coopers Hill Society as a "live institution"[116] over so many years must go personally to Arthur Hicks. While the society's room and the events, dinners, and garden parties he arranged were important, they were attended by only a minority of the membership. The truly vital link that held the society together at home and abroad was the *Coopers Hill Magazine*, which made its way to every corner of the empire and was read with interest, even if its recipient never attended an event. An article reprinted from *Indian Engineering* considered that the magazine possessed "a certain charm from the atmosphere of the bond of union that it expresses," the "intimate record of a family circle."[117]

Another part of the longevity was certainly due to Coopers Hill's much-vaunted esprit de corps. The same *Indian Engineering* article illustrated this with a metaphor involving the college colours: "The special attributes of the chameleon are said to be that the creature turns the colour of anything you put it on … but the Coopers Hill man, whatever colour you may put him on, turns purple and gold."[118]

For the first two decades there was a strong sense of responsibility towards the men still in service: "We may anticipate that for another 30 years [the youngest generation of all] will be worthily maintaining the traditions that have been handed down to them. So that for all this long time to come Coopers Hill men will be in evidence and Coopers Hill will be a living and moving force. Even when the day shall arrive for the very last Coopers Hill man of all to cease from his labours and turn his head homeward, the good work that has been done through a period of 65 years or more will remain."[119]

The First World War offered a rallying cry at a time when the society might otherwise have started to lose momentum. With so many CH men and their sons eligible to serve and with retirees still able to contribute meaningfully to the war effort, the war brought out and focused members' natural inclinations to serve their country.

Between the wars, younger members retired from overseas and returned to refresh the relatively small group of active society organizers, swelling the fraction of members at home. But it never really recovered from its hiatus during the Second World War; even though there were still 300 regular members, attendance at events was rarely more than 20.

In much the same way as the Haileyburians had seen their world change a century before, the Coopers Hill men saw themselves becoming the last representatives of a disappearing way of life. As their members retired, the public works, telegraphs, and railways they had made their own were administered by others, increasingly by the Indians themselves. The society's duty, formerly towards future colleagues, therefore shifted to one of honouring departed comrades by maintaining traditions and memories, standing steady to the end, and not letting the side down. Such ideals were at the very core of British imperialism at the time of their education.[120] In this context, the return of Alexander Taylor's statue to Coopers Hill would have represented a last fulfilment of this duty.

There were two other reasons for a sense of completeness in 1960. The first was that the Coopers Hill estate was once again home to an educational establishment, one that had embraced the legacy of the Coopers Hill Society and offered continuity for the trophies and traditions of *their* Coopers Hill. The second was the completion of two records of Coopers Hill and its graduates. From the first suggestion in 1910 of a register of OCH men and their achievements, proposals for annals, histories, and registries had been made in 1920, 1925, 1939, and 1957, but nothing had come of them. Now John Cameron's *A Short History of the Royal Indian Engineering College, Coopers Hill* was circulated in 1960 and Arthur Astbury completed a register of the careers of Coopers Hill men.[121]

With the completion of these projects, their stories and identities had been preserved. Duty had been done and it was finally acceptable to "stand down."

10
Conclusions and Reflections

Conclusions

This is the first complete history of the Royal Indian Engineering College at Coopers Hill, an institution that was for a brief period the pre-eminent engineering school of the British Empire.

Unusually for an educational establishment, the story of Coopers Hill has a clearly defined start, middle, and end. This gives a unique opportunity to study the entire trajectory of such an institution. Accordingly, this chapter first draws together common themes from earlier chapters into a conclusion, before offering some personal reflections on the Royal Indian Engineering College. Recall that the two overall goals of the book from chapter 1 were: (1) to understand the culture, education, and achievements of its students; and (2) to assess the significance of Coopers Hill in the picture of British engineering education. These will be summarized here in reverse order.

After the Crown assumed responsibility for India, the East India Company's colleges at Haileybury and Addiscombe were closed in 1858 and 1861, respectively. Entrance to the Indian Civil Service was by open competition, and all Royal Engineers were trained at Chatham. During its existence, Coopers Hill was therefore the sole British institution dedicated to training young British men for work in India.

Partly as a result, the college was plagued by contradictions and competing influences. Perhaps the most significant was that, despite all indications to the contrary – such as the word "college" in its title and the presence of numerous students with no intention of going to India – the Secretary of State for India in Council did not consider Coopers Hill to be an educational establishment at all, but a government department. The college was therefore an anomaly, not wholly part of the government, like Woolwich, nor belonging entirely to the academic world of other colleges. As a result, it faced criticisms and challenges from both sources.

From the Government of India, Coopers Hill received conflicting messages: after being not wanted for more than half its existence, it became indispensable after it was too late. The initial reservations arose largely from the vested interests of the Royal Engineers and the Government of India's pique at not being consulted by London. The situation persisted for two decades, through the 1879 call for the college's abolition, until the hegemony of the Royal Engineers was gradually replaced by that of Coopers Hill. The continued agitation by the Coopers Hill engineers over their conditions and pensions was also regarded negatively by the establishment in India.

Oversight of the college's operations by the India Office was primarily fiscal. The ongoing demands for financial self-sufficiency, together with the unfortunate (from the Coopers Hill perspective) decline in demand for PWD engineers, created a cloud over the college that never fully dissipated. While such pressure could be seen as keeping the college accountable, these periodic poison pills of uncertainty had little educational value and served mostly to undermine confidence in the college and its graduates. One way in which this may have been felt was in the difficulty of recruiting applicants. By adjusting the Practical Course, starting the Telegraph and Forest Programs, and taking "extra" and non-PWD students, the annual operating costs were brought nearly into balance. But, as the college authorities concluded more than once, it was never realistically going to cover the interest on the capital investment in its facilities.

The latter requirement was spurious in two ways. First, although it provided the excuse to close Coopers Hill, the ultimate sale of the estate recouped just 12 per cent of the India Office's capital expenditure. Covering the interest on *that* amount from the college's revenues would have been much more feasible. Second, the India Office was being two-faced. When it was convenient for them, Coopers Hill was treated as a government department whose "professors" were just employees with no role in collegial governance. But, as George Chesney pointed out in the college's early days, other governmental training establishments such as Woolwich or Chatham had no similar requirement to be self-sufficient.

The seeds of this problem were sown in the parliamentary debates even before the college opened, and successive secretaries of state, with encouragement from the Government of India, interpreted the commitment to covering costs quite strictly. This highlights the political underpinning of Coopers Hill. Although it never had to pay for its buildings and facilities, the college was beholden to the India Office in ways that prevented it from achieving full control of its academic affairs – despite fewer than half of its engineering graduates actually joining the PWD (although 70 per cent of all graduates went to India in some capacity). To add insult to injury, the public universities and colleges resented what they perceived to be unfair subsidization of Coopers Hill. They were unanimous in believing that the funding would have been better spent expanding their facilities than founding a new school. When the final enquiry asked for their input in 1903, the other institutions eagerly added their weight to calls for the college's abolition.

When Coopers Hill was established, the curriculum was broadly based on that at Roorkee, adapted according to Chesney's own vision and experience and with input from those he consulted. Chesney was one of the earliest proponents of combining theoretical and practical approaches to engineering education. Coopers Hill was therefore a pioneer of what is now an accepted principle. At the time, however, the college was criticized for this innovation by the pupilage-trained engineering establishment.

But once Chesney left, who was responsible for oversight of the college's academic affairs? Sometimes engineers in India were consulted, such as on the abolition of the Practical Course, but, for example, it took several years for the complaints of Reynolds and Pitman about the Telegraph curriculum to be heard. It effectively fell to the Board of Visitors to represent the technical needs of the PWD in absentia, but their effectiveness in this regard was questionable.

Little changed in the Engineering curriculum at Coopers Hill for the middle twenty years of its existence. Possibly little adjustment was deemed necessary because the nature of the work in India did not itself change greatly. If so, this was an oversight, because the wider context of engineering education in Britain *did* change significantly during this period. With Coopers Hill as a catalyst, other institutions caught up and surpassed it, as stated repeatedly by witnesses at the 1903 enquiry. Such an omission was important when, again, the majority of engineering students were there for a general education that was now becoming widely available elsewhere (and for whom, one could argue, both the fees and the specialized Indian content at Coopers Hill were disadvantages).

The belief that other institutions had surpassed Coopers Hill proved to be a critical factor in determining the college's fate. In 1900–1 it fell to President John Ottley to implement overdue curricular reforms. The resulting furore was largely due to his authoritarian approach (encouraged by the India Office's attitude that Coopers Hill was a government department), rather than the academic substance of Ottley's proposals. The post-1901 curriculum was a great improvement, especially in electrical engineering. Unfortunately, the highly visible controversy resulting from Ottley's associated staff dismissals, together with the expense of upgrading the college facilities, made the political cost of maintaining the college prohibitive. When combined with the inevitable, but this time credible, assurances of other institutions that they could educate adequate numbers of engineers better than Coopers Hill, these factors led to the ultimate abolition of the college.

In this context it is worthwhile looking again at the nature and composition of the Board of Visitors. Although the general impression is that it was a group of experts to advise the college, it was in reality selected and appointed by the government to ensure that activities at Coopers Hill met the government's requirements. For much of the college's existence, this meant an attention to finances rather than education. When it came to the final few years, it is also noteworthy that all of the men who signed the 1901 report advocating closure were still on the Board of Visitors during the 1903 enquiry. Moreover, of the twenty-four witnesses who appeared before the enquiry, four (William Preece, Alexander Rendel, William Bisset, and John Wolfe-Barry), as 1901 signatories, had already decided on closure, and a fifth (Charles Crosthwaite) was the enquiry chairman.

Few of the men involved in 1903 were impartial. Of the board, two were members of the India Office (Crosthwaite, Bisset), three had strong ties to the government in India (Ilbert, Pearson, Glass), and three (Edward Busk, Andrew Forsyth, and G.C. Bourne) were from rival universities. The remainder (James Mansergh, Preece, Rendel, Wolfe-Barry) were members of what has been called the "Great George Street Clique."[1] In addition to these three, witnesses with connections to the clique included Benjamin Baker, George Whitehouse, Francis O'Callaghan, and possibly William Unwin (who was later president of the ICE).

The Great George Street Clique comprised the close network of senior members of the ICE and Consulting Engineers whose offices were located near its headquarters at 1 Great George Street, London. This location was close to the major government and colonial offices. Such men would have had their roots in the old pupilage system, but while this may have factored somewhat into their opinions, that debate was fast losing its relevance. However, they did rely on government contract work in Britain, India, and throughout the empire for their businesses and reputations. As we saw in chapter 7, bridge designer Alexander Rendel was one example. At the same time, the ICE was the major determinant of hierarchy and prestige within the engineering profession. During this period, therefore, "a high position in the ICE enhanced the ability to obtain imperial projects while imperial projects provided a path to the highest positions in the ICE."[2] Additionally, the awarding of knighthoods and other civil honours often resulted from building successful high-profile engineering works in the empire.

Many of the men deciding the fate of the college were therefore members of this network of power, finance, prestige, and patronage focused on Westminster.[3] Their dependence on lucrative government contracts placed them in a delicate position, if not an outright conflict of interest, by subjecting them to actual or perceived pressure to implement the government's wishes on the future of Coopers Hill.

Lacking the capacity for self-determination, and with political masters who saw their mandate regarding the college in narrow terms, Coopers Hill had no opportunity to adapt. It was, in effect, immobilized in a web of competing, incompatible, and changeable interests. Had the PWD demand for engineers remained steady at the 1871 levels, the history of Coopers Hill might have been very different. With a clear and defensible raison d'être, the political and fiscal challenges would have been greatly reduced. The dilution of its primary mandate by the admission of non-PWD students (needed to balance the books), would have been unnecessary. Coopers Hill's curriculum could have focused exclusively on engineering requirements for India relatively independent of the development and complaints of other institutions.

Coopers Hill was a prestigious institution. By Chesney's design, it perfectly supplemented the established pathways that led from the English public schools to either a university (usually Oxbridge or a handful of others) or a military academy. Coopers Hill's culture, environment, and sporting prowess were firmly in the same tradition, making the college a comfortable option for well-off families. This was aided by the lure of a manly, patriotic career in the service of the empire. Although high pupilage fees had long limited access to the upper echelons of civil engineering to the wealthy, Coopers Hill offered an alternative, perhaps more socially acceptable, pathway into engineering.

The culture, and indeed the very existence, of Coopers Hill was rooted in the colonial thinking of the time, requiring public works in India to be led by Europeans, with European training, and a strong esprit de corps. The education and the culture imparted by the college were seen to be of similar importance, and both were determined by the nature of the work in India.

Under the British model of PWD administration, the engineer needed a high degree of self-reliance for day-to-day operations. Earlier chapters looked in detail at the experiences of Coopers Hill graduates such as Ernest Hudson, Eustace Kenyon, Lionel and Bertram Osmaston, James Best, Alfred Newcombe, and others. After a variable but often short period of on-the-job training in India, these young men were placed in charge of technical projects, personal servants, contractors, Indian subordinates, accounts, and reporting. Life on construction or tours of inspection could be isolated, arduous, and nomadic, not to mention dangerous in some locations. Periods in the office allowed for the development of engineering designs, forest working plans, and technical discussions, but were equally often dominated by tedious paperwork.

Each officer possessed specialized technical knowledge according to his branch of engineering or forestry. But even the technical work shared a great deal in common across disciplines. Through courses in accounting, surveying, drawing, estimating, mathematics, and descriptive engineering, as well as practising report preparation, Coopers Hill aimed to prepare their graduates for these common duties. Students at Coopers Hill were "saturated with Indian ideas,"[4] as much from the environment as the course content.

Nevertheless, the fresh engineer in India had a great deal to learn, causing established engineers in India to criticize the first batches of Coopers Hill graduates. The Telegraph Department had a systematic approach to bringing new graduates up to speed, necessitated between about 1885 and 1900 by an increasingly inadequate preparation at Coopers Hill. In the Public Works and Forest Departments, the training consisted of an ad hoc apprenticeship to a working officer.

To complement the technical training, Coopers Hill promoted a particular culture that characterized the "Coopers Hill Man," which consisted of two parts: esprit de corps or group loyalty, and individual "character." The first was analogous to the military concept of regimental honour, a social pressure to bring credit to the group rather than shame (according to the subjective standards of the group). This was manifested in an official way at Coopers Hill through the British tradition of sports such as rugby, rowing, and cricket, as well as the volunteer regiment. Socially, cohesion was enhanced by living together in the college[5] and by shared activities such as the Bal Masqué, smoking concerts, and picnics by the river. Student pranks such as ruxing, breaking curfew, and removing the famous signpost all served to establish a set of shared traditions and mythologies that cemented loyalty within the group.

Characteristics of personal honour and integrity were highly valued. This was achieved in part by recruiting students from the appropriate social background, usually with a public school education that specifically taught boys to become leaders. At Coopers Hill, one of the president's duties was to "eliminate the bad men" before they could go to India; here, Ottley was referring to their character, not their technical abilities. This concept of character, often summarized as being a "gentleman," was complex but was primarily used by the engineers to mean immunity to corruption. It also encompassed "efficiency," delivering a project as economically as possible, and the self-reliance necessary to wield power appropriately in remote areas.

Based on the nature of the jobs performed by Coopers Hill graduates described in this book, the need for engineers in the British system to be "gentleman generalists," while criticized today, is understandable. Indeed, these concepts of esprit de corps and efficiency were effectively precursors of the engineering professional codes of practice drawn up in the 1920s, when loyalty to fellow members of the profession and the welfare of the client were amongst the first items to be codified. But before the development of these broader professional ideals, the narrowly defined British "gentleman generalist" approach discriminated against Indian engineers and held back engineering education in the subcontinent.

As part of Britain's colonial machinery, Coopers Hill played its part in the well-documented legacies of the Raj. Its graduates' domination of public works and forestry extended well into the twentieth century and only gradually gave way as independence approached.

Much of India's current infrastructure was built by Indian labour under British leadership. It has been argued that the British motivations were self-serving, that the infrastructure would have been built even in the absence of the British, and that the costs to the Indian population of the empire were egregious. Regardless of the merits of these positions, a substantial part of this infrastructure was built under the management of Coopers Hill engineers. The college's impact on the Indian infrastructure, on education, and the engineering profession was therefore immense.

In England, the establishment of the college in 1871 stepped up the quality of engineering education. It offered the first complete program that systematically combined theory and practice. The Practical Course foreshadowed modern co-op engineering programs. By providing a high-profile target for other institutions, Coopers Hill arguably catalyzed the transition of engineering education from the pupilage system to a university discipline. Even though other institutions eroded the college's leading position, it retained its prominence until the end. Ultimately, as a result of its political constraints, the Coopers Hill model was unsustainable. To some degree, the college's spirit still lingers at the University of Oxford, in the form of the transferred Forestry School and, indirectly, its Engineering Department.

Coopers Hill also influenced British engineering through its people. Professors and graduates alike contributed to the technical discourse by presenting papers at the ICE and other professional societies, as well as by publications in scientific journals. They also wrote technical books and textbooks, a number of which were standard reading at the time. It is also important to remember that around 30 per cent of Coopers Hill graduates had careers outside the imperial services, many in British engineering companies. Some engineers, retiring back to Britain after service overseas, joined engineering consulting companies or, like Lionel Osmaston, took up related work at home. At the Coopers Hill Society dinner on 7 December 1920, Harold Blake Taylor (CH 1879–82)[6] listed thirty-five different careers that, from his knowledge, the college's retirees had taken up.[7] To the extent that the education at Coopers Hill reflected Indian conditions and needs, it may be said that India did, via Coopers Hill, influence the development of British engineering.

One major way in which imperial engineering affected the British profession was by both professional periodicals and the popular press. Articles describing imperial engineering triumphs, increasingly illustrated by photographs, fuelled national pride in the engineering itself and the civilizing influence of the empire. Authors such as Rudyard Kipling portrayed imperial engineers as the spiritual successors to the explorers of a previous generation.[8]

As the primary public intermediaries in this process, the consultants of the Great George Street Clique exerted a strong influence on the communication of engineering. Consultant engineers were not permitted to advertise, so articles lauding colonial engineering works were important to them in order to publicize their achievements, generate public support, and maintain government financing.[9]

Consultant engineers in London dominated the profession in Britain through a combination of their own positions of prestige and by their control of the ICE, the country's pre-eminent engineering institution. Much of this influence stemmed from their imperial construction works. However, at the turn of the twentieth century the fact that 10 per cent of the ICE's membership resided in India and another 14 per cent lived in other parts of the empire, gave ordinary engineers abroad a strong claim for having greater influence in the ICE's affairs.[10] Indeed, in response to the domination of London consultants, colonial engineers pushed for changes in organization of the institution to reflect better the colonial membership's interests. Although changes to this effect were enacted in 1896 under the presidency of Benjamin Baker, the Great George Street Clique retained its dominance over the affairs of the ICE. The combined imperial influence on the ICE, and hence on the engineering profession in Britain, was therefore substantial.

Reflections

Plus Ça Change

What has changed in engineering education since Coopers Hill? While many principles remain the same, there have been huge advances in engineering knowledge over the past century, resulting in a plethora of new subdisciplines of engineering and a rapid increase in the pace of diversification discussed in chapter 8. But, in other ways, the fundamental challenges of educating engineers in colleges and universities have changed little.

Your author having spent fifteen years of his professional career in a leadership role establishing a new engineering school, several fascinating parallels have emerged with the Royal Indian Engineering College in the areas of balancing theory versus practice; including practical experience and work-integrated learning; achieving financial sustainability; maintaining curricular relevance; and the competing demands of different stakeholders.

At its heart, engineering is a practical discipline, and the balance between theory and practice is an inherent question in educating engineers. This is reflected even in the choice of words: training versus education. The transition from the pupilage system of training engineers to education by a course of study at universities and colleges was a key issue of debate when Coopers Hill was established. Even as late as 1917, only 15 per cent of prominent British engineering firms valued a degree as a qualification for entry.[11]

That core division has generally disappeared, because the complexity of modern engineering is founded on using theory to predict the performance of designs before they are constructed. But there remains an expectation from students and employers that graduates are immediately ready for the workplace. Earlier examples showed that this was also true for the Coopers Hill graduates.

The college addressed this issue by introducing the Practical Course in which final-year students spent time in the workplace learning practical skills, before receiving their diplomas. Now called work-integrated learning, modern engineering programs do exactly the same, by incorporating "professional experience years" or interspersing time at the university with time at work (which may be called co-operative education or sandwich courses).

Throughout its lifetime, Coopers Hill struggled with the issue of financial sustainability. From the initial, optimistic promises to Parliament on its creation in 1871 to Ottley's cost-saving measures in 1901, the question of how to educate an engineer cost-effectively and with minimal government outlay remained central to the college's administration. This reality has not changed, of course, and significant investment is required to start a new engineering school, often in the form of philanthropic donations together with matching funding from government and the institution. In a fiscal environment where sustainability of the undergraduate program is an expectation, there must be an optimization between program delivery (class sizes, facilities, student-to-faculty ratios, student experience) and revenues from student tuition fees and grants.

Because of its high tuition fees, Coopers Hill was a "rich man's college" where students were drawn from the "limited class which can afford to spend a large sum on education."[12] To break even, it also needed to take in more students than there were places guaranteed in the Indian Services. The same economics apply in today's engineering schools where the tuition and government grants (which may or may not be regulated), scope of services, and domestic/international student mix must be adjusted to balance the budget. The significant difference between Coopers Hill and engineering at a public Canadian university is, however, the modern emphasis on accessibility to education, diversity, student well-being, and the public good.

At its inception, the curriculum at Coopers Hill included non-engineering subjects such as languages, Indian history and geography, accounting, and freehand drawing that were important for an engineer's career. In his opening speech at Coopers Hill in 1871, George Chesney touched on several themes that are beginning to reoccur in discussions on the future of engineering education in the twenty-first century – the importance of the liberal arts, life-long learning, and the passionate pursuit of knowledge:

> Above all do not give any weight to the pestilent doctrine that culture interferes with the acquisition of expertness in business. Depend upon it the two things are in no sense antagonistic, and practical ability is compatible with any degree of culture.
>
> There is no such thing as remaining stationary with respect to knowledge. If it be true that motion is the condition of the physical universe, it is not the less true of the mental state of man. You may go forward, or you fall backward, but you cannot stand still.

> I may remind you that the pursuit of knowledge is the one pleasure that never palls.... By laying yourself out to be studious men ... you are securing for your-selves a certain fund of happiness, independent of any caprices fortune may have in store.[13]

With an ever-increasing body of engineering knowledge, engineering programs have been faced with the problem of how to fit the broad set of academic requirements into a reasonable length of time. The historical tendency has been to focus on fundamental mathematics, science, and core principles of the discipline, at the expense of open-ended design, professional skills, and cross-disciplinary breadth. This is, however, changing, with a return to valuing engineers' understanding of the broader context of their work.[14]

These touch on age-old issues for engineers; to what extent do their responsibilities extend beyond meeting purely technical specifications to the ethical, societal, moral, and environmental aspects of their work? Responsibility for climate change could be attributed to politicians, consumers, or business leaders. However, the opinion is growing amongst engineers that they should play a more active role in the decision-making. A strong code of ethics has long been central to the engineering profession, and modern engineering education is increasingly preparing young engineers to tackle these broader questions.

Over 18,000 engineers graduated from accredited Canadian engineering programs in 2019, less than half of whom become licensed engineers.[15] The other half finds employment in finance, business, research, and other enterprises. This indicates that an engineering education is valued beyond the profession itself. But it also raises the question of how best to educate each of the two halves. To become accredited, a program must meet an extensive set of criteria, covering the curriculum and other matters, designed to prepare graduating engineers to take their places in a legally regulated profession. It is not clear, however, that a program designed for accreditation provides the best education for those graduates who do not become licensed engineers. This is similar to the dilemma faced by Coopers Hill, that by taking students not destined for India it introduced two audiences with distinct needs and thus opened itself to criticism from both.

Where Ye Muses Sport on Cooper's Hill

Impressive though the engineering achievements mentioned above may be, they are the work of a relatively few eminent engineers.[16] Even taking into account that an astonishing one in eight Coopers Hill alumni received civil awards (including thirty-five knighthoods, and numerous commanders and officers of the Order of the British Empire, and companions and knight commanders of the Order of the Indian Empire),[17] the vast majority of alumni did not rise to positions of particular prominence. They were the ordinary men and their families, like the Hudsons and Osmastons. These are the ones who struggled to build bridge piers on shifting riverbed sands, constructed telegraph lines through the mountains in winter, and endured lonely months in the forests. They may equally have been those who remained in Britain as engineers, patent agents, and businessmen. Or those who explored the world in other ways such as George Reynolds and David Carnegie, prospectors for petroleum and gold, respectively.[18]

Major engineering achievements are the tangible monuments to the sons of Coopers Hill, visible today, but the true impact of their work is less apparent, especially with the passage of time; this is the social impact. Consider the 6,500,000 public telegrams that were sent in 1903–4,[19] and the impact they had on private, social, and business decisions of ordinary Indians and Europeans alike. Or think of the 200,000,000 third-class passengers, mostly Indians, who rode the railways in 1904 and what their journeys meant to them.[20] Or the economic impact of the 52,000,000 tons of goods traffic those same railways carried that year.[21] Or the changes brought to the 44,000,000 acres of land irrigated by dams and canals.[22]

Although largely unseen, the combined decisions and transactions facilitated by the telegraphs and railways are an integral and significant part of the great tide of events that made the world as it is today. The construction, operation, and maintenance of this kind of infrastructure was the bread and butter of Coopers Hill men in India, and indeed elsewhere, and so their contribution to shaping today's world is truly part of their legacy.

In the almost sixty years between Ottley's remark about the monumental works and the dissolution of the Coopers Hill Society quoted in chapter 1, ideas had changed about the role of the college. In 1906 only the first few batches of Coopers Hill men had retired, and the men graduating from those final years would have gone on to experience a very different India. Cameron concluded his *Short History* with a note that, in essence, summarizes this broader view of Coopers Hill's legacy:

> In India the period between the opening of the RIEC College and the retirement of its last graduates was one of great political and economic expansion and progress. During the same period, throughout the world, new theories and techniques in all branches of engineering were evolved and these were applied in India, after modifications as necessary to suit conditions in that country. As a result, when the sub-continent attained independence, all Departments of the Engineering and Forest Services were handed over to the new rulers in efficient running order and the task of the Royal Indian Engineering College was successfully completed.[23]

It is also interesting to reflect on just how few graduates there were from Coopers Hill – only 1623 over a span of thirty years.[24] The clearly defined nature of this group is one reason why the study of Coopers Hill is so fascinating.

It helps us to get to know them, their strengths and weaknesses, their ideas and jokes, and the workings of their community. Our ongoing interaction with that community continues the Coopers Hill narrative and thus forms an integral, perhaps essential, part of their legacy. Those interactions continue to shape our understanding of the world, our appreciation of the challenges overcome, and our ability to re-evaluate past actions.

The End of an Era

The last Coopers Hill man alive was probably Anthony Wimbush, the Forest student whose memoirs formed part of chapter 6. He attended the college briefly in 1904–5 before transferring to Oxford with the Forestry School. While on furlough in 1935, Wimbush presided over the Coopers Hill dinner and commented in his speech on the "hopeful signs" he saw in India, such as the increasing prosperity of villagers, greater female education, and a widespread interest at all levels of society in playing cricket.[25]

Although he was short of the official retirement age of fifty-five, the prospect of leaving his family to return to India after that furlough was "not at all a happy one." So, after a friend lent him a book on fruit trees, Wimbush decided to remain in England and apply his forestry skills to commercial fruit growing.[26]

Anthony Wimbush died on the last day of 1980 in Somerset at the age of ninety-six.[27] The era of the Royal Indian Engineering College had come to a close.

Notes

1. Dastardly Murder

1 "Edmund Colvile Elliot," *Minutes of Proceedings of the Institution of Civil Engineers* 116 (1894): 377–9.

2 *Pioneer* (Allahabad, India), 15 November 1893, 4–5.

3 *Pioneer* (Allahabad, India), 5 September 1893, 6.

4 *Pioneer* (Allahabad, India), 15 November 1893, 4–5.

5 Writers are inconsistent about using "Cooper's Hill," with an apostrophe, or "Coopers Hill," without. Although many external publications use the former, the college calendar, magazines, and society omit the apostrophe. Official college documents adopted this convention in July 1871 in what appears to be a conscious decision. For example, the printed copy of the president's opening address in August 1871 does not use the apostrophe (British Library, India Office Records and Private Papers [hereafter BL IOR] MSS EUR F239/106, 10), and a corrected proof for the 1872 prospectus specifically indicates that the apostrophe should be deleted (BL IOR L/PWD/8/7, 173). John Denham's famous seventeenth-century poem is titled "Coopers Hill" or "Coopers-Hill," with no apostrophe. Geographically, the region next to the college grounds appears on ordnance survey maps as "Cooper's Hill." The edition surveyed 1865–70 (Ordnance Survey Six-Inch series, Berkshire Sheet XL) identifies the college building, then Baron Albert Grant's mansion, as "Coopershill." The college style, with no apostrophe, is used here.

6 Because the RIEC was not a university it could not grant degrees, so successful students were awarded a diploma. For convenience, however, the word "graduate" will sometimes be used here to describe students who were awarded the diploma.

7 In later years the college also supplied a small number of men for the Accounts Branch of the Indian Public Works Department. These were seen by the students as a consolation prize for those who performed poorly in engineering.

8 *Coopers Hill Magazine* (hereafter *CHM*) 5, no. 7, June 1904, 115.

9   Although students at Coopers Hill came from a variety of countries (typically around fifteen in a given year), they were all men. Alice Perry is generally considered to be the first woman to graduate from an engineering program in Ireland or Great Britain, and possibly the world. She graduated in 1906, the same year that Coopers Hill closed, from Queen's College, Galway. See Marie Coleman, "Perry (Shaw), Alice Jacqueline," *Dictionary of Irish Biography*, https://www.dib.ie, last revised October 2009.

10 "The New Indian Service," *Spectator*, 10 December 1870, 1472–3.

11 Difference between 5,074 miles in 1871 and 28,054 in 1905, from *Imperial Gazetteer of India* (Oxford: Clarendon, 1908), 3:411.

12 40,950 miles open by 1928–9 from: Nalinaksha Sanyal, *Development of Indian Railways* (University of Calcutta, 1930), 376. Similar data are also presented in these recent works: Ian Kerr, *Building the Railways of the Raj* (Delhi: Oxford, 1995), 211–12; Kartar Lalvani, *The Making of India* (London: Bloomsbury, 2016), 221–2.

13 Lalvani, *Making of India,* 253.

14 *Report of the Indian Irrigation Commission 1901–1903, Part I: General*, UK parliamentary command paper (hereafter PP Cd.) 1851 (1903), 11.

15 *Imperial Gazetteer of India*, 3:445. Also Roland Wenzlhuemer, *Connecting the Nineteenth-Century World: The Telegraph and Globalization* (Cambridge: Cambridge University Press, 2013), chap. 8.

16 R.S. Troup, *The Work of the Forest Department in India* (Calcutta: Superintendent Government Printing, 1917), 4. Troup was an Old Coopers Hill man (hereafter OCH).

17 Ian Hay, preface to *Tiger Tiger* by W. Hogarth Todd (London: Heath Cranton, 1927), 8. This passage is misquoted without full attribution by both Maud Diver in *The Unsung* (London: William Blackwood & Sons, 1945) and Rosamund Lawrence in *Indian Embers* (Oxford: George Ronald, 1949).

18 Shashi Tharoor, *Inglorious Empire: What the British Did to India* (London: Hurst, 2016), 180.

19 Roy MacLeod and Deepak Kumar, eds., *Technology and the Raj* (New Delhi: Sage, 1995); David Arnold, *Science, Technology and Medicine in Colonial India*, The New Cambridge History of India (Cambridge: Cambridge University Press, 2000).

20 Kerr, *Building the Railways of the Raj*. Also Ian Kerr, *Engines of Change: The Railroads That Made India* (Westport, CT: Praeger, 2007).

21 David Arnold and Ramachandra Guha, eds., *Nature, Culture, Imperialism* (Delhi: Oxford University Press, 1995); William Beinart and Lotte Hughes, *Environment and Empire* (Oxford: Oxford University Press, 2007); Deepak Kumar, Vinita Damodaran, and Rohan D'Souza, eds., *The British Empire and the Natural World: Environmental Encounters in South Asia* (Oxford: Oxford University Press, 2011); Madhav Gadgil and Ramachandra Guha, *This Fissured Land: An Ecological History of India* (Oxford: Oxford University Press, 2012).

22 David Gilmour, *The Ruling Caste: Imperial Lives in the Victorian Raj* (New York: Farrar, Straus and Giroux, 2006); Gilmour, *The British in India: Three Centuries of Ambition and Experience* (UK: Allen Lane, 2018).

23 Aparajith Ramnath, *Birth of an Indian Profession: Engineers, India, and the State, 1900–47* (Delhi: Oxford University Press, 2017).

24 J.G.P. Cameron, *A Short History of the Royal Indian Engineering College* ([Richmond]: Coopers Hill Society, 1960). A copy is in BL IOR MSS EUR F239/133.

25 Brendan Cuddy, "Royal Indian Engineering College, Coopers Hill (1871–1906): A Case Study of State Involvement in Professional Civil Engineering Education" (PhD diss., London University, 1980).

26 Several initiatives were proposed by Coopers Hill Society members in 1909–10 to record the legacy of the college graduates, including photographs, a register of careers, and a map of their great works (*CHM* 8, no. 14, 1909, 200, 209; *CHM* 8, no. 15, January 1910, 229–30, 241). Some 800 questionnaires were sent to members to seek information for a registry, but only 20 per cent were returned, and the initiative was abandoned (*CHM* 11, no. 10, November 1920, 145–6).

27 For example, Coopers Hill is mentioned only in passing and with some inaccuracy in R.A. Buchanan, *The Engineers: The History of the Engineering Profession in Britain 1750–1914* (London: Jessica Kingsley Publishers, 1989).

28 The college was established on the understanding that it would be self-supporting from its tuition fees, which were £150 per annum when it opened (approximately one-third of the annual starting salary for an engineer in India).

29 This is not to say that engineers were considered socially equal to these other professions in India. One author wrote that his grandfather had resigned from his club "when he heard of a proposal to admit engineers and forest-officers" (Dennis Kincaid, *British Social Life in India* (London: George Routledge & Sons, 1938), 215.

30 For example, E.H. Stone's evidence to the 1903 enquiry into the fate of Coopers Hill. PP Cd. 2056 (1904), 91.

31 A substantial proportion of the membership of the Institution of Civil Engineers (hereafter ICE) came from families of higher social status. Almost half of those born between 1770 and 1859 had attended public schools, and the fraction was still high in the early 1900s (Brian Harper, "Civil Engineering: A New Profession for Gentlemen in Nineteenth-Century Britain?" *Icon* 2 [1996]: 59–82; and Shin Hirose, "Two Classes of British Engineers: An Analysis of Their Education and Training, 1880s–1930s," *Technology and Culture* 51, no. 2 [2010]: 388–402). However, the extent to which the membership of the ICE was representative of the profession as a whole is not clear. Buchanan argued that the ICE had been established precisely to increase the prestige of the profession (R.A. Buchanan, "Gentlemen Engineers: The Making of a Profession," *Victorian Studies* 26, no. 4 [1983]: 407–29).

32 For example, H.S. Hele Shaw, "The Royal Indian Engineering College," *Engineer* 58 (21 November 1884): 392.

33 Reported in *CHM* 7, no. 8, September 1906, 107.

34 Tharoor, *Inglorious Empire*.

35 For example, Daniel Headrick, *The Tentacles of Progress: Technology Transfer in the Age of Imperialism 1850–1940* (New York: Oxford University Press, 1988); Arnold and Guha, *Nature, Culture, Imperialism*, 237–59; Beinart and Hughes, *Environment and Empire*; Kerr, *Engines of Change*; Christopher Hill, *South Asia: An Environmental History* (Santa Barbara, CA: ABC-CLIO, 2008).

36 Beinart and Hughes, *Environment and Empire*, 130–47.

37 Pallavi Das, "Colonialism and the Environment in India: Railways and Deforestation in 19th Century Punjab," *Journal of Asian and African Studies* 46, no. 1 (2011): 38–53; Das, "Railway Fuel and Its Impact on the Forests in Colonial India: The Case of the Punjab, 1860–1884," *Modern Asian Studies* 47, no. 4 (2013): 1283–1309.

38 Kerr, *Engines of Change*; Deep Kanta Lahiri Chaudhury, *Telegraphic Imperialism: Crisis and Panic in the Indian Empire c. 1830* (New York: Palgrave Macmillan, 2010).

39 Elizabeth Whitcombe, "The Environmental Costs of Irrigation in British India: Waterlogging, Salinity, Malaria," in *Nature, Culture, Imperialism,* ed. David Arnold and Ramachandra Guha, 237–59 (Delhi: Oxford University Press). A connection between canals and malaria had been known since the 1860s from the work of T.E. Dempster, a surgeon, and W.E. Baker, an engineer (W.E. Baker, T.E. Dempster and H. Yule, *The Prevalence of Organic Disease of the Spleen* [Calcutta: Office of the Superintendent of Government Printing, 1868]).

40 Dipak Sarmah, *Forestry in Karnataka* (Chennai: Notion, 2019), chap. 2.

41 David Gilmartin, "Scientific Empire and Imperial Science: Colonialism and Irrigation Technology in the Indus Basin," *Journal of Asian Studies* 53, no. 4 (November 1994): 1127–49.

42 For the social impact of the railways, including the spread of disease, see Ritika Prasad, *Tracks of Change: Railways and Everyday Life in Colonial India* (Delhi: Cambridge University Press, 2015); Hill, *South Asia*, chap. 7.

43 On, for example, the issues of waterlogging and increased salinity, see J.G. Medley, ed., *Roorkee Treatise on Civil Engineering in India* (Roorkee: Thomason College Press, 1867), 2:390. Other discussions are in A. Ross, J. S. Beresford, S. Preston, H. Marsh, E. Benedict, L.W. Dane, and J. Benton, "Discussion: Punjab Triple Canal System," *Minutes of Proceedings of the Institution of Civil Engineers* 201 (January 1916): 48–62; and Dr. Center, "A Note on Reh or Alkali Soils and Saline Well Waters," *Indian Forester* 7 (1882): 61–83.

2. "This College Has Been Established at Cooper's-Hill"

1 The trial was reported in the *Pioneer*'s issues of 15, 16, and 19 November 1893.

2 Latitude 28.789, longitude 78.985.

3 *Khalasi*s were traditionally port workers specializing in winches and pulley systems for shipyards. Their skills at manoeuvring large, unwieldy objects made them useful in the construction of bridges.

4 This account from 15 November 1893. Paragraph breaks have been added.

5 *Pioneer* (Allahabad, India), 5 September 1893, 6.

6 See, for example, PP Cd. 2056 (1904), 2, where the witness, John Ottley, was asked to clarify what he meant by these terms.

7 Elizabeth Buettner, *Empire Families* (Oxford: Oxford University Press, 2004).

8 James W. Best, *Forest Life in India* (London: John Murray, 1935), 138.

9 The issue of Hawkins's sanity was raised several times in the *Pioneer* over the succeeding weeks, generally with a more constructive attitude to the question than was expressed at the trial. Other news included the revelations that Hawkins had been acquitted on another count of murder and that his grandfather had also been hanged for murder (*Amrita Bazar Patrika*, 19 November 1893, 7).

10 The event was mentioned in passing in an article in the *Overland Ceylon Observer*, 6 January 1894, 12.

11 Robert Fenelon Hawkins was born in 1859 and married Marian (or Mary Anne) Vincent Hatcher in 1884 at Parembore, Madras. Marian bore at least three children, two of whom died in infancy. She died in 1940 in Madras at the age of seventy-eight.

12 George Chesney, *Indian Polity,* 2nd ed. (London: Longmans, Green, 1870), 361–5.

13 In larger provinces, matters relating to irrigation were usually conducted separately under a parallel structure.

14 Kerr, *Building the Railways of the Raj*, 39.

15 "The New Indian Service," *Spectator*, 10 December 1870, 1472–3.

16 Sally Mitchell, *Daily Life in Victorian England* (Westport, CT: ABC-CLIO, 2008), 58.

17 Hirose, "Two Classes of British Engineers."

18 As H.G. Wells commented in his autobiography, "The great group of schools at South Kensington which is now known as the Imperial College of Science and Technology, grew therefore out of an entirely technical school, born of the base panic evoked in England by the revelation of continental industrial revival at the Great Exhibition of 1851" (H.G. Wells, *Experiment in Autobiography Discoveries and Conclusions of a Very Ordinary Brain [since 1866]* [London: Gollancz, 1934], 1:208).

19 J. Scott Russell, *Systematic Technical Education for the English People* (London: Bradbury, Evans, 1869), 86.

20 "Dr. Lyon Playfair to Lord Taunton, Chairman of the Schools Inquiry Commission," *Engineer*, 19 July 1867, 60.

21 Warington W. Smyth, lecturer on mining and mineralogy, Royal School of Mines, *Engineer*, 26 July 1867, 68.

22 *The Education and Status of Civil Engineers in the United Kingdom and in Foreign Countries* (London: Institution of Civil Engineers, 1870).

23 A fee, often of several hundred pounds.

24 *Education and Status of Civil Engineers*, viii–ix.

25 Designs were traditionally achieved by measuring the performance of scale models. Models, both physical and computer-generated, are still used to verify, supplement, and extend theoretical predictions.

26 In reality, the association of "shop culture" with pure practice and "school culture" with book learning was neither rigid nor the only difference between the two cultures. For example, shop culture was based on personal mentoring, whereas school culture involved educational institutions, lecture rooms, and examinations. Both approaches evolved. Monte A. Calvert, *The Mechanical Engineer in America* (Baltimore, MD: Johns Hopkins University Press, 1967).

27 Fleeming Jenkin, *A Lecture on the Education of Civil and Mechanical Engineers in Great Britain and Abroad* (Edinburgh: Edmonston & Douglas, 1868), 10. The figure of £500 seems to have been about the going rate, because Henry Conybeare mentions the same amount in his letter in *The Education and Status of Civil Engineers*, 181.

28 J.M. Heppel, in *The Education and Status of Civil Engineers*, 185.

29 R.A. Buchanan, "The Rise of Scientific Engineering in Britain," *British Journal for the History of Science* 18, no. 2 (July 1985): 218–33.

30 Dates from Peter Lundgreen, "Engineering Education in Europe and the U.S.A., 1750–1930: The Rise to Dominance of School Culture and the Engineering Professions," *Annals of Science* 47, no. 1 (1990): 33–75.

31 Hirose, "Two Classes of British Engineers."

32 Henry Conybeare made sensible suggestions about how the ICE itself could be proactive in improving engineering training: "(1) That the Institution might influence for good the scientific education given in the Engineering departments of the colleges and schools. (2) That no Engineer should take a pupil without adequate scientific training. (3) That Engineers in different departments of the profession should exchange pupils for a time."

33 Authors have usually focused on particular aspects of Chesney's career, so the full scope of his overall achievements has not been highlighted. This is beginning to change, e.g., Patrick M. Kirkwood, "The Impact of Fiction on Public Debate in Late Victorian Britain: The Battle of Dorking and the 'Lost Career' of Sir George Tomkyns Chesney," *Graduate History Review* 4, no. 1 (Fall 2012): 1–16.

34 George Chesney, *Memorandum on the Employment of the Corps of Royal Engineers in India* (London: Spottiswoode, 1868); Chesney, *Indian Polity*.

35 Roger T. Stearn, "Chesney, Sir George Tomkyns (1830–1895)," *Oxford Dictionary of National Biography* (Oxford: Oxford University Press, 2004).

36 See H.M. Vibart, *Addiscombe: Its Heroes and Men of Note* (London: Archibald Constable, 1894); and W. Broadfoot, "Addiscombe: The East India Company's Military College," *Blackwood's Edinburgh Magazine*, May 1893, 647–57.

37 "Examination at Addiscombe," *Morning Chronicle* (London, England), 9 December 1848.

38 *Thomason Civil Engineering College Calendar, 1871–72* (Roorkee: Thomason Civil Engineering College Press, 1871).

39 Frederick Sleigh Roberts (Field-Marshall Lord Roberts of Kandahar), *Forty-One Years in India, from Subaltern to Commander-in-Chief* (London: Macmillan, 1901), 514.

40 "Parliamentary Intelligence," *Jackson's Oxford Journal*, 19 August 1871.

41 George Chesney to John Blackwood, 11 March 1871, in Mrs. Gerald Porter, *Annals of a Publishing House: John Blackwood* (Edinburgh: William Blackwood & Sons, 1898), 3:300.

42 George Chesney to John Blackwood, 11 March 1871, in Porter, *Annals of a Publishing House*, 299.

43 A. Michael Matin, "Scrutinizing the Battle of Dorking: The Royal United Service Institution and the Mid-Victorian Invasion Controversy," *Victorian Literature and Culture* 39 (2011): 385–407; I.F. Clarke, "Before and after 'The Battle of Dorking,'" *Science Fiction Studies* 24, no. 1 (1997): 33–46.

44 Chesney wrote two more works of fiction – *The True Reformer* (1873) and *The Dilemma* (1876) – but neither captured the public imagination like *The Battle of Dorking*.

45 A. Cameron Taylor, *General Sir Alex Taylor C.G.B., R.E.: His Times, His Friends, and His Work* (London: Williams & Norgate, 1913), 2:244.

46 John Blackwood to William Blackwood, 15 May 1871, in Porter, *Annals of a Publishing House*, 3:302.

47 Chesney, *Memorandum*, 4.

48 Chesney, *Memorandum*, 5.

49 Chesney, *Memorandum*, 6.

50 "East India (Civil Engineers' College): Further Papers Relating to the Indian Civil Engineering College," PP 148 (1871), 4.

51 PP 148 (1871), 3.

52 Indian Public Works Department (hereafter PWD) dispatch, 11 April 1862, quoted in Cuddy, "Royal Indian Engineering College," 66.

53 Chesney, *Memorandum*, 14.

54 PP 148 (1871), 13.

55 Chesney, *Memorandum*, 8.

56 Chesney is listed as officiating ("offg.") in the capacity as principal of Fort William in the *Quarterly Army List of Her Majesty's British and Indian Forces on the Bengal Establishment* (Calcutta: Military Orphan, 1859), 90. Other institutions offering an engineering education included the University of Madras (1858), College of Engineering at Poona (1856), and the Bengal Engineering College at Seebpore, where the Fort William College was relocated in 1856 (Subrata Dasgupta, *Awakening* [Noida: Random House India, 2010]. The supply of engineers from those

institutions was very small, rarely exceeding four per year (Cuddy, "Royal Indian Engineering College," 49).

57 PP 148 (1871), 2. Roorkee is 150 kilometres northeast of Delhi, on the banks of the Ganges Canal and home of the Bengal Engineers.

58 For the history of Thomason Civil Engineering College, refer to K.V. Mital, *History of the Thomason College of Engineering (1847–1949)* (Roorkee: Indian Institute of Technology, 2008); and "History," Indian Institute of Technology Roorkee, http://www.iitr.ac.in/institute/uploads/File/history10022009.pdf.

59 *Account of Roorkee College Established for the Instruction of Civil Engineers, with a Scheme for Its Enlargement* (Agra: Secundra Orphan, 1851), 5–6.

60 *General Report on Public Instruction in the North Western Provinces of the Bengal Presidency for 1843–44* (Agra: Agra Ukhbar, 1845), appendix D, vi.

61 Mital, *History of the Thomason College*, 61.

62 The details that follow are from the *Thomason Civil Engineering College Calendar, 1871–72.* Many of those with European-sounding names would have been Indian nationals.

63 Engineer Class course of study at Thomason Civil Engineering College, Roorkee, 1871–72.

64 Based on the names of passing students listed in the 1871–72 *Thomason Civil Engineering College Calendar.*

65 Chesney, *Memorandum*, 13.

66 Cuddy, "Royal Indian Engineering College," 61.

67 A refrain familiar to administrators of modern-day engineering schools – "Principal's Annual Report 1870–71" in *Thomason Civil Engineering College Calendar, 1871–72*, 268. Lang took over as principal from Lieut.-Col. Julius George Medley RE (Royal Engineer), who had done a great deal during his tenure to raise the profile of the college. The introductory part of Medley's 1873 *India and Indian Engineering* lectures at the Royal Engineer Institute in Chatham contains his interesting views of the country and its people, and the challenges and rewards of being an engineer in India (before moving on to technical issues).

68 "Principal's Annual Report 1870–71," *Thomason Civil Engineering College Calendar, 1871–2*, 269.

69 PP 148 (1871), 11.

70 Martin Moir, *A General Guide to the India Office Records* (London: British Library, 1988), 61–124. This page range indicates that the present description covers only the very fundamental structure of British rule in India.

71 For a fascinating look at the resort life of the British Raj, see Vikram Bhatt, *Resorts of the Raj* (Middletown, NJ: Grantha, 1998).

72 Based on H.C.G. Matthew, "Campbell, George Douglas, Eighth Duke of Argyll in the Peerage of Scotland, and First Duke of Argyll in the Peerage of the United Kingdom (1823–1900)," 23 September 2004, *Oxford Dictionary of National Biography* (Oxford University Press, https://www.oxforddnb.com).

73 John Lowe Duthie, "Pressure from Within: The 'Forward' Group in the India Office during Gladstone's First Ministry," *Journal of Asian History* 15, no. 1 (1981): 36–72.

74 In later life Argyll continued forming his own views, independent of fashion or politics. Kirsteen Mairi Mulhern, 'The Intellectual Duke: George Douglas Campbell, 8th Duke of Argyll, 1823–1900" (PhD diss., University of Edinburgh, 2006).

75 Argyll, letter to the editor, *Times*, 6 April 1895, 14. Argyll was objecting to the *Times*'s obituary for Chesney, which he felt gave the latter all the credit for founding Coopers Hill. From PP 115 (1871) it seems that William Thornton (secretary, Public Works Department, India Office) had been in discussions with Chesney about a possible college and presumably introduced Chesney to Argyll. Argyll may not have known Chesney at that time, but an anonymous writer to the *Pall Mall Gazette* accused Argyll of plagiarizing Chesney's *Indian Polity* in a speech on the Indian budget (*Pall Mall Gazette*, 24 September 1869).

76 The following details are from Cuddy, "Royal India Engineering College," 98–110, which provides an in-depth account of the comments of council members and Argyll's actions, based on Argyll's private papers. See also Brendan Cuddy and Tony Mansell, "Engineers for India: The Royal Indian Engineering College at Cooper's Hill," *History of Education* 23, no. 1 (1994): 107–23.

77 PP 115 (1871), 9.

78 Cameron Taylor, *General Sir Alex Taylor*, 2:244. This biography of the man who replaced Chesney as president of Coopers Hill was written long after the events and after Chesney's death. While the broad strokes agree with the official record, the details should be treated with caution.

79 PP 115 (1871), 19.

80 PP 115 (1871), 22–3.

81 It presents the question of whether Argyll used the chair's privilege of setting meeting dates to ensure that these members could not attend.

82 Argyll, "Sir G. Chesney and Coopers-Hill."

83 Argyll, "Sir G. Chesney and Coopers-Hill."

84 *Engineering*, 27 May 1870, 376.

85 Hansard, House of Commons Debate 5 August 1870, vol. 203, col. 1665.

86 Hansard, House of Commons Debate 9 August 1870, vol. 203, col. 1733.

87 G.C. Foster, "The Intended Engineering College," *Nature*, 18 August 1870, 316–17.

88 Cuddy, "Royal India Engineering College," 128.

89 Grant Duff stated this in his response to University College. Cameron Taylor attributed this argument to Chesney, but states that it related to the University of Glasgow.

90 Letter 12 January 1871, PP 148 (1871), 23.

91 Hansard, House of Commons Debate 24 February 1871, vol. 204, cols. 868–919.

92 Hansard, House of Commons Debate 3 March 1871, vol. 204, col. 1272.

93 Hansard, House of Commons Debate 18 May 1871, vol. 206, col. 954.

94 Henry Colebrook was listed as thirty-seventh in the list of examination results published by the *Times* on 11 July 1871. A profile of Colebrook appeared in *Indian Engineering*, 28 June 1919, 351 (BL MSS EUR F239/108) confirming his success at both the entrance and final examinations, but that he never attended the college.

95 "Engineers for India," *Madras Mail*, 24 January 1871, 2.

96 "The Indian Civil Engineering College," *Amrita Bazar Patrika*, 13 January 1871, 4.

97 Evans Bell and Frederick Tyrrell, *Public Works and the Public Service in India*, (London: Trübner, 1871), 24.

98 *Thomason Civil Engineering College Calendar, 1871–72*, 308.

99 Quoted in Mital, *History of the Thomason College of Engineering*, 73–4.

100 Mital, *History of the Thomason College of Engineering*, 75.

101 *Engineer*, 2 December 1870, 377.

102 Cameron, *Short History of the Royal Indian Engineering College*, 6. This story owes some elements to the 1913 version in Alicia Cameron Taylor's account, which adds a picturesque detail: "When boating down the Thames, he saw a large building crest the wooded ridge overhanging the classic reaches which connect old Windsor with Runnymede." The college buildings are not visible from the river today.

103 Each was described, along with its advantages and disadvantages, by George Chesney on 16 July 1870 (BL IOR L/PWD/8/7, 69–76). Many of these documents are copied in both IOR L/PWD/8/7 and IOR L/PWD/8/8.

104 "Proposed Purchase of Cooper's Hill Estate for the Purposes of the Proposed Civil Engineering College," 9 November 1870 (BL IOR L/PWD/8/8, 151).

105 George Chesney, "Confidential: The Coopers Hill Estate," 6 October 1870 (BL IOR L/PWD/8/7, 109–12).

106 Chesney, "Confidential," 6 October 1870, 111.

107 Or, as Mital disparagingly described it, a "ridge of sand 44 metres high, on the south side of the river Thames about 30km west of London." Mital, *History of the Thomason College of Engineering*, 69n1.

108 Chesney, "Confidential," 6 October 1870, 111.

109 Coopers Hill was sequestrated from John Denham at the end of the English Civil War.

110 John Denham, *Coopers Hill. A Poeme*, 2nd ed. with additions (London: Humphrey Moseley, 1650), University of Toronto Library, B-11 03457.

111 27 October 1870, BL IOR L/PWD/8/7, 113–27.

112 BL IOR L/PWD/8/8, 152.

113 George Chesney, "Accommodation for Additional Students," 27 December 1872 (BL IOR L/PWD/8/7, 361–3). The full proposal included a quarantine area, extra classrooms, equipment storage, and a chapel, in addition to the accommodation.

This "New Block" ran perpendicularly to the 1871 building, joining it at the bottom right in the plan.

114 "Royal Indian Engineering Civil Engineering College, Cooper's Hill, near Staines," *Builder*, 5 August 1871, 597.

115 Colonel Chesney was informed of the change in a letter from the India Office, 4 January 1875. Cameron, *Short History*, 9. The approval was announced in the newspapers a few days later, e.g., *Pall Mall Gazette*, 7 January 1875, 9.

116 Chesney, *Memorandum*, note on p.14.

117 PP 148 (1870), 12.

118 Cuddy, "Royal Indian Engineering College," 99–102.

119 "The New Indian Service," *Spectator*, 10 December 1870, 8.

120 Hansard, House of Commons Debate 3 March 1871, vol. 204, cols. 1326–55.

121 Under the reforms of Lord Salisbury, the comparable issue of maintaining esprit de corps in the Indian Civil Service was addressed between 1879 and 1892 by restricting the two years of university study taken by probationary members of the service to approved institutions. The majority attended the University of Oxford. (David Gilmour, *The Ruling Caste: Imperial Lives in the Victorian Raj* [New York: Farrar, Straus and Giroux, 2006], 46–68.)

122 Peter J. Cain, "Empire and the Languages of Character and Virtue in Later Victorian and Edwardian Britain," *Modern Intellectual History* 4, no. 2 (2007): 249–73.

123 P.J. Cain, "Character and Imperialism: The British Financial Administration of Egypt, 1878–1914," *Journal of Imperial and Commonwealth History* 34, no. 2 (June 2006): 177–200.

124 W.E.H. Lecky, *The Empire: Its Value and Its Growth* (London: Longmans, Green, 1893), 14.

125 Lord Curzon of Kedleston, *The Romanes Lecture 1907: Frontiers*, 2nd ed. (Oxford: Clarendon, 1908), 56.

126 David Cannadine has argued that British attitudes towards social ranking and race were of similar importance in the empire at the turn of the twentieth century. Ornamentalism – involving royalty, elaborate ceremonies, and grand palaces – was a major way in which social hierarchy was displayed. Although Cannadine did not include large-scale engineering achievements, they too were demonstrations of hierarchy, albeit technical rather than social. David Cannadine, *Ornamentalism: How the British Saw Their Empire* (Oxford: Oxford University Press, 2001).

127 At the time, *Engineering* attributed it to the "miserable jealousy" of an individual or individuals. Later, when the notification was retracted, the magazine blamed the "personal animus" of Col. Strachey himself towards the civil engineers. "The Strachey Manifesto," *Engineering* 8 (19 November 1869): 340. Possibly it simply reflected an ill-thought-out concern that civil engineers would turn to alternative means of compensation when their legitimate attempts to improve their government salaries were thwarted.

128 "The Simla Calumny," *Engineering* 8 (29 October 1869): 291. The full text of PWD Notification No. 242 was reproduced in *Engineering* 8 (5 November 1869): 308.

129 "Strachey Manifesto."

130 William Thornton, "A Note for the Duke of Argyll," PP 115 (1871), 4.

131 The version with the additional clause was typeset in a format identical to the published version and appears in both BL IOR L/PWD/8/8, 15 and BL IOR L/PWD/8/7, 46 (also numbered 91). Page 58v (or 116) of this folio contains the published version of this sentence.

132 "The Civil Engineering College, Cooper's Hill," *Engineer* (11 August 1871): 94.

133 Testimony of J.G.H. Glass in PP Cd. 2056 (1904), 84. Somewhat poignantly, Glass's obituary stated, "He served on the board of visitors of Cooper's Hill Indian Engineering College until it ceased to exist." *Journal of the Royal Society of Arts* 59, no. 3049 (28 April 1911): 592.

134 Harold Laski, "The Danger of Being a Gentleman: Reflections on the Ruling Class in Britain," in *The Danger of Being a Gentleman and Other Essays*, chap. 1 (London: Allen and Unwin, 1939).

135 See Michael Brander, *The Victorian Gentleman* (London: Gordon Cremonesi Publishers, 1975), especially chap. 2 on the public school education.

136 Robin Gilmour, *The Idea of the Gentleman in the Victorian Novel* (London: George Allen & Unwin, 1981), 8.

137 PP Cd. 2056 (1904), 2–3.

138 C.J. Dewey, "The Education of a Ruling Caste: The Indian Civil Service in the Era of Competitive Examination," *The English Historical Review* 88, no. 347 (1973): 262–85.

139 For studies on the membership of the Institution of Civil Engineers, see Harper, "Civil Engineering"; and Hirose, "Two Classes of British Engineers."

140 Laski, "Danger of Being a Gentleman," 22.

141 Obituaries for Coopers Hill alumni occasionally referred to the deceased as a gentleman, but they were more likely to describe him as a sportsman: "first and last a sportsman and a gentleman," "a thorough sportsman," "a sportsman in every sense of the word." In the context of hunting and games, "sportsman" embodied the ideals of loyalty, courage, determination, team spirit, and competence. It was therefore one possible component of being a gentleman. See also chapter 4.

142 "Cooper's Hill College," *Engineer* 53 (28 April 1882), 307.

143 "Cooper's Hill College," *Engineer* 52 (22 July 1881), 63.

144 "Indian Civil Engineering College," *Times* (London), 7 August 1871, 10.

145 "Engineer," "The Abolition of the Royal Indian Engineering College, Cooper's Hill," *Spectator*, 30 July 1904, 152.

146 Cameron Taylor, *General Sir Alex Taylor*, 2:248.

147 Steven Patterson, *The Cult of Imperial Honor in British India* (New York: Palgrave Macmillan, 2009).

148 *CHM* 8, no. 10, November 1908, 158. The engineer in question, not a Coopers Hill man, accepted bribes of Rs15,000 to allow a contractor to charge the government at a higher rate. Is the use of "*almost* unheard of" significant?

149 William Unwin to the Coopers Hill Society Dinner, 19 June 1907 (*CHM* 8, no. 4, July 1907, 49).

150 See Aparajith Ramnath, *Birth of an Indian Profession: Engineers, India, and the States, 1900–47* (Delhi: Oxford University Press, 2017), chap. 3.

151 Carl Mitcham, "A Historico-Ethical Perspective on Engineering Education: From Use and Convenience to Policy Engagement," *Engineering Studies* 1, no. 1 (2009): 35–53.

152 *CHM* 8, no. 1, December 1906, 4.

153 It has been argued that the nature of engineering professionalism and its ideals developed in response to the engineers' divided loyalties between the profession and their employers, distinguishing technical knowledge from its business application. Edwin T. Layton, *The Revolt of the Engineers: Social Responsibility and the American Engineering Profession* (Baltimore, MD: Johns Hopkins University Press, 1986).

154 This section is based on several accounts: "Indian Civil Engineering College," *Times,* 7 August 1871, 10; "The Civil Engineering College, Cooper's Hill," *Engineer* (11 August 1871): 94; "Cooper's Hill College," *Engineer* (6 October 1871): 240–1.

155 Two members absent at the final council vote did attend the opening (Sir H. Montgomery and Mr. R.D. Mangles).

156 Diaries of Herbert McLeod, Archives of Imperial College, London. In this text McLeod's abbreviations of common words have been spelled out fully.

157 *Engineer* (11 August 1871): 94.

158 *Times,* 7 August 1871, 10.

159 While being editorially pro-education, the *Engineer* took pains not to offend the establishment by being overly critical of traditional methods of education and training.

160 *Engineer* (6 October 1871): 241. A booklet containing Chesney's opening address is also in BL IOR MSS EUR F239/106.

3. The Prime Years: 1871–1896

1 Leading article in the *Oracle* 4, no. 32, October 1877, 381.

2 PP Cd. 831 (1901), 3.

3 PP Cd. 831 (1901), 13.

4 Alfred. C. Newcombe, *Village, Town and Jungle Life in India* (Edinburgh: Blackwood & Sons, 1905), 411.

5 Motivations for joining the Indian Civil Service also included the early responsibility, the outdoor life, and plenty of shooting (Gilmour, *Ruling Caste*, 51).

6   Engineers Canada, "2019 Membership Information," 4.

7   George Walter MacGeorge, *Ways and Works in India* (Westminster: Archibald Constable, 1894), 1.

8   Julius George Medley, *India and Indian Engineering: Three Lectures Delivered at the Royal Engineer Institute, Chatham in July 1872* (London: E. & F.N. Spon, 1873), 16–17. The opposite was also true; the reduction of status that came with retirement could come as a let-down (chapter 9).

9   BL IOR L/PWD/8/8, 189, 26 November 1870. The college entrance examination was still advertised extensively through 1872 and 1873 in more than fifty British newspapers and *Allen's Indian Mail* (BL IOR L/PWD/8/7, 441–2).

10 "Indian Civil Engineering College, Open Competition," June 1871 (BL IOR L/PWD/8/8, 513–17).

11 Arrangements for the open competition of 1871 (BL IOR L/PWD/8/8, 455–6).

12 For example: "Indian Civil Engineering College, Cooper's-Hill," *Times*, 23 February 1871, 8.

13 "Indian Civil Engineering College, Open Competition," June 1871.

14 John Watkin died in May 1876 and was memorialized in one of the College Chapel windows (*Oracle* 4, no. 31, July 1877; and *CHM* 6, no. 4, February 1905).

15 The advance prospectus in the *Engineer* (2 December 1870: 374, gave the original number of 750. An official advertisement for the college in 1871 (e.g., *Pall Mall Gazette*, 6 April 1871) showed the 2000 figure. See also Cuddy, "Royal Indian Engineering College," 147n3.

16 W. Baptiste Scoones, "The Civil Service of India," *Macmillan's Magazine* 30, May–October 1874, 365–76.

17 Report by George Chesney, 1 February 1872 (BL IOR L/PWD/8/7, 239v, 246).

18 Report by Chesney, 6 January 1873 (BL IOR L/PWD/8/7, 347–52).

19 The names are given in the obituary to Russell, the last member of the quartet to die, on 24 November 1932 (see *CHM* 12, no. 27, April 1933, 428). Maclean died at an early age in 1879 of "remittent fever" (*Oracle* 6, no. 51, December 1879).

20 "Tales of Old Coopers Hill," *CHM* 1, no. 11, May 1899, 147–8.

21 PP Cd. 2056 (1904), 28.

22 Hirose, "Two Classes of British Engineers."

23 George Chesney prepared an outline of the academic, administrative, and financial operations of the college in "Note on Civil Engineering College," 10 June 1870 (BL IOR L/PWD/8/7, 77–94). In the public papers, Chesney wrote more simply: "I would propose to enlarge the field of candidates by abolishing all special tests. The examination should be conducted in the ordinary subjects of English education, thus affording to all well-educated young Englishmen the same equal chance of success" (PP 115 [1871], 7).

24 "Royal Indian Engineering College," *Times*, 14 July 1875, 10.

25 *CHM* 9, no. 4, December 1913, 50. In the English sense, "public school" refers to a privately operated, fee-paying school rather than to a state school.

26 Cuddy, "Royal Indian Engineering College," 180. Many if not most of these students would have attended the crammer after the traditional public school education.
27 Hudson diary for 1887.
28 Donald P. Leinster-Mackay, "The English Private School 1830–1914, with Special Reference to the Private Proprietary School" (PhD diss., Durham University, 1971), quoting Henry Wolffram's 1885 pamphlet *The Private Tutor's Raison D'être.*
29 Cramming to get into Haileybury College: G.R. Elsmie, *Thirty-Five Years in the Punjab, 1858–1893* (Edinburgh: David Douglas, 1908), 3.
30 As stated in the *Indian Civil Engineering College Calendar*, 1874–5, 10.
31 *CHM* 8, no. 16, May 1910, 249.
32 There is no indication that Chesney knew about this, but it is easy to imagine him having a quiet word with some of the better students about the issue.
33 See Minchin's evidence in PP Cd. 2056 (1904), 31.
34 "The 1875–78 Year: A Retrospect and Review," *CHM* 3, no. 6, May 1901.
35 W.E. Bagot, "R.I.E.C.," *CHM* 13, no. 7, May 1940, 171–2.
36 *Indian Civil Engineering College Calendar*, 1874–5, 9.
37 George Chesney, "Mode of Classifying the Students of the C.E. College," 29 July 1870 (BL IOR L/PWD/8/7, 132–4).
38 PP Cd. 2056 (1904), 29.
39 From "Indian Civil Engineering College, Coopers Hill Regulation," 5 (BL IOR L/PWD/8/7, 316). In 1900, Ottley found that students spent thirty-four, thirty-one, and fourteen hours per week in class in years 1, 2, and 3, respectively and was concerned that the workload had fallen since Chesney's time (PP Cd. 490 [1900], 22). However, this may have reflected an increased proportion of time spent outside of class in laboratories, workshops, and fieldwork.
40 PP Cd. 539 (1901), 70.
41 Chesney, "Note on Civil Engineering College," para. 19.
42 Baptisms for babies of people associated with the college were occasionally performed in the chapel – see "Baptisms at the Royal Indian Engineering College," Untold Lives Blog, 24 September 2019, https://blogs.bl.uk/untoldlives/2019/09/baptisms-at-the-royal-indian-engineering-college.html. See College Chapel Baptismal Register 1897–1904 (BL IOR/L/PWD/8/404).
43 *Indian Civil Engineering College Calendar*, 1874–5, 11.
44 The nickname "Snippy" comes from a reminiscence of the 1875–8 year in *CHM* 3, no. 6, May 1901. Egerton was a classmate of Edmund Elliot, who was murdered in 1893. He also a member of the editorial staff of the *Oracle* (obituary for Norman McLeod, *CHM* 11, no. 12, June 1921, 181).
45 PP Cd. 2056 (1904), 44.
46 *Engineer* (6 October 1871): 241.
47 After Chesney left the college in 1880, J.C. Hurst taught the Accounts course based on Chesney's syllabus. PP Cd. 539 (1901), 18.

48 The presence of William Stocker, professor of physics and electrotechnology, who was appointed at Coopers Hill in 1883, suggests that the photograph was in fact taken in that year, not 1882. Professors George Minchin (Applied Mathematics) and Herbert McLeod (Chemistry) are absent from the photograph.

49 Details of Reilly's earlier life are to be found in his obituary in *Minutes of Proceedings of the Institution of Civil Engineers* 142 (1900): 376–9. See also Cameron-Taylor, *General Sir Alex Taylor*, 2:253.

50 Reilly obituary, *Minutes of Proceedings of the Institution of Civil Engineers*.

51 Calcott Reilly, "On the Longitudinal Stress of the Wrought Iron Plate Girder," *Civil Engineer and Architect's Journal* 23 (September 1860): 261–4; Reilly, "On Uniform Stress in Girder Work, Illustrated by Reference to Two Bridges Recently Built," *Minutes of Proceedings of the Institution of Civil Engineers* 24 (1865): 391–425; Reilly, "Studies of Iron Girder Bridges, Recently Executed, Illustrating Some Applications of the Modern Theory of the Elastic Resistance of Materials," *Minutes of Proceedings of the Institution of Civil Engineers* 29 (1870): 403–500.

52 Ernest Hudson's diary entry for 22 July 1900 records him sending 10/6 to the fund for the Callcott Reilly Medal "in memory of Pomph." Hudson's diary records Sunday lunch with Pomph and some other students on 13 February 1887 and visiting Mrs. Pomph on 26 February 1888.

53 Reilly obituary, *Minutes of Proceedings of the Institution of Civil Engineers* 142 (1900): 376–9.

54 From the 1881 UK census.

55 *CHM* 8, no. 23, December 1911, 356.

56 *CHM* 15, no. 5, October 1950, 79.

57 Obituary for Professor Joseph Wolstenholme, *Times*, 23 November 1891, 6. Also University of Cambridge online alumni database, http://venn.lib.cam.ac.uk/. Wolstenholme was born at Eccles, near Manchester, on 30 September 1829 and was educated at Wellesley College School, Sheffield.

58 Leslie Stephen, *Sir Leslie Stephen's Mausoleum Book* (Oxford: Clarendon, 1977), 79.

59 The first edition published in 1867 contained 1628 problems; it was enlarged to 2815 problems for the 1878 second edition.

60 Dr. Forsyth, Sadlerian Professor of Pure Mathematics at Cambridge, in the 1900 edition of the *Oxford Dictionary of National Biography*. The *DNB* was edited by Leslie Stephen, but there is no hint in Wolstenholme's entry of their friendship. The entry acknowledges information from his sister, Elizabeth Wolstenholme Elmy, a prominent suffragist.

61 Cameron Taylor, "General Sir Alex Taylor," 2:255.

62 *CHM* 15, no. 5, October 1950, 79.

63 Cameron Taylor, "General Sir Alex Taylor," 2:255.

64 Ellen Tremper, "'The Earth of Our Earliest Life': Mr. Carmichael in *To the Lighthouse*," *Journal of Modern Literature* 19, no. 1 (Summer 1994): 163–71.

65 Obituaries in *Nature* 112, no. 2817 (October 1923): 628; *Journal of the Chemical Society, Transactions* 125 (1924): 990–2; *CHM* 12, no. 1, January 1924, 1–3.

66 See the excellent series of papers by Hannah Gay: "A Questionable Project: Herbert McLeod and the Making of the Fourth Series of the Royal Society Catalogue of Scientific Papers, 1901–25," *Annals Of Science* 70, no. 2 (2013): 149–74; "Science and Opportunity in London, 1871–85: The Diary of Herbert McLeod," *History of Science* 41 (2003): 427–58; "Science, Scientific Careers and Social Exchanges in London: The Diary of Herbert McLeod, 1885–1900," *History of Science* 46 (2008): 457–96.

67 *The Declaration of Students of the Natural and Physical Sciences* (London: Simpkin, Marshall, 1865).

68 6–21 April 1871. McLeod's memo and plans regarding the laboratory, dated 14 April 1871, are preserved in BL IOR L/PWD/8/8, 375–87. One of the rooms is shown in the photograph in chapter 7.

69 *CHM* 8, no. 19, January 1911, 293.

70 *CHM* 15, no. 5, October 1950, 80.

71 Herbert McLeod, "Apparatus for Measurement of Low Pressures of Gas," *Proceedings of the Physical Society* 1 (1874): 30–4.

72 George S. Clarke and Herbert McLeod, "The Telephone," *Nature* 18 (2 May 1878): 11.

73 *CHM* 3, no. 6, May 1901.

74 Sydenham of Combe, *My Working Life* (London: J. Murray, 1927), 15.

75 Lewis Carroll, *The Hunting of the Snark* (London: Macmillan, 1876).

76 Sydenham of Combe, *My Working Life*, 12.

77 George Sydenham Clarke, *Practical Geometry and Engineering Drawing* (London: Spon, 1875), iii–iv.

78 Sydenham of Combe, *My Working Life*, 17.

79 Jason Tomes, "Clarke, George Sydenham, Baron Sydenham of Combe (1848–1933)," 23 September 2004, *Oxford Dictionary of National Biography* (Oxford University Press, https://www.oxforddnb.com).

80 E.G. Walker, *The Life and Work of William Cawthorne Unwin* (London: Unwin Memorial Committee,1938). Unwin's obituary is in *Minutes of Proceedings of the Institution of Civil Engineers* 236 (1933): 514–19.

81 J.S. Wilson, "William Cawthorne Unwin 1838–1933," *Proceedings of the Royal Society* 1, no. 3 (December 1934): 167–78.

82 Gay, "Science and Opportunity in London," 437.

83 *CHM* 8, no. 23, December 1911, 358.

84 "Obituary Sir George Greenhill," *Engineer* 143 (18 February 1927): 189.

85 Linde Lunney, Enda Leaney, and Patricia M. Byrne, "Minchin, George Minchin," *Dictionary of Irish Biography*, rev. March 2013, https://www.dib.ie/.

86 Stephen N. Elrington, *Election of Fellows at Trinity College: Report of the Proceedings at a Visitation* (Dublin: Hodges, Foster, 1871).

87 Richard A. Gregory, "Prof. G.M. Minchin, F.R.S.," *Nature* 93, no. 2318 (2 April 1914): 115–16.

88 "Major-General William Henry Edgcome, R.E. Retired," *Minutes of Proceedings of the Institution of Civil Engineers* 168 (1907): 364.

89 "Major-General E.H. Courtney, R.E.," *Times*, 21 June 1913, 11.

90 Surveys were typically performed in teams of eight to twelve under the supervision of a non-commissioned officer. Courtney was therefore probably responsible for overseeing and monitoring the work of several teams in the field.

91 Cameron Taylor, *General Sir Alex Taylor*, 2:260.

92 The later character of Q in the James Bond stories was named as an abbreviation for quartermaster, the officer responsible for accommodation and supplies.

93 UK National Archives Admiralty Records ADM/196/11/318, ADM/196/76/26, and ADM/196/79/7.

94 George Minchin as president of the Coopers Hill Society paid tribute to Pasco at the 1913 Annual Summer Dinner, *CHM* 9, no. 3, August 1913, 34–5. Assuming they were referring to the bombardment of Odessa that took place in April 1854, one of them must have been mistaken. Pasco's continuous Navy service record does not list him as serving on the HMS *Encounter* at any time, and nor does that ship appear to have been involved in the bombardment of Odessa. In 1854, according to his officer pay record, Pasco was on HMS *Indefatigable*, which was serving in the South Atlantic (http://www.pdavis.nl/ShowShip.php?id=1608).

95 *CHM* 9, no. 2, June 1913,19.

96 George Chesney, Public Works letter no. 168, 16 March 1877, nine pages plus appendix (BL IOR L/PWD/8/7 518–23).

97 "Mr. John George Whiffin," *Times*, 11 January 1892, 9.

98 Walter J. Boyes, "Life in the XIIth Regiment: Fifty-Six Years Personal Reminiscences," *Suffolk Regimental Gazette*, 203–10, 1907/8 (Suffolk Archives). See also *CHM* 11, no. 3, September 1918, 41–2.

99 "List of Staff and Establishment, Indian C.E. College, Coopers Hill," 1871 (BL IOR L/PWD/8/8, 578–9). That table did not include, for example, the grounds staff such as the watchman, lodgekeeper, two groundsmen, two gardeners, dairyman, carter, and later the boatsman.

100 "Tales of Old Coopers Hill," *CHM* 1, no. 11, May 1899, 147.

101 "Reminiscences of Coopers Hill Staff, 1888 to 1891," *CHM* 15, no. 5, October 1950, 81.

102 W.L.S.L. Cameron (CH 1874–7), *CHM* 8, no. 23, December 1911, 356.

103 PP Cd. 490 (1901), 12; PP Cd. 2055 (1904), 42.

104 PP Cd. 2055 (1904), 42.

105 Cuddy, "Royal Indian Engineering College," 203–4.

106 Chesney, "Note on Civil Engineering College," 77–8.

107 House of Lords Hansard 263 (15 July 1881), cols. 997–9. Argyll's statement offended the Institution of Civil Engineers, which argued that the barracks failures

he cited in support of his statement were constructed by military, not civil, engineers. Argyll ultimately agreed he had been mistaken and apologized. *His Grace the Duke of Argyll on Indian Public Works and Cooper's Hill College: A Reply with Correspondence between His Grace and the President of the Institution of Civil Engineers* (London: E. & F.N. Spon, 1882).

108 Louis Mallet, extract letter to G. Chesney, 11 October 1876 (BL IOR L/PWD/8/7, 516).

109 George Chesney, Public Works letter no. 168, 6. Chesney observed with some frustration that self-sufficiency of an institution like Coopers Hill had "never been recognized as a necessary aim or condition in any country or in any branch of the Service."

110 George Chesney, Public Works letter no. 168.

111 Minute by R. Strachey, 2 May 1877, BL IOR L/PWD/8/7, 525.

112 Chesney, "Note on Civil Engineering College," 86v.

113 George Chesney, "Coopers Hill and the Supply of Engineers for India," 30 May 1879 (BL IOR/C/142, 395–9).

114 RIEC Calendar for 1876–7, 7.

115 RIEC Calendar for 1877–8, 7.

116 See the *Oracle* 5, no. 37, April 1878, 451; and 5, no. 40, July 1878, 478.

117 *Oracle* 4, no. 32, October 1877, 381.

118 E.C. Baker, *Sir William Preece, F.R.S.: Victorian Engineer Extraordinary* (London: Hutchinson, 1976), 83–7.

119 Charles Pitman, one of Preece's Lambs, described the origin to a Telegraph Department dinner in London on 3 June 1926. "Indian Telegraph Department," *CHM* 12, no. 10, July 1926, 154.

120 PP Cd. 2055 (1904), 40.

121 "Appointment of Lt. Col. Chesney as Principal of New Civil Engineering College," 21 July 1870 (BL IOR L/PWD/8/8, 145–6).

122 Roger T. Stearn, "General Sir George Chesney," *Journal of the Society for Army Historical Research* 75, no. 302 (Summer 1997): 106–18.

123 Telegram 3 June 1879, as stated in a recapitulation of events in PP Cd. 2055 (1904), 62. In their follow-up dispatch the reasons for closing Coopers Hill were given as "1st, that with a fluctuating expenditure of Public Works it was impossible to employ the large number of men regularly being sent out; 2nd, that is was a disadvantage to have all the engineers trained in the same school; 3rd, that it was the desire of the Government of India to utilise natives of India more largely." Coincidently, a false rumour circulated at the time that the College was in fact closing: "The report that Cooper's-hill College is to be closed is entirely without foundation. There would, indeed, be much cause for regret were so admirable an institution abolished" (*Engineer*, 27 June 1879, 472). The rumour was caused by the sale of an estate of the same name in Ireland.

124 PP Cd. 831 (1901), 10. Chesney had argued that because the president lived in the college, there were expectations that he should entertain at his own expense

men coming home from India, the staff, and the students. Noting that the salaries of the heads of Sandhurst and Woolwich had been increased to £1500 at about the time of his appointment, Chesney requested the same consideration. It was approved as a "personal allowance" rather than a salary.

125 Argyll, "Sir G. Chesney and Coopers-Hill," *Times*, 6 April 1895, 14.

126 Information and quotes from "General Sir Alexander Taylor," *Times*, 27 February 1912, 11.

127 Cameron Taylor, *General Sir Alex Taylor*, 2:251.

128 *CHM* 11, no. 11, January 1921, 164.

129 Cameron Taylor, *General Sir Alex Taylor*, 2:283–4.

130 RIEC calendar 1882–3.

131 RIEC calendar 1880–1, 8.

132 "Cooper's Hill College: Return of the Working of the New Scheme," PP 166 (1884).

133 Fifteenth Report of the Special Committee on Home Charges, 24 April 1890, in "East India Home Charges," PP 327, 13 July 1893, 138.

134 Hansard House of Commons Debate, 21 April 1882, vol. 268, cols. 1111–30.

135 Mr. Pugh later asked for information on the supply of engineers from Roorkee College. Hartington responded that in 1880, 1881, and 1882 *admissions* of students to Roorkee's engineer class were eight, nine, and ten, respectively. There were eight guaranteed positions in the Public Works Department (PWD). Allowing for student attrition, Roorkee was therefore supplying about half as many engineers to the PWD as Coopers Hill (Hansard House of Commons Debate, 4 May 1882, vol. 269, col. 99).

136 *Engineer*, 28 April 1882, 307.

137 "Engineers for India," *Engineer*, 14 October 1881, 279–80.

138 Such a board had in fact been part of George Chesney's early plans ("Note of Proposal for Admission, Regulations, &c.," BL IOR L/PWD/8/7, 106). There was also a college board made up of the professors of the principal branches of engineering for the discussion of academic matters. The frequency of those meetings varied widely according to the year and the president but was typically every two or three months (PP Cd. 831 [1901], 11).

139 Quoted in the 12 July 1901 report of a Committee of the Board of Visitors in PP Cd. 831 (1901), 9.

140 G.F. Pearson, "The Teaching of Forestry," *Journal of the Society of Arts* 30 (3 March 1882): 422–33.

141 Later, plantations were established specifically to provide wood for burning.

142 Pallavi Das, "Hugh Cleghorn and Forest Conservancy in India," *Environment and History* 11, no. 1 (2005): 55–82.

143 Hugh Cleghorn, "Address Delivered at the Nineteenth Annual Meeting," *Transactions of the Scottish Arboricultural Society* 7 (1875): 1–9.

144 F.J. Bramwell and H. Truman Wood, "Education in Forestry," *Journal of the Society of Arts* 30 (14 July 1882): 879.

145 PP Cd. 490 (1901), 12.

146 "The Forestry School at Cooper's Hill," *Nature* 37 (5 April 1888): 529–31.

147 Born Wilhelm Philipp Daniel Schlich in 1840 in Flonheim, Germany. See R.S.
     Troup, revised by Andrew Grout, "Schlich, Sir William Philipp Daniel (1840–
     1925)," 23 September 2004, *Oxford Dictionary of National Biography* (Oxford
     University Press, https://www.oxforddnb.com). See also E. Mammen, M.S.
     Tomar, and N. Parmeswaran, "A Salute to William Schlich, Forestry Pioneer,"
     Journal of the Forestry Commission 34 (1965): 170–5; R.S. Troup, "Sir William
     Schlich, K.C.I.E., F.R.S.," *Nature* 116 (25 October 1925): 617–18.

148 PP Cd. 490 incorrectly gave Schlich's starting date as 1889 instead of 1885, an er-
     ror that has been propagated elsewhere.

149 Peter G. Ayres, *Harry Marshall Ward and the Fungal Thread of Death* (St. Paul,
     MN: American Phytopathological Society, 2005).

150 Chapter by Ward's long-time friend and colleague William Thiselton-Dyer,
     "Harry Marshall Ward 1854–1906," in *Makers of British Botany*, ed. F.W. Oliver,
     261–79 (Cambridge: Cambridge University Press, 1913). Ward is buried in the
     picturesque Ascension Parish Burial Ground in Cambridge, with his tombstone
     all but engulfed by bushes.

151 PP Cd. 2055 (1904), 40.

152 Other costs included those of mail and telegraphs, interest on India Stock, deben-
     tures and loans, and the contribution to naval defence.

153 Fifth report of the Special Committee on Home Charges, 26 October, 1888, in
     "East India Home Charges," PP 327 (13 July 1893), 29–35.

154 "East India Home Charges," PP 327, 35.

155 Indian Public Works Department despatch no. 24, 11 March 1890, as quoted in a
     recapitulation of events in PP Cd. 2055 (1904), 62.

156 Report to the undersecretary of state for India, 27 June 1895, reprinted in PP Cd.
     2055 (1904), 3–7.

157 Report to the undersecretary of state for India, 27 June 1895, 5.

158 Report to the undersecretary of state for India, 27 June 1895, 7.

159 Cameron Taylor, *General Sir Alex Taylor*, vol. 2, chap. 25.

160 *CHM* 8, no. 24, March 1912, 371–4.

161 Professor Minchin was not so charitable in his view of "superannuated generals"
     as presidents – see chapter 8.

4. Student Life

1 This statement refers specifically to the Royal Indian Engineering College; more
   recent generations have been educated on the Coopers Hill estate while it housed
   Shoreditch College of Education (1951–80) and the Runnymede Campus of
   Brunel University (1980–2007).

2 *Oracle* 5, no. 41, October 1878, 490.

3 The *Oracle* was followed in the mid-1880s by the short-lived *Charta*, of which no copies survive. The *Oracle*'s motto was a quote in Greek from Euripides, "*mantis d'aristos hostis eikazei kalôs*," roughly translated as "The best prophet is he who conjectures well." This may have referred to the manner in which the Classical Scholar approached his engineering examinations (see chapter 3). The *Coopers Hill Magazine* adopted a Latin quotation from Horace: *Et semel emissum volat irrevocabile verbum* (Once spoken, a word cannot be recalled).

4 *Oracle* 6, no. 50, November 1879, 603.

5 *CHM* 3, no. 1, October 1900.

6 *CHM* 1, no. 3, November 1897, 41.

7 Obituary of Robert "Bob" "Snippy" Egerton, *CHM* 12, no. 29, June 1934, 463.

8 Pennycuick's letter was printed in the October 1899 issue of the *Coopers Hill Magazine* 2, no. 14, 195–6. That the publication in the magazine was at Pennycuick's instigation, and that the editor later regretted doing so, was in Arthur Hick's evidence to the Board of Visitors (PP Cd. 539 [1901], 82–3).

9 For example, the obituary for the college pony in 1899 suggests that even the "docile temper" of that "noble quadruped" had been sorely tested by the students' fancy-dress antics (*CHM* 1, no. 10, February 1899, 138).

10 Steuart C. Bayley, "College Literature," in *Memorials of Old Haileybury College*, ed. Frederick Charles Danvers, M. Monier-Williams, Steuart Colvin Bayley, Percy Wigram, the late Brand Sapte, and many contributors, 257–305 (London: Archibald Constable, 1894). The *Observer* was turgid reading, and after going through its issues, Bayley commented, "I cannot say that the literary charm has been throughout such as to repay my labour."

11 F.G. Guggisberg, *The Shop* (London: Cassell, 1900), 179.

12 Hudson's names appear in this order on his birth record, the 1871 census, the probate of his father, and his own probate. On his Indian Telegraph Service record, his name is misspelled as "Ernest James Bonnel Hudson." He labels his diary in handwriting simply "E. Hudson," although an inked stamp "E.B. Hudson" was used occasionally. As late as 1905, Hudson noted in his diary, "Names as given to India Office: James Bonnell Ernest Hudson; Usual signature: Ernest Hudson."

13 Hudson celebrated his twenty-first birthday on this date in 1887.

14 Thomas Hudson (1836–1922) saw action in the Second Opium War in the Canton River. He was promoted to captain on his retirement in 1873 (Thomas Keith Hudson Royal Navy service records, National Archives ADM 196/70/400 and ADM 196/14/119).

15 John Hudson served during the Indian uprisings of 1857–8 and was commander of the Quetta division of the Indian Amy at the time Ernest left Coopers Hill. He was killed by a fall from his horse at Poona on 19 June 1893 ("Sir John Hudson," *Times*, 10 June 1893, 12).

16 Tottenham, Montague Hill, (later Sir) Ottway Cuffe, John Ridout, J.C.D. Couper, W.E. Bagot.

17 Cameron Taylor, *General Sir Alex Taylor*, 2:282.

18 Cameron, *Short History*, 16.

19 Ian F.W. Beckett, *Riflemen Form: A Study of the Rifle Volunteer Movement 1859–1908* (Aldershot, UK: Ogilvy Trusts, 1982).

20 Sydenham of Combe, *My Working Life*, 15.

21 "Thomas Henry Eagles," *Minutes of Proceedings of the Institution of Civil Engineers* 110 (1892): 392–3. "He was buried with military honours at Englefield Green, being followed to the grave by the Students' Volunteer Company, by the President and Staff of the college, and by many friends."

22 Beckett, *Riflemen Form*, 104. The volunteers became part of the Territorial Force in 1908.

23 RIEC calendar 1877–8, 12.

24 A large photograph in the British Library (MSS EUR F239/142) shows the massed student volunteer corps in formation during the "farewell inspection and parade by Sir Alex Taylor." Photographs of the volunteers building trestle bridges and observation towers are preserved in the photograph albums of Frederick Canning in the archive of the Centre of South Asian Studies (CSAS), University of Cambridge.

25 The *Oracle* could be scathing in its criticisms of individuals, particularly members of the college's sports teams. For example, in commenting on the performance of each of the eight individuals in the 1877 rowing boat, it described the faults of Edmund Elliot (the engineer murdered by Robert Hawkins) as the stroke rower: "Stroke does not get, we think, as much work out of his crew as he might do; we do not mean for one moment that he spares himself in the least, but without much increased exertions we fancy greater distances might be covered in lesser times" (*Oracle* 4, no. 30, June 1877).

26 *Oracle* 5, no. 42, November 1878, 501–2.

27 *Oracle* 6, no. 49, October 1879, 590.

28 *Oracle* 5, no. 45, March 1879, 548–9.

29 Referring to signatures on notes of permission to leave the college.

30 Mufti – the wearing of civilian clothes by someone who normally wears a uniform.

31 "'O' Company, 1st V.B. Royal Berks Regiment," *CHM* 2, no. 13, July 1899, 186.

32 "'O' Company, 1st V.B. Royal Berks Regiment," 186.

33 *CHM* 1, no. 8, July 1898, 107.

34 Ralph Strachey (1868–1923) was second son of Lieut.-General Sir Richard Strachey, member of the Board of Visitors for Coopers Hill. He spent his entire career from 1889 until 1919 as an engineer for the East India Railway, becoming its chief engineer in 1913, responsible for 2,700 miles of track. Ian J. Kerr, "Strachey, Ralph [1868–1923]," 23 September 2004, *Oxford Dictionary of National Biography* (Oxford University Press, https://www.oxforddnb.com).

35 *With the Cooper's Hill Volunteer Company at Aldershot by One of the Awkward Squad* (Beccles: Caxton, 1890[?]), BL 011913205.

36 Referring to Thomas Eagles.

37 Cameron, *Short History*, 27.

38 *CHM* 4, no. 3, October 1902, 54.

39 *CHM* 7, no. 5, March 1906, 66.

40 E.H.H. Collen, "The Volunteer Force of India: Its Present and Future," United Service Institution of India prize essay, 1883.

41 This large number of units proved difficult to coordinate when they were called into service during the First World War, and the Volunteer Force was replaced first by the Indian Defence Force in 1917 and then by the Auxiliary Force (India) in 1920.

42 "Indian Volunteer Corps List Corrected to 31 December 1912," in *The Quarterly Army List* (October–December 1913), 2:1904–17.

43 Edwin Collen, "The Volunteer Forces of India," *Empire Review* 5, no. 27 (April 1903): 221–32.

44 In chapter 1 of *Among the Railway Folk* Rudyard Kipling briefly described the East India Railway Volunteers drilling on Tuesdays and Fridays in their unlovely uniforms and wondered when the government would fully value their contribution.

45 *The Bengal-Nagpur Railway: Its History and Development, Parts I and II, Reprinted from the January and February Issues of "The Indian Railway Gazette,"* ca. 1910, 12, DeGolyer Library, Southern Methodist University, http://digitalcollections.smu.edu/cdm/ref/collection/eaa/id/1511.

46 S. Martin Leake, "Early Recollections of the B-N R (continued)," *Bengal-Nagpur Railway Magazine*, August 1929, 30–4 (Hertfordshire County Archives Acc. 599, under DE/MI Family Papers of the Martin Leake family).

47 Ernest Hudson diary, 7 March 1890.

48 "Coopers Hill Engineering College," *Times*, 28 July 1898, 8.

49 PP Cd. 2056 (1904), 12.

50 Sir Hugh Keeling wrote to the magazine in July 1944 (*CHM* 13, no. 16, October 1944, 427) that the Bells of Ouseley had been flattened by a "flying bomb ... which will be regretted by every Coopers Hill man."

51 *CHM* 2, no. 14, October 1899, 203.

52 A punt paddled like a Canadian canoe by a team of six.

53 Hilary Lushington Holman Holman-Hunt (1879–1949) spent his career in Burma Public Works, becoming chief engineer and secretary to the Government of Burma (*Times*, 14 September 1949, 7). In 1911, Holman-Hunt's polo pony was struck by lightning on the Rangoon racecourse (*CHM* 8, no. 21, August 1911, 335).

54 Corridors of student rooms were named after prominent figures in British India, mostly viceroys. The original corridors in 1871 were Canning, Clive, Cornwallis, Dalhousie, and Hastings. Lawrence and Mayo were added later.

55 Gay, "Science and Opportunity," 450n45 and 453n95.

56 Several of these brief notes indicating the students' interest in being considered for a telegraph appointment can be found in the BL IOR L/PWD/8/105.

57 Hudson and Henderson were not alone in doing this. A.J. Cradock, A.G. Romilly, and W.J. Howley also appeared in the first-year exam results for both years. This also explains why, in his second year at the college, Hudson was elected as first-year representative on the Boat Club Committee (diary 4 February 1887).

58 While questioning Prof. Stocker, Sir William Preece, the pre-eminent British telegraph engineer of the day, opined, "There have been some very great duffers sent out." PP Cd. 539 (1901), 34.

59 In 1888, thirty-one students completed the Cooper Hill course, but only fourteen were appointed to India, including the Telegraph Department (PP Cd. 2055 [1904], 41).

60 These events, or others like them, were probably what Taylor referred to as "marked and disagreeable occurrences" in his speech at the 1887 annual Prize Day (*Indian Forester* 13 [1887]: 479–88).

61 See, for example, Eric Partridge and Jacqueline Simpson, *A Dictionary of Historical Slang* (London: Penguin Books, 1972).

62 McLeod diary, 3 and 16 October 1871.

63 Based on Rollo Appleyard's copy of the College Rules dating from about 1885 (BL IOR MSS EUR F239/106).

64 George Chesney, "College Regulations: Memorandum," 21 July 1873 (BL IOR L/PWD/8/7, 415–16).

65 Lodge was appointed assistant professor of mathematics in 1884 to support Wolstenholme. He had been educated at Oxford. He was "an enlightening teacher" and became a master at Charterhouse when Coopers Hill was closed. He was particularly noted for his "computing" – the calculations used in the compilation of mathematical reference tables – primarily for his industry rather than his ability: "If he did little to improve the technique of computing, he laboured incessantly in the practice" (*Nature* 938, no. 3561 [29 January 1938]: 191).

66 Hudson diary, 12 and 14 November 1887. A smoking concert was an event with live music, and the all-male audience would discuss politics and the events of the day over alcohol and tobacco.

67 Hudson diary, 24–8 November 1887.

68 Hudson diary, 4 May 1888.

69 Hudson had been elected Boat Club captain in September 1887 and notes in his diary on several occasions meeting with Pomph about buying new boats and "squared off boat club matters with Pomph" (12 May 1888).

70 In 1885, the House Committee consisted of the captains of the Rugby Fifteen, Boats, Cricket Eleven, and Lawn Tennis Club, as well as two of the college scholars (1885 College Rules, BL IOR MSS EUR F239/106). By 1900, the committee also included the captain of the Association Football Eleven and any other

students nominated by the president (Standing Orders, effective 25 September 1900, 9 [BL IOR L/PWD/8/9]).

71 The 1891 census shows that Lodge was living with his young wife on St. Jude's Road, Englefield Green.

72 1885 College Rules.

73 The leading article in the *Oracle* 5, no. 38, May 1878, discusses the proper distribution of fine money between the various clubs, arguing that clubs that "really represent the College" – rugby, cricket, rowing – should be supported over the lawn tennis and music clubs.

74 *CHM* 2, no. 15, November 1899, 216.

75 James Pycroft, *Oxford Memories: A Retrospect after Fifty Years* (London: Richard Bentley & Son, 1886), 1:36.

76 Cameron, *Short History*, 14.

77 *CHM* 9, no. 3, August 1913, 36–7.

78 Hudson diary, 23 January 1889. Horace Turner (1867–1930), Hudson wrote, "won almost everything" at the 1888 Sports Day. Turner received a direct commission in the Royal Engineers and joined the Survey of India (*CHM* 12, no. 22, March 1931, 350).

79 Graham Dennis, *Englefield Green in Pictures* (Englefield Green Village Association, 1994), 6.

80 For example, the *Coopers Hill Magazine* of 1897 commented on the excellence of that year's Bal Masqué program's caricatures, and that the editors heard "that some of the originals are desirous of disowning such dastardly designed drawings and deem them decidedly deserving of destruction" (*CHM* 1, no. 3, November 1897, 38).

81 Photographs of a later incarnation of the signpost in 1909 and 1912 are shown in Graham Dennis and Richard Williams, *The Englefield Green Picture Book* (Egham-by-Runnymede Historical Society, 2000), 3; and Dennis, *Englefield Green in Pictures*, 6.

82 "Royal Indian Engineering College," *Times*, 27 July 1888, 4.

83 William Patrick Henderson was born in Livorno, Italy, but educated in England. He served his whole career in the Telegraph service, and when he retired in 1915, he was director of the Telegraphic Service in Assam. He retired to Florence, where he became heavily involved in astronomy, accurately calculating the orbits of comets and asteroids (UK Probate Office records; *History of Services of the Officers of the Indian Telegraph Department corrected to 1 July 1907* (Calcutta: Government of India Department of Commerce and Industry, 1907), 22–3; G. Abetti, *Memorie della Società Astronomia Italiana* 20 [1949]: 345–6, in Italian).

84 Two years later the ship ran aground off Plymouth and was wrecked.

85 See 1881 and 1891 UK census data, and "Medical News," *Medical Times and Gazette*, 28 April 1877, 461.

86 PP Cd. 490 (1901), 15.

87 "Arsenical Tissues," *British Medical Journal* 1 (23 June 1888): 1347–8

88 James C. Whorton, *The Arsenic Century* (Oxford: Oxford University Press, 2010), chap. 8, "Walls of Death."

89 Herbert McLeod diary, 19 and 28 November 1887.

90 "Arsenical Tissues," 1348.

91 "Arsenical Tissues," 1348.

92 Peter F. Bladin, "Julius Althaus (1833–1900): Neurologist and Cultural Polymath; Founder of Maida Vale Hospital," *Journal of Clinical Neuroscience* 15 (2008): 495–501.

93 Julius Althaus, *On Sclerosis of the Spinal Cord* (New York: G.P. Putnam & Sons, 1885), 360–1.

94 "Football: The Killed and Wounded," *British Medical Journal* 1, no. 1214 (5 April 1884), 689. Lewis's plaque, now relocated to the wall of Egham Parish Church, records his date of death as 31 March 1884.

95 *Oracle* 4, no. 36, March 1878, 440; and 5, no. 37, April 1878, 449.

96 *CHM* 8, no. 1, December 1906, 1.

97 *Oracle* 5, no. 37, April 1878, 449.

98 *Oracle* 5, no. 45, March 1879, 552.

99 *CHM* 8, no. 1, December 1906, 1 and 9.

100 *CHM* 1, no. 3, November 1897, 37.

101 This was probably one of Callcott Reilly's scale models.

102 "The Ball," *Oracle* 4, no. 35, February 1878, 422–3.

103 *CHM* 1, no. 8, October 1898, 107.

104 *Oracle* 4, no. 33, November 1877, 401–2.

105 "The Bal Masque," *CHM* 1, no. 9, December 1898, 124–5. All three of these young men died relatively early. Norman Ramsay joined the Royal Artillery and was killed in action in France on 4 November 1915. Arthur S. Appleby died of enteric fever (typhoid) at Mussoorie in 1902, aged twenty-three. Herbert C. Walker, a forester, died at Maymyo also of typhoid in 1925.

106 Elizabeth Knowles, ed., *Oxford Dictionary of Quotations*, 7th ed. (Oxford: Oxford University Press, 2009).

107 *Oracle* 4, no. 36, March 1878, 438.

108 *CHM* 6, no. 5, March 1905, 66–7, quoting *Sporting Life* magazine.

109 *Oracle* 6, no. 51, December 1879, 613 and 622. The correspondent wrote that "billiards is the root of all loafing," referring to an emerging habit of some students loafing in the billiard room rather than participating in sports, or worse still, being indifferent to the college's success in sports.

110 An ulster is an overcoat with a cloak and sleeves. Smoking caps were used to prevent the gentleman's hair from smelling of tobacco smoke.

111 "Vestitus" (Latin for "clothed"), letter to the editor, *Oracle* 6, no. 50, November 1879, 611.

112 *CHM* 6, no. 5, March 1905, 66.

113 *CHM* 6, no. 5, March 1905, 67.

114 Oracle 5, no. 38, May 1878, 453–4.

115 Minchin to George Francis Fitzgerald, n.d., probably from December 1879 by its context (Archive of Royal Dublin Society, reference 10/41).

116 Oliver Lodge, "George Minchin Minchin (1845–1914)," *Proceedings of the Royal Society London*, ser. A, 92, no. 645 (2 October 1916): xlvi–l.

117 This version of the story is from *CHM* 6, no. 5, March 1905, 67. The cup was first awarded for the England-Scotland competition on 10 March 1879.

118 *CHM* 2, no. 12, June 1899, 170. Datchet is a community about seven kilometres north of Englefield Green.

119 *CHM* 7, no. 5, March 1906, 71.

120 *Oracle* 6, no. 46, April; and May 1879, 553.

121 "Cooperzillian," letter to the editor, *CHM* 7, no. 6, May 1906, 87.

122 There was considerable concern in the pages of the magazine over the fate of the college buildings and estate after the closure (see chapter 9).

123 Charles William Hodson (1851–1910) felt that one of the scholarships he had been awarded should have gone to another student. He therefore remained seated when the Duke of Argyll called his name at Prize Day (*CHM* 8, no. 16, May 1910, 246–7). Hodson served as engineer-in-charge of the Mushkaf-Bolan Railway in the 1890s when Ernest Hudson was constructing the telegraph lines there, so the two interacted frequently (see chapter 5). In November 1894, Hudson complained that he did not go to the viceroy's levee in Quetta because his invitation had mistakenly been sent to Hodson.

124 *CHM* 8, no. 4, July 1907, 47.

125 *CHM* 8, no. 4, July 1907, 49.

126 A 500-foot pontoon bridge across the Kelani River constructed of wooden boats lashed together that supported a wooden deck.

127 John Bertram Mais Ridout (1866–1944).

128 *Overland Ceylon Observer*, 24 July 1890, 743.

129 Hudson diary, 31 December 1888.

130 Newcombe, *Village, Town and Jungle Life*, 1.

131 Newcombe, *Village, Town and Jungle Life*, 10–11.

5. An Officer of the Indian Telegraph Department

1 For example, Hudson's trip to Kan, Burma, November 1890.

2 In November 1890, for example, Hudson's spelling of the Myanmar city evolved from Pokoko, to Pokoku, to Pakokku.

3 William O'Shaughnessy, "Memoranda Relative to Experiments on the Communication of Telegraphic Signals by Induced Electricity," *Journal of the Asiatic Society of Bengal* 8, no. 93 (1839): 714–31.

4 Charles G. Adley, *The Story of the Telegraph in India* (London: E. & F.N. Spon, 1866), 4.

5 Chaudhury, *Telegraphic Imperialism*, 23.

6 Chaudhury, *Telegraphic Imperialism*, 40. Chaudhury also argues that the notion of the telegraph saving India is a myth rather than a reality. A fascinating compilation of telegrams sent during the latter part of the rebellion was published in S.N. Sinha, ed., *Mutiny Telegrams* ([Lucknow]: Department of Cultural Affairs, U.P., 1988).

7 *Imperial Gazetteer of India*, 3:445.

8 "Particulars Regarding the Indian Telegraph Department," RIEC calendar 1882–3, 17–18.

9 This was a point of contention for early Coopers Hill engineers, as discussed in chapter 9.

10 These were termed "substantive positions," the base positions held by each man. On top of that, officers could be "temporary" or "officiating" at a more senior level.

11 David Gilmour, *British in India*, 293. On the occasion when a colleague announced his engagement, Eustace Kenyon commented that the colleague's "superiors will not like his getting married" (Kenyon to his mother, 12 October 1883, Kenyon papers, CSAS, University of Cambridge).

12 In 1914, Sir Ganendra Roy calculated that he had been transferred twenty-two times in his nineteen years with the Telegraph Department, which he thought was only slightly above average. *Amrita Bazar Patrika*, 21 January 1914, 9.

13 *History of Services of the Officers of the Indian Telegraph Department, 1907*.

14 Kenyon to his mother, 13 January 1898, CSAS.

15 Kenyon to his mother, 13 January and 19 May 1898, CSAS.

16 In his article on the history of the Indian telegraph, P.V. Luke wrote about Henry Archibald Mallock (1834–1923): "To his talents and energy, combined with a thorough practical knowledge of Indian requirements, the Department is indebted for the high class of posts, brackets, wire, and insulators that are now generally in use." P.V. Luke, "Early History of the Telegraph in India," *Journal of the Institution of Electrical Engineers* 20 (1891): 102–22.

17 An Indian clerk who spoke English. An indispensable guide to the meanings of Anglo-Indian words is Henry Yule and A.C. Burnell, *Hobson-Jobson* (London: John Murray, 1886).

18 The exchange rate Hudson received in 1892 was Rs16/9 to £1. Murray's *Handbook for Travellers in India* states that the rate in 1911 was fixed at £1 = Rs15. Hudson's monthly starting salary was Rs300.

19 This is consistent with remarks made in 1899 by Charles Pitman, director general of telegraphs, over curricular reform at Coopers Hill. PP Cd. 539 (1901), 133.

20 PP Cd. 2056 (1904), 76.

21 Hudson had to prepare numerous other kinds of reports – on making over a subdivision or division to another officer, weekly progress reports during construction, a completion report when construction was finished, annual confidential reports of his own service and that of his European and Indian subordinates, monthly reports on interruptions, an annual "administration" report, financial reports, and notifications when going on or returning from leave. Eustace Kenyon also remarked on the number of reports he needed to prepare: "Awfully busy ... sending in accounts completion, reports of works, returns of stores, progress reports of other works, etc. some the result of the end of a month, some of a quarter and some of the official year." Kenyon letter to his mother, 13 April 1882, CSAS.

22 Sydney Hutchinson, then officiating as temporary superintendent in the Bombay Office. Hutchinson was appointed director general of telegraphs in 1903 and was knighted at Karachi during the tour of India by the Prince and Princess of Wales in 1906.

23 Henry Edward Thompson, superintendent in charge of Nagpur Division, who arrived in India in the year of the Indian Rebellion (1857) and retired in 1892.

24 Hudson diary, 6 December 1889.

25 RIEC calendar for 1884–85, 22–5.

26 Mrs. M.A. Handley, *Roughing It in Southern India* (London: Edward Arnold, 1911), 7.

27 Hudson diary, 18 December 1889.

28 Newcombe described the process: "Four coolies, or trollymen, run behind, pushing it along the line. Two sit on the trolly, while two push. After a few minutes of pushing, if on the level or a decline, they jump on the trolly, and it may run half a mile or more before they have to get off again to give it another push to keep up the pace." Newcombe, *Village, Town and Jungle Life,* 345.

29 In his diary for 15 March 1893, Hudson specifically mentions learning about where to shoot snipe and other game from the line book. Other entries in the diary about accommodations and other details of the day's journey are clearly prepared for use elsewhere (for example his trip to Fort Sandeman in May 1895).

30 The objective was good relations, but challenges to British authority were severely punished, typically by fines but in some instances by burning villages (although that was generally considered to be counterproductive). Detailed accounts were collected of the retaliations against villages "causing trouble"; see, for example, J.H. Parsons, *The History of the Third Burmese War; 1890–91 Period VI. The Winter Campaign of 1890–91* (Simla: Government Central Printing Office, 1893).

31 Parsons, *History of the Third Burmese War,* 28.

32 A covered litter carried by two or four men.

33 Hudson diary, 26 December 1890.

34 Under the nom de plume "Tar Wallah," Indian slang for a telegram messenger, Hudson wrote up his reminiscences of his time in Burma for the *Coopers Hill*

Magazine, apparently prompted by news in the *Times* of "offensive patrols" along the road from Pakokku to Gangaw. *CHM* 13, no. 15, May 1944, 388–91.

35 Hudson diary, 20 March 1891.

36 W.O. Horne, *Work and Sport in the Old I.C.S.* (Edinburgh: Blackwood, 1928), 81.

37 *Imperial Gazetteer of India*, 3:436.

38 A Bengali word for an enclosure constructed for trapping elephants.

39 Ronald Ross, *Memoirs, with a Full Account of the Great Malaria Problem and Its Solution* (London: Murray, 1923), 95.

40 G.H.F.N, "Sir Ronald Ross (1857–1932)," *Obituary Notices of Fellows of the Royal Society* 1, no. 2 (December 1933): 108–15.

41 Hudson diary, 12 October 1896. On 4 April 1896 the director general had offered Hudson the Rajputana Division, but Hudson requested to stay at Quetta and let the next man on the seniority list take the division. His opinion of Quetta was in marked contrast to his relief at departing from Akyab on 20 November 1898, about which he wrote, "Up anchor 10:30 and in a couple of hours saw the last of Akyab."

42 *A Handbook for Travellers in India and Ceylon* (London: John Murray, 1892), 229.

43 James Ramsay, "The Mushkaf-Bolan Railway, Baluchistan, India," *Minutes of Proceedings of the Institution of Civil Engineers* 128 (1897): 232–56.

44 On the train he met the prominent palaeontologist Carl Griesbach travelling to Mud Gorge to collect ammonites.

45 Hudson diary, 5 February 1894.

46 Construction of the railway was also progressing from the north, with the railhead located just south of Mach at this time.

47 Hudson diary, 12 February 1894.

48 Hudson diary, 16 February 1894.

49 Hudson diary, 18 February 1894.

50 T. Gracey, *Administration Report of the Railways in India for 1896–97* (London: Her Majesty's Stationery Office, 1897), 20.

51 Hudson diary, 30 September 1895.

52 Hudson diary, 8 August 1895.

53 Hudson diary, 10 and 11 August 1895.

54 Measured from the railway junction at Rahuki.

55 Hudson field notes, 19 September 1895.

56 Hudson field notes, 2 August 1895.

57 Hudson field notes, 6 August, 6 September, and 12 September 1895.

58 Hudson field notes, 20 August and 31 August 1895.

59 Hudson field notes, 28 July 1895. Value of a maund: *Imperial Gazetteer of India*, 3:vii.

60 Hudson field notes, 5 August 1895.

61 Hudson field notes 13 and 28 August 1895.

62 Hudson field notes, 26 August 1895.

63 Hudson diary, 2 September 1895.

64 Hudson field notes, 16–24 September 1895.

65 Hudson diary, 4 October 1895.

66 Hudson diary, 12 October 1895; field notes for 11 October state 1600, which was presumably an estimate.

67 In the Hudson Papers, CSAS. Having failed the entrance examinations for the army, Herbert had had difficulty settling on a career, so Ernest and their father had been exploring the idea that Herbert might run a coffee estate in India. Herbert had therefore come to India to visit a coffee plantation to see if the life might suit him.

68 Reflecting the order of names on Hudson's official file with the India Office.

69 *Administration Report of the Baluchistan Agency for 1895–96* (Calcutta: Office of the Superintendent of Government Printing India, 1896), 155.

70 Hudson diary, 6 February 1896.

71 Hudson received a letter from Elfie Alington on 21 August 1897 in which she thanked him for the wrench.

72 Hudson diary, 6 October 1901.

73 "Marriage of Miss S.E.E. Alington and Mr. J.B.E. Hudson," *Lincoln, Rutland and Stamford Mercury*, 15 August 1902, 6.

74 For descriptions of these and other hill stations, see Bhatt, *Resorts of the Raj.*

75 Hudson diary, 6 March 1904.

76 Hudson diary, 13 September 1907.

77 Hudson diary, 20 January 1905.

78 Hudson diary, 16 April 1904.

79 Hudson diary, 7 March 1905.

80 Hudson diary, 30 January 1892. Hudson "concocted a reply" to the director general acknowledging his mistake but explaining that he had visited the Coimbatore office not to inspect it (therefore that he had not missed the faulty battery) but to instruct the postmaster on new procedures for record-keeping. He "promised to be a good boy in future."

81 Hudson diary, 28 December 1901.

82 Hudson field notes, 17–18 January 1905. Campbell had built 16 miles in one month, whereas Hudson expected 1½ miles per day.

83 Hudson diary, 22 February 1908.

84 Hudson diary, 19 and 20 August 1904. Hudson notes that on that morning he had to report himself to the local hospital for plague inspection as part of the government's efforts to combat a significant outbreak at the time.

85 Hudson field notes, 12 and 14 March 1905.

86 Hudson diary, 23 November 1904: "The nuts [at Tiptur] are of exceptional size + flavour and are shipped not only up north but to England and the continent."

87 Hudson diary, 17 December 1908.

88 Hudson diary, 3 January 1908.

89 *Imperial Gazetteer of India*, 3:438 and 445.

90 Nearly 6.5 million of the 7.3 million telegrams sent in 1903–4 were by the public. *Imperial Gazetteer of India*, 3:443.

91 *Report of the Telegraph Committee 1906–07* (Calcutta: Superintendent of Government Printing, India, 1907), i.

92 *Report of the Telegraph Committee*, 123–6. The large-scale changes recommended in the report were not implemented until 1910.

93 Just a few years later in 1913, the abolition of the subdivisions was criticized in evidence to the Royal Commission on the Public Services in India because it removed a valuable training ground for young telegraph officers.

94 *Summary of the Administration of the Earl of Minto, Viceroy and Governor-General of India in the Department of Commerce and Industry, November 1905 to July 1910* (Simla: Government Central Branch Press, 1910), 73. Britain and North America had employed female telegraphists since the 1870s – see Thomas C. Jepsen, *My Sisters Telegraphic: Women in the Telegraph Office, 1846–1950* (Athens: Ohio University Press, 2000).

95 Also see Jepsen, *My Sisters Telegraphic*, 31.

96 For a detailed account of the Telegraph General Strike of 1908, see Choudhury, *Telegraphic Imperialism*, chap. 7.

97 Kenyon to Tizzie, 8 July 1909, CSAS.

98 "Mr. John Newlands: From Delivery Boy to Telegraph Chief," *Times*, 3 March 1937, 16. A clipping of this obituary is enclosed with Hudson's 1908 diary.

99 "Improvement in the Telegraph Department. Mr. Newlands' Lecture," *Tribune* (Lahore), 3 November 1907, 3.

100 "Meeting of the Telegraph Unions. Detailed Proceedings," *Tribune* (Lahore), 11 February 1908, 4.

101 "The Indian Telegraph Operators. Messages Wilfully Delayed," *Times*, 4 April 1908, 7.

102 Hudson diary, 1 February 1908.

103 Hudson diary, 2 April 1908.

104 Hudson diary, 4 April 1908.

105 Hudson diary, 6 April 1908.

106 Hudson diary, 7 April 1908.

107 "Telegraph Crisis. Drastic Measures. Extraordinary Circular. Alarming Developments," *Amrita Bazar Patrika*, 10 April 1908, 6.

108 D.K.L. Chaudhury, "First Virtual Community and the Telegraph General Strike of 1908," in "Uncovering Labour in Information Revolutions, 1750–2000," supplement 11, *International Review of Social History* 48 (2003): 45–71.

109 Hudson diary, 13 April 1908. See also "The Indian Telegraphists' Strike," *Times*, 13 April 1908, 9.

110 This according to a wire Hudson received from Berrington on 16 April.

111 "The Indian Telegraph Strike Ended," *Times*, 21 April 1908, 3.

112 Hudson diary, 26–31 December 1908.

113 "The New Indian Service," *Spectator* 43, 10 December 1870, 1472–3.

114 "Edmund du Cane Smithe (1853–1908)," *Minutes of Proceedings of the Institution of Civil Engineers* 178 (1909): 369.

115 "Frederick Benbow Hebbert (1854–1905)," *Minutes of Proceedings of the Institution of Civil Engineers* 166 (1906): 387–8.

116 PP Cd. 2056 (1904), 52 (du Cane Smithe), and 72 (Hebbert). Coincidentally, both men died within a few years of their testimony at this enquiry.

117 See college calendars and *Indian Civil Engineering College, Coopers Hill. Syllabus of the Course of Study*" (London: Wm. H. Allen, 1871), 39–41. This is contrary to the statements of some authors that the accounting curriculum commenced in 1882. John Black, "A More Vigorous Prosecution of Public Works: The Reforms of the Indian Public Works Department," *Accounting History* 6, no. 2 (2001): 91–119; John Black and John Richard Edwards, "Accounting Careers Traversing the Separate Spheres of Business and Government in Victorian Britain," *Accounting History* 21, no. 2–3 (2016): 306–28.

118 PP Cd. 539 (1901), 18. Hurst was paid five guineas per lecture.

119 PP Cd. 539 (1901), 21.

120 Hudson diary, 30 October 1892.

121 Hudson diary, 26 March 1891.

122 Hudson diary, 15 February 1894.

123 These techniques allowed the simultaneous transmission of two and four messages, respectively, along a single telegraph wire.

124 Duffey was appointed on 11 December 1878 to teach students on the mechanical aspects of telegraph lines, as well as to maintain the instruments. Soon afterwards he suffered a spinal injury while attaching an insulator to a tree when the branch he was standing on broke, causing him to fall twenty-five feet. McLeod later recorded that Dr. Giffard was treating Duffey for a broken leg after his earlier injury led him to tumble down the steep bank at Egham Wick in January 1887. Duffey resigned from his post in 1896. (Letters from June to August 1896 by President Taylor, Duffey, McLeod and Giffard concerning Duffey's resignation and pension are in BL IOR L/PWD/8/105.)

125 In 1904, Reynolds said that they were satisfied at first with the Coopers Hill education but then "the thing became slack." PP Cd. 2056 (1904), 76.

126 PP Cd. 539 (1901), 131–2.

127 Two works of particular relevance to the British-operated lines were W.H. Preece and J. Sivewright, *Telegraphy*, which went through numerous editions between 1876 and 1905 (London: Longmans, Green); and John Christie Douglas, *Manual of Telegraph Construction* (London: Griffin, 1877). Both books were used at Coopers Hill, but Pitman and Reynolds advocated replacing Douglas by the Indian Telegraph Department Code. Williams's *Manual of Telegraphy* was

commissioned specifically to be a practical guide to operating and testing Indian telegraph lines. W. Williams, *Manual of Telegraphy* (London: Longmans, Green, 1885). These books are available online for the reader interested in telegraph technology. Sir William Preece (1834–1913) was a pioneer of British telegraphy, working with the railways and the Post Office, and replaced Carl W. Siemens (who also made significant contributions to telegraphy) on the Board of Visitors for the college upon the latter's death.

128 From the detailed requirements for telegraphy students in the 1902–3 RIEC Calendar.
129 See chapter 4 about reasons for Hudson's choice of career. See also Gilmour, *British in India*, 383; and Kincaid, *British Social Life in India*, 215.
130 PP Cd. 2056 (1904), 78 and 81.
131 Hudson diary, 28 February 1889.
132 Hudson diary, 25 February 1890.
133 Kenyon to his mother, 11 April 1898.
134 For example, Hudson diary, 1 September 1908.
135 *Times*, 14 January 1952, 1.

6. Jungle-Wallahs

1 Robert Greenhalgh Albion, *Forests and Sea Power: The Problem of the Royal Navy 1652–1862* (Cambridge, MA: Harvard University Press, 1926), chap. 9, "Searching the World for Timber."
2 William Milburn, *Oriental Commerce* (London: Black, Parry, 1813), 1:328.
3 E.P. Stebbing, *The Forests of India* (London: John Lane, The Bodley Head, 1922), 1:37.
4 G.D. Holmes, "History of Forestry and Forest Management," *Philosophical Transactions of the Royal Society of London*. Series B, *Biological Sciences* 271, no. 911 (10 July 1975): 69–80; Albion, *Forests and Sea Power*, chap. 3.
5 Stebbing, *Forests of India*, 1:107. For an account of Cleghorn's career, see Das, "Hugh Cleghorn and Forest Conservancy in India." The *kumri* system of agriculture had been practised sustainably for generations, but the value it placed on the clearance differed from the largely economic perspective of the Europeans.
6 H. Cleghorn, "Second Annual Report, 1858–59," in *The Forests and Gardens of South India* (London: W.H. Allen, 1861), 32–49. For an in-depth modern analysis of this issue, refer to Pallavi V. Das, *Colonialism, Development, and the Environment: Railways and Deforestation in British India, 1860–1884* (New York: Palgrave Macmillan, 2015).
7 Troup, *Work of the Forest Department in India*, 2–3. See also "Dietrich Brandis, the Founder of Forestry in India," *Indian Forester* 10, no. 8 (August 1884): 343–57.
8 "Sir Dietrich Brandis, 1824–1907," *Indian Forester* 33 (December 1907): 569–74.
9 "Sir Dietrich Brandis," *Indian Forester*. B. Ribbentrop, *Forestry in British India* (Calcutta: Office of the Superintendent of Government Printing, India, 1900), 228.

10 A system of harvesting where a ring was first cut all around the trunk to kill the tree, which was left standing for up to two years before felling.

11 Quoted in Stebbing, *Forests of India*, 2:9.

12 B.H. Baden-Powell and J. Sykes Gamble, eds., *Report of the Proceedings of the Forest Conference 1873–74* (Calcutta: Office of the Superintendent of Government Printing, 1874); Ramachandra Guha, "Forestry in British and Post-British India: A Historical Analysis," *Economic and Political Weekly* 18, no. 45/46 (5–12 November 1983): 1940–7; Guha, "An Early Environmental Debate: The Making of the 1878 Forest Act," *Indian Economic and Social History Review* 27, no. 1 (1990): 65–84.

13 Unsurprisingly many protected forests were in fact converted to reserved forests. In 1889–90 there were 56,000 square miles of reserved forests and 30,000 square miles of protected forests; by 1900 these numbers had changed to 81,4000 and 8,800, respectively (Ribbentrop, *Forestry in British India,* 111).

14 Ribbentrop, *Forestry in British India,* 110–11.

15 See, for example, Gadgil and Guha, *This Fissured Land.*

16 There is some question whether the Forest Service continued to promote conservationism. Authors such as Gopa Joshi ("Forests and Forest Policy in India," *Social Scientist* 11, no. 1 [January 1983]: 43–52) and Ramachandra Guha and Madhav Gadgil ("State Forestry and Social Conflict in British India," *Past & Present* 123 [May 1989]: 141–77) argue that commercialism became the dominant force in Indian forestry. This view is countered by authors such as Gregory Barton and Brett Bennett ("Environmental Conservation and Deforestation in British India 1855–1947," *Itinerario* 32, no. 2 [2008]: 83–104).

17 Government of India circular no. 22-F, reprinted in *Punjab Record* 30 (Lahore: Civil and Military Gazette, 1896): 40–4. It is striking how many of these topics are still represented nearly a century later in India's 1988 National Forest Policy (reprinted in *Social Change* 33, no. 2 and 3 [2003]: 192–203).

18 S. Abdul Thana, "Forest Policy and Ecological Change," in Kumar, Damodaran, and D'Souza, *Environmental Encounters in South Asia*, 262–80.

19 Anthony Wimbush, "Life in the Indian Forest Service," unpublished, May 1964, Archives of the Centre of South Asian Studies, University of Cambridge, UK, 54.

20 Rudyard Kipling, "In the Rukh," in Kipling, *Many Inventions* (New York: Appleton, 1893), 222–64.

21 Best, *Forest Life in India*, 74.

22 Best, *Forest Life in India*, 81.

23 Gregory Barton and Brett Bennett, "'There Is a Pleasure in the Pathless Woods': The Culture of Forestry in British India," *British Scholar* 3, no. 2 (September 2010): 219–34.

24 Kevin Hannam, "Utilitarianism and the Identity of the Indian Forest Service," *Environment and History* 6, no. 2 (May 2000): 205–28.

25 Best, *Forest Life in India.*

26 E.P. Stebbing, *The Diary of a Sportsman Naturalist in India* (London: John Lane, the Bodley Head, 1920).

27 Wimbush, *Life in the Indian Forest Service*. Wimbush was in the last group of students admitted to the Forest course at Coopers Hill in 1904. He spent one year there before transferring to Oriel College, Oxford.

28 From the Osmaston Papers, Archives of the Centre of South Asian Studies, University of Cambridge, UK.

29 Lionel Sherbrooke Osmaston (1870–1969), Bertram Beresford Osmaston (1867–1961), and Arthur Edward Osmaston (1885–1972).

30 B.B. Osmaston, Henry Osmaston, and B.E. Smythies, *Wild Life and Adventures in Indian Forests* (Ulverston, UK: Henry Osmaston, 1999).

31 RIEC calendar for 1901–2, 213.

32 Henry Osmaston, "The Osmaston Family, Foresters & Imperial Servants," *Commonwealth Forestry Review* 68, no. 1 (March 1989): 77–87.

33 E.P. Stebbing, *Forests of India*, 2:54.

34 *The India Office List for 1916* (London: Harrison & Sons, 1916), 520.

35 Osmaston Papers, CSAS.

36 A deputy revenue officer.

37 A Marathi word meaning "village accountant," usually for a group of villages.

38 Osmaston diary, 7 April 1892.

39 Allen Thornton Shuttleworth directed Bombay's famine relief operations after the short monsoon of 1896. As deputy conservator, Shuttleworth did much to reform and organize the Bombay Forest Department. However, Stebbing criticized Shuttleworth's "unfortunate attitude" concerning grazing rights and free grant of wood to villagers, particularly important during the famine, which led to more land being taken under the Forest Act (Stebbing, *Forests of India*, 3:32). Shuttleworth and his wife are mentioned frequently by L.S. Osmaston in both personal and professional contexts.

40 Best, *Forest Life in India*, 68.

41 Osmaston, *Wildlife and Adventures in Indian Forests*, 2.

42 Osmaston diary, 29 November 1896.

43 Osmaston had himself returned from leave in England only the day before and resumed charge of his division the next day.

44 Wimbush recalled of his time at the college in 1904–5, "Learning to ride had been compulsory at Coopers Hill and the forestry students used to go to Windsor for instruction at the Barracks of the Royal Horse Guards 'The Blues.'" Wimbush, *Life in the Indian Forest Service*, 22.

45 Osmaston diary, 7 December 1896.

46 Best, *Forest Life in India*, 182.

47 Osmaston diary, 5 December 1896.

48 *The India Office List for 1928* (London: Harrison & Sons, 1928), 651. His brother Charles Hodgson retired at the rank of conservator in 1923.

49 "Forest Administration Report of the Northern Circle, Bombay Presidency, for the Official Year 1891–92," 67, in *Administration Reports of the Forest Department in the Bombay Presidency Including Sind for the Year 1891–92* (Bombay: Government Central Press, 1893).

50 W.E. D'Arcy, *Preparation of Forest Working-Plans in India* (Calcutta: Office of the Superintendent of Government Printing, India, 1895), 6.

51 Ribbentrop, *Forestry in British India*, 122, gives the total forest area, while page 138 gives the areas with working plans and in preparation (2,484, and 3,514 square miles, respectively).

52 "Forest Administration Report on the Central Circle, Bombay Presidency, for the Forest Year 1893–94," 77, in *Administration Reports of the Forest Department in the Bombay Presidency Including Sind for the Year 1893–94* (Bombay: Government Central Press, 1895).

53 "Forest Administration Report of the Central Circle, Bombay Presidency, for the Forest Year 1894–95," 2, in *Administration Reports of the Forest Department in the Bombay Presidency including Sind for the Year 1894–95* (Bombay: Government Central Press, 1896).

54 Best, *Forest Life in India*, 108.

55 "Forest Administration Report of the Central Circle, Bombay Presidency, for the Forest Year 1893–94," 14.

56 "Forest Administration Report of the Central Circle, Bombay Presidency, for the Forest Year 1894–95," 37.

57 Born 8 August 1867 at Goodnestone, Kent, died 14 June 1949. Lionel and Selina shared a grandfather, Francis Wright, who later changed his name to Osmaston.

58 Osmaston diary, 12 November 1896.

59 Osmaston diary, 4 March 1897.

60 Selina's description of camp, Osmaston diary, 9 January 1909.

61 Lionel Osmaston describing the view from the top of Yeola Hill, 28 December 1893.

62 Osmaston diary, 18 and 20 February, and 3 and 4 March 1896.

63 Best, *Forest Life in India*, 67.

64 Osmaston diary, 12 January, 23 February, and 7 April 1910.

65 Best, *Forest Life in India*, 75.

66 Osmaston, *Wildlife and Adventures in Indian Forests*, 83.

67 Hugo Wood (1870–1933) joined the Service in 1893 and made conservator in 1919 (*India Office List for 1928*, 871).

68 Wimbush, *Life in the Indian Forest Service*, 129.

69 Diana Currie's recollections, from Mary MacDonald Ledzion's excellent little book, *Forest Families*, containing her family's and other tales of life in the Forest Service between the Wars (Putney: British Association for Cemeteries in South Asia, 1991), 41.

70   Wilson Thomas, "Forest Department to Pay Homage to Hugo Wood," Hindu, 12 December 2017, https://www.thehindu.com/todays-paper/tp-national/tp-tamil-nadu/forest-department-to-pay-homage-to-hugo-wood/article21457571.ece.

71   Osmaston diary, 9 March 1897.

72   Wimbush, *Life in the Indian Forest Service*, 112.

73   Best, *Forest Life in India*, 88.

74   Best, *Forest Life in India*, 94–5.

75   Osmaston diary, 3 March 1902.

76   Charles Gilbert Rogers, *A Manual of Forest Engineering for India*, 3 vols. (Calcutta: Office of the Superintendent of Government Printing, India, 1900, 1901, and 1902). Rogers "was a good man in every sense of the word" (*CHM* 13, no. 3, May 1938, 59–60).

77   Lionel also mentions spending an afternoon in Ahmednagar reading Shuttleworth's administration report and making corrections, 8 January 1894.

78   Osmaston diary, 23 April 1892.

79   Best, *Forest Life in India*, 78.

80   Best, *Forest Life in India*, 77.

81   Stebbing, *Diary of Sportsman Naturalist*, 24–5.

82   Osmaston diary, 27 February 1892.

83   Here, Osmaston mixed terms from the Muslim ("mussulman") and Hindu ("brahmin") faiths. After several years in India, he would have known the distinction, so he may have used the latter word as a general term for a leader.

84   Osmaston diary, 20 April 1896.

85   Osmaston diary, 24 March 1892.

86   Osmaston diary, 31 May 1892.

87   "Forest Administration Report of the Central Circle, Bombay Presidency, for the Forest Year 1896–97," 2, in *Administration Reports of the Forest Department in the Bombay Presidency including Sind for the Year 1896–97* (Bombay: Government Central Press, 1898).

88   Mike Davis, *Late Victorian Holocausts: El Niño Famines and the Making of the Third World* (London: Verso, 2002), 19.

89   In 1900, Lionel and Selina recorded paying their servants an extra allowance on account of the famine.

90   Frederick Canning, "Uttar Pradesh 1903–1937," in *100 Years of Indian Forestry* (Dehra Dun: Forest Research Institute, 1961), 1:37–40.

91   Osmaston diary, 14 January 1897.

92   Osmaston diary, 21 April 1897.

93   Osmaston diary, 1 April 1897.

94   Extracted from *Administration Reports of Forest Department in the Bombay Presidency 1896–97*.

95   Official letters and testimonials, Osmaston Papers, Cambridge CSAS.

96   Best, *Forest Life in India*, 255.

97 Osmaston diary, 3 January 1899.

98 Best, *Forest Life in India,* 109.

99 See, for example, Alfred Newcombe quoted in chapter 7 and Robert Egerton in chapter 8.

100 John Peake Wildeblood to his mother, 4 February 1910 from "Camp, Almora Dist," Wildeblood Papers, CSAS, University of Cambridge, 13–16.

101 Fowl.

102 E.A. Kenyon to his mother, 12 March 1884 (CSAS).

103 Selina's experience was not unique; marriages straight off the boat were officially sanctioned by exemption from normal residence requirements by Act of Parliament. Olivia Hamilton, who married forester Arthur Hamilton in 1925, had a lengthy journey to reach her new husband's primitive bungalow, with his equally basic eating habits. Charles Allen ed., *Plain Tales from the Raj* (London: Futura, 1976), chap. 6.

104 The page is headed "Bombay," but that is in Lionel's writing.

105 Their formal honeymoon holiday was at the hill station of Mahabaleshwar in mid-March 1898.

106 Osmaston diary, 4 April 1894.

107 Best, *Forest Life in India,* 109–10.

108 Newcombe, *Village, Town and Jungle Life in India,* 70.

109 Osmaston diary, 11 May 1895.

110 Best, *Forest Life in India,* 236.

111 Handley, *Roughing It in Southern India,* 1.

112 For three excellent books on the roles of women in the British Raj, refer to Margaret MacMillan, *Women of the Raj* (New York: Thames & Hudson, 1988); Pat Barr, *The Memsahibs: The Women of Victorian India* (London: Secker & Warburg, 1976); Mary A. Procida, *Married to the Empire* (Manchester: Manchester University Press, 2014).

113 Maud Diver, *The Englishwoman in India* (London: William Blackwood & Sons, 1909), 21.

114 She was not very unusual in this – see Mary A. Procida, "Good Sports and Right Sorts: Guns, Gender, and Imperialism in British India," *Journal of British Studies* 40, no. 4 (October 2001): 454–88.

115 MacMillan, *Women of the Raj,* 11.

116 Lawrence, *Indian Embers,* 29–30.

117 Sara Jeannette Duncan, *Simple Adventures of a Memsahib* (New York: D. Appleton, 1893), 6. Duncan (1861–1922) was a Canadian-born journalist and author who married a civil servant in India.

118 Osmaston diary, 7 March 1897.

119 Flora Annie Steel and Grace Gardiner, *Complete Indian Housekeeper & Cook,* 7th ed. (London: William Heinemann, 1909), chap. 12.

120 Referring to Indian representatives of the divisional revenue office. The *chuprassi* was the official message carrier. "Such an official makes the Civil Officer a terror to the villagers, who look upon his camp as a flight of locusts" (Steel and Gardiner, *Complete Indian Housekeeper*, 158).

121 Steel and Gardiner, *Complete Indian Housekeeper & Cook*, 151.

122 Steel and Gardiner, *Complete Indian Housekeeper & Cook*, 152.

123 MacMillan, *Women of the Raj*, 15.

124 Procida, *Married to the Empire*, 60.

125 Chota Mem, *The English Bride in India*, 2nd ed. (Madras: Higginbotham, 1909).

126 Chota Mem, *English Bride in India*, 55.

127 Wimbush, *Life in the Indian Forest Service*, 20.

128 Diver, *Englishwoman in India*, 34.

129 Not only was the *ayah* often the only female servant, but she usually also spoke good English, understood British ways, and had authority over other servants (Procida, *Married to the Empire*, 99–100). For a view from the children's perspective, see Allen, *Plain Tales from the Raj*, chap. 1.

130 Chota Mem, *English Bride in India*, 61–2.

131 Steel and Gardiner, *Complete Indian Housekeeper & Cook*, 166.

132 Osmaston diary, 17 August 1898.

133 And two months after giving birth.

134 Osmaston diary, 10 April 1899.

135 Steel and Gardiner, *Complete Indian Housekeeper & Cook*, 166.

136 Diver, *Englishwoman in India*, 44.

137 The passion of the ardent sportsman of the time can be appreciated only by reading the original works. Fortunately, many of them are now available in scanned form on the internet.

138 "My camp tomorrow is in a place where a tiger roams about so I may get a change again there. You see there are plenty of tigers to spare." Forester R.S. Pearson quoted in E.W. March and A.E. Osmaston, "Coopers Hill," *Commonwealth Forestry Review* 50, no. 3 (September 1971): 243–6.

139 E.P. Stebbing, *Jungle By-Ways in India* (London: John Lane, The Bodley Head, 1911), xiv–xv.

140 Stebbing, *Jungle By-Ways*, xix.

141 Osmaston, *Wildlife and Adventures in Indian Forests*, plate 2.

142 Wimbush, *Life in the Indian Forest Service*, 117–18.

143 Bertram Osmaston made similar comments about the poor state of the teeth of another man-eating tiger, suggesting that it had turned to eating people when it found difficulty catching its normal prey. That tiger's stomach contained twelve human fingernails and toenails (Osmaston, *Wildlife and Adventures in Indian Forests*, 214).

144 Quotes from *Pioneer*, 17 May 1889, 1; and Osmaston, *Wildlife and Adventures in Indian Forests*, 4–6.

145 Stebbing, *Diary of a Sportsman Naturalist*, 285.

146 Osmaston diary, 5 December 1902.

147 L.S. Osmaston, "A Man-Eating Panther," *Journal of the Bombay Natural History Society* 15 (1903): 135–8. Lionel sent a less polished but more immediate account to his mother that same day. L.S. Osmaston to his mother, 5 December 1902 (Osmaston papers, CSAS).

148 Stebbing, *Diary of a Sportsman Naturalist*, 11.

149 Stebbing, *Jungle By-Ways*, viii.

150 Stebbing, *Diary of a Sportsman Naturalist*, 241.

151 Stebbing, *Diary of a Sportsman Naturalist*, 264.

152 Stebbing, *Diary of a Sportsman Naturalist*, 288–9.

153 Originally from P. Chalmers Mitchell, "Zoological Gardens and the Preservation of Fauna," *Science* n.s. 36, no. 925 (20 September 1912): 353–65.

154 PP Cd. 2056 (1904), 6.

155 In 1889 Blandford replaced Mr. A. Shipley from Cambridge University, who had been teaching entomology since 1887.

156 Based on W.R. Fisher, "The Forestry Branch at Coopers Hill," *Indian Forester* 31 (December 1905): 679–86.

157 PP Cd. 2055, appendix IV, 40. Beginning in 1890, students not destined for India were admitted to the college for the entire forest course, or just some specific subjects, and included students from South Africa, Mauritius, Spain, and Malaysia (Fisher, "Forestry Branch," 683).

158 Based on "The Forestry School at Cooper's Hill," *Nature* 37 (1888): 529–31. The same article was published in *Indian Forester* 14 (1888): 277–82.

159 Ribbentrop, *Forestry in British India*, 229.

160 Wimbush, *Life in the Indian Forest Service*, 10.

161 Wimbush, *Life in the Indian Forest Service*, 8–15.

162 B.H. Baden-Powell, *Forest Law* (London: Bradbury, Agnew, 1893). Badon Henry Baden-Powell was the half-brother of Robert Badon-Powell, founder of the scouting movement and author of *Pig-Sticking or Hoghunting*, "the premier sport of India" (London: Harrison & Sons, 1889). Recall that Selina Osmaston tried her hand at this sport.

163 Best, *Forest Life in India*, 68.

164 William Peter Sangster diary 16 December 1895 (BL IOR MSS EUR F245).

165 W.R. Fisher, "Practical Forest Training Grounds for the Coopers Hill College," *Indian Forester* 18 (1892): 12–18. Fisher was defending the Coopers Hill curriculum against accusations that it provided practical training only in the third year. The general issue of the appropriate level of practical training for forestry students occupied the pages of the *Indian Forester* on several occasions.

166 Fisher, "Forestry Branch," 680.

167 William Schlich, "Forestry Education," *Indian Forester* 24 (1898): 228–44. There had been a lectureship in forestry at the University of Edinburgh since 1887, but

it was not until after the closure of Coopers Hill that a degree was offered in the subject ("The Forestry Department of Edinburgh University," *Nature* 106 [1921]: 706–7). Edward Stebbing was appointed chair of Forestry at Edinburgh in 1910.

168 This section is based on a paper read in 1885 by director of the Forest School, F. Bailey, to the Aberdeen meeting of the British Association ("The Indian Forest School," *Transactions of the Scottish Arboricultural Society* 11 pt. 2 [1886]: 155–61). Bailey later became lecturer in forestry at the University of Edinburgh.

169 *Indian Forester* 32 (1906): 315–17.

170 *CHM* 13, no. 15, May 1944, 391.

171 Wimbush, *Life in the Indian Forest Service*, 53–4.

172 Stebbing, *Forests of India*, 3:436.

173 Gadgil and Guha, *This Fissured Land*, chap. 4.

174 In *Seeing Like a State: How Certain Schemes to Improve the Human Condition Have Failed* (New Haven, CT: Yale University Press, 2005), James C. Scott uses scientific forestry as an example of the "constriction of vision" necessary to control the complex natural environments.

175 See, for example, Barton and Bennett, "Environmental Conservation and Deforestation."

176 Benjamin Weil, "Conservation, Exploitation, and Cultural Change in the Indian Forest Service, 1875–1927," *Environmental History* 11, no. 2 (April 2006): 319–43.

177 Data and quote from Stebbing, *Forests of India*, 3:618.

178 Hanqin Tian, Kamaljit Banger, Tao Bo, and Vinay K. Dadhwal, "History of Land Use in India during 1880–2010: Large-scale Land Transformations Reconstructed from Satellite Data and Historical Archives," *Global and Planetary Change* 121 (2014): 78–88.

179 The India Office List for 1928 includes service records of no fewer than four Osmastons: Lionel's brothers Bertram and Arthur, Lionel's son Fitzwalter Camplyon, and Bertram's son Bertram Hutchinson Osmaston.

180 In January 1894 Lionel notes that he had just finished reading through the Bible from cover to cover, having started in 1891 at Thana. The following day, he indicated his intention to start again from the beginning.

181 "Lionel S. Osmaston," *Commonwealth Forestry Review* 48, no. 2 (1969): 97–8.

7. Engineers in India

1 "Engineers in India," *Fraser's Magazine* 18, no. 107 n.s. (November 1878): 559–65.

2 Hogarth Todd was mauled by a tigress during one of his hunting trips, causing an injury that rendered him an invalid in later years, during which time he wrote his books. Obituary, *CHM* 13, no. 9, May 1941, 225–6.

3 George Chesney, "Note on Civil Engineering College," 10 June 1870 (BL IOR L/PWD/8/7, 81). This dilemma was also recognized by *Engineering*, which wrote, "In such a college as Coopers Hill is intended to be, it is by no means easy to draw the

line between the subjects which should be pursued, and those which may be neglected. It is obvious, however, that time would not permit, neither would it be desirable, even if practical, to cram too many subjects into the heads of the students, lest the result should be that they obtained a smattering of many, and a real proficiency in few, if any." "Education at Coopers Hill," *Engineering* 12 (25 August 1871): 126–7.

4 "Memoranda by Colonel Ottley on the Educational Course," PP Cd. 490 (1901), 22–34.

5 Initially, a student achieving this minimum mark qualified for public service. Later, the limited number of available public service positions was allocated on a competitive basis, so the minimum mark became the requirement for the diploma, independent of whether a student was posted to India.

6 Students could also "Pass," meaning that they qualified at a ranking below Third Class, or be "Not Placed," meaning that they did not qualify.

7 Stated in Thomas Hearson's letter to the *Engineer*, 15 March 1901, 274. Earlier, the qualifications were titled Diploma of Associate of Coopers Hill (for first-class students) and the ordinary Diploma of Graduate (for second-class and telegraph students). For example, RIEC calendar 1880–1, 10.

8 "Memoranda by Colonel Ottley," 30.

9 These included future Coopers Hill professor Joseph Wolstenholme. Public Works Department Reference Paper, 10 June 1872, in BL IOR L/PWD/8/7, 147.

10 Wimbush, *Life in the Indian Forest Service*, 16. John Benton (see later this chapter) was the only man awarded a Fellowship of Coopers Hill in 1873.

11 "Testing Laboratory, Cooper's Hill Engineering College," *Engineer* 58 (25 July 1884): 70.

12 Evidence of A.W. Brightmore, PP Cd. 2056 (1904), 27.

13 PP Cd. 490 (1901), 24–5.

14 The allocation of marks was not necessarily consistent with the instructional time spent on a subject. In 1900, 39.3 per cent of the total instructional hours were devoted to Branch I, 19.2 per cent to Branch II, and 16.9 per cent to sciences. The total times of 58.5 per cent for engineering, broadly defined, and 75.4 per cent for engineering + science are both lower than the mark weighting would suggest. Instructional hours from PP Cd. 490 (1901), 36.

15 "Memoranda by Colonel Ottley," 23.

16 "Coopers Hill College," *Engineering* 12 (1 December 1871): 358.

17 Evidence of R.W Egerton, PP Cd. 2056 (1904), 44.

18 Evidence of F.B. Hebbert, PP Cd. 2056 (1904), 66. This labour resource meant that mechanization was much less common on engineering works in India than in Britain. See also Medley, *India and Indian Engineering*, 61–4.

19 According to the evidence of R.W. Egerton, PP Cd. 2056 (1904), 44.

20 For example, "Appendix 1," PP Cd. 539 (1901), 87.

21 Obituary in *CHM* 12, no. 9, March 1926, 137–8, echoing comments written in the magazine upon his enforced retirement in 1901.

22 For example, "Mid-Term Examination in Descriptive Engineering: Materials, 27 October 1871," in BL IOR L/PWD/8/7, 206. Also "Examination in Materials, January 1872," in BL IOR L/PWD/8/7, 251.

23 "Syllabus of Courses of Study for Engineer Students: Epitome of Work," RIEC Calendar for 1901–2, 63–7.

24 To pack down the earth to make it watertight and reduce settling.

25 W. Hogarth Todd, *Work Sport and Play* (London: Heath Cranton, 1928), chap. 5, "The Small Tank."

26 Todd, *Work Sport and Play*, chap. 17, "The Big Tank."

27 Todd, *Work Sport and Play*, 177–9.

28 Evidence given in PP Cd. 2056 (1904), 11–12.

29 From the 1891 and 1901 UK census returns for the residents of Coopers Hill College.

30 These figures are from the RIEC Calendar for 1880–1, 11.

31 Information on whether a PWD appointee was going directly to India or taking the practical course (and their employer) was listed in the RIEC calendar and sometimes in the college magazine (e.g., *Oracle* 6, no. 49, October 1879, 596–7).

32 Appendix VII of PP Cd. 2055 (1904), 43.

33 *East India (Home Charges)*, PP 327, 13 July 1893, 34.

34 This calculation accounts for the fact that the fiscal and academic years do not align; for example, the costs associated with students starting the one-year practical course in September 1880 come from both the 1880–1 and 1881–2 fiscal years.

35 Chesney, "Public Works Letter No. 168," 518–22. In reality, the student salaries plus prizes alone amounted to nearly £200, without including the pupillage fee (around £80) or the salary of the practical course superintendent.

36 "Government of India Public Works Department Resolution No. 2352G," Simla, 1 September 1897, printed in *Indian Engineering*, 18 September 1897, 186.

37 "Some Disappointed Ones," *CHM* 1, no. 8, October 1898, 112.

38 *CHM* 1, no. 9, December 1898, 113–14.

39 These restrictions were specified in the Resolution 2352G; the RIEC calendar for 1901–2 just said "a few Assistant Engineers" would stay in England and that the special training would not normally exceed one year.

40 This is based on the short descriptions of the PWD structure provided in the RIEC calendars.

41 *Government of India Public Works Department Code* (Calcutta: Office of the Superintendent of Government Printing, India, 1892), 1:90.

42 Each rank of engineer was further subdivided. The chief engineer and superintending engineer ranks each contained three classes, while the executive engineer and assistant engineer ranks each contained four (later reduced to three) and three grades, respectively.

43 In the 1890s this was changed to third grade because of a revision to the PWD grades.

44 1892 PWD Code, 99.
45 1892 PWD Code, 100.
46 Unlike navigation canals with their relatively still water maintained by locks, irrigation canals needed controlled flow so the water would reach the fields. The effective gradient of the canal could be controlled by the use of weirs or, in this case, rapids. The idea of the rapids was to eliminate the scouring due to water action under the weir by achieving the same fall in height using a gradual slope rather than a vertical drop. The reduced maintenance achieved by using rapids came at the expense of using more stone in their initial construction.
47 From the 1913 diary of William Peter Sangster, BL IOR MSS:EUR F245.
48 Newcombe, *Village, Town, and Jungle Life*, 122.
49 Newcombe, *Village, Town, and Jungle Life*, 80.
50 "High Court, N.-W.P.," *Pioneer*, 15 November 1893, 4–5.
51 Based on the accounts in the *Pioneer*, 15 November 1893, 4–5, and 16 November 1893, 1, 4, and 5.
52 *Pioneer*, 19 November 1893, 5.
53 Benton, Brodie, and Kennedy were the students Herbert McLeod described in his diary as being "Scotch and a little rough" when he met them on 7 August 1871.
54 Information on Brokenshaw and Farquharson comes from the 1945 India Office and Burma Office List. Ahsan entered the government in the late 1930s. His entry in the Coopers Hill Society membership list identified him as "serving" until 1948, but this may be because he, like several others who had retired earlier, had not updated the society on his status. Brokenshaw was not a member of the Coopers Hill Society and is absent from its records.
55 Ramnath, *Birth of an Indian Profession*, 115–17.
56 "George Moyle M.Inst.C.E.," *Indian Engineering*, 14 February 1920, 85–7. In 1886 Moyle was appointed superintendent of the Coopers Hill practical course, despite Alexander Taylor's reservations about his youth. Moyle spent his career on the Indian railways, becoming engineer-in-chief of the Eastern Bengal State Railway, expanding Bell's techniques of bridge design to crossings previously thought unfeasible.
57 Quotes are from George Moyle, "The Platelaying of the Jacobabad or Broad Gauge Section of the Kandahar Railway," *Minutes of Proceedings of the Institution of Civil Engineers* 61 (1880): 286–94. Other information comes from the companion paper: James Richard Bell, "The First Section of the Kandahar Railway," *Minutes of Proceedings of the Institution of Civil Engineers* 61 (1880): 274–85.
58 The Indus was bridged at Attock and Sukkur in the 1880s.
59 W.H. Cole, *Notes on Permanent-Way Material, Platelaying and Points and Crossings* (London: E. & F.N. Spon, 1890), 28.
60 Moyle, "Platelaying of the Jacobabad or Broad Gauge Section," 289.
61 Kerr, *Building the Railways of the Raj*, contains a detailed examination of the sources, conditions, and unrest of the labourers employed on the constructions

of India's railways. See also Ian Derbyshire, "The Building of India's Railways," in *Technology and the Raj*, ed. Roy MacLeod and Deepak Kumar, 177–215 (New Delhi: Sage Publications, 1995).

62　RIEC Calendar for 1884–5, 68.

63　Sometimes spelled "Hurnai."

64　Sometimes spelled Peshin." For roughly contemporary articles on the railways to Quetta, see Scott-Moncrieff, "The Frontier Railways of India," *Professional Papers of the Corps of Royal Engineers* 11 (1885): 213–56, written before the completion of the Sind-Pishin line; E.W.C. Sandes, *The Military Engineer in India*, vol. 2 (Chatham: Institution of Royal Engineers, 1935), chap. 9 on Frontier Railways; and G.K. Scott-Moncrieff, "Sir James Browne and the Harnai Railway," *Blackwood's Magazine* 177 (May 1905), 608–21. In addition, P.S.A. Berridge gives a readable account of the Sind-Pishin and other lines in the region, with excellent photographs, in *Couplings to the Khyber* (Newton Abbot, UK: David & Charles, 1969).

65　Sandes, *Military Engineer*, 2:146.

66　Epitaph quoted in Sandes, *Military Engineer*, 2:143.

67　Moncrieff, "Sir James Browne," 618.

68　"Mr. E.I. Shadbolt" in two parts, *Indian Engineering*, 21 February 1925, 99–100 and 28 February 1925, 113–14. The editor of *Indian Engineering* clearly had an axe to grind because he took several opportunities in this sequence of profiles to criticize Browne, such as in the profiles of William Drew (26 March 1921, 169–70) and F.D. Fowler (15 August 1925, 85–7).

69　Scott-Moncrieff, "Frontier Railways of India," 241.

70　Berridge, *Couplings to the Khyber*, 143–5.

71　The breaches are summarized in Berridge, *Couplings to the Khyber*, 150–1.

72　Hudson diary, 29 March 1894.

73　"Major-General Sir James Browne, RE, KCSI, 1839–1896," *Minutes of Proceedings of the Institution of Civil Engineers* 125 (January 1896): 428–30.

74　Hudson diary, 13 June 1896.

75　Hudson diary, 26 January 1894 and 8 February 1894.

76　In his 1885 lectures, "Frontier Railways of India," Moncrieff complained that the press had drawn "invidious comparisons" between the speed of the civil engineers and that of the Royal Engineers. While stating justifiably that there was "no comparison between a permanent line, independent of floods and a temporary line in the bed of a river," he continued defensively that "all the difficult works on the line, without exception have been either partly or altogether under Royal Engineer officers" while "the only part that is under civilian engineers is the easiest part of all."

77　Before this Chaman was "a desolate sort of country of brown earth, with scrubby bushes resembling lavender." It was then irrigated using the stream of water that emerged from the Khojak tunnel, allowing the growth of fruit trees brought from

Kandahar by Richard Woods (CH 1875–8). Woods was one of those who laid out the railway line into Chaman. "Indian Frontier Reminiscences," *CHM* 13, no. 11, May 1942, 267–73. In his first tour of duty, Woods was responsible for a difficult section of the Sind-Pishin line, constructing the Mudgorge tunnel, and survey work for the Chaman Extension. He later joined the staff of Coopers Hill ("Major R.J. Woods, O.B.E., F.C.H.," *Indian Engineering*, 11 June 1921, 323–4).

78 Weightman completed his practical course on the railways in England and served on the Sind Pishin Railway construction. He went on to design the Nilgiri Mountain Railway and others. He retired in 1905 and served in several government ministries during the First World War. He died in 1942 aged eighty-three years (obituary *CHM* 13, no. 12, October 1942, 307).

79 Walter James Weightman, "The Khojak Rope-Inclines," *Minutes of Proceedings of the Institution of Civil Engineers* 112 (1893): 310–20.

80 Weightman, "Khojak Rope-Inclines," 311.

81 Hodson was appointed secretary of a committee of experts to report on the "mud gorge problem." The committee recommended works to stabilize the gorge, which Hodson implemented, as well as the establishment of an alternative route, which led to the Mushkaf-Bolan Railway.

82 C.J. Cole (CH 1876–9), W.A. Johns (CH 1877–80), T.E. Curry (CH 1872–5), E.A.C. Lister (CH 1889–92), H.R. Walton (CH 1888–91).

83 Ramsay, "Mushkaf-Bolan Railway."

84 This story, repeated and elaborated on the internet, seems to stem in recent times from Berridge's *Couplings to the Khyber*, 172. In a similar story, the Barog Tunnel on Herbert Harington's (CH 1872–5) Kalka-Simla line is supposed to be haunted by the spirit of Col. Barog, an engineer who, as a result of his tunnel headings not meeting, took his own life. Contemporary guides speak of the tunnel passing beneath the Barogh Hill or Ridge, with nary a mention of an incompetent engineer. Edward J. Buck, *Simla Past & Present* (Calcutta: Thacker, Spink, 1904), 19; and Murray's *Handbook for Travellers in India, Burma and Ceylon*, 1911, 218.

85 "Sir Francis Langford O'Callaghan, KCMG, CSI, CIE, 1839–1909," *Minutes of Proceedings of the Institution of Civil Engineers* 179 (1910): 364–5.

86 Thomas Elmitt Curry (CH 1872–5), "Some Personal Recollections," *CHM* 13, no. 15, May 1944, 397. Curry also officiated as instructor of surveying at Coopers Hill in 1901.

87 Hudson's course notes mentioned that the "standard work is Simm's Tunnelling": Frederick Walter Simms and D. Kinnear Clark, *Practical Tunnelling*, 4th ed. (London: Crosby Lockwood & Son, 1896), first published in 1844.

88 William Arthur Johns, "The Tunnels on the First Division of the Mushkaf-Bolan Railway, India," *Minutes of Proceedings of the Institution of Civil Engineers* 128 (1897): 257–64; Charles John Cole, "The Tunnels on the Second Division of the Mushkaf-Bolan Railway, India," *Minutes of Proceedings of the Institution of Civil Engineers* 128 (1897): 265–77.

89 Johns, "Tunnels on the First Division," 259.

90 Cole, "Tunnels on the Second Division," 265. Hudson walked through the tunnel in February 1894 and commented that the tunnel was not nearly finished, "there being only the heading to go through in places."

91 S. Martin-Leake, "Early Recollections of the B.-N. R.," *Bengal-Nagpur Railway Magazine*, July 1929, 51.

92 Martin-Leake, "Early Recollections of the B.-N. R.," 51.

93 J.G. Medley and A.M. Lang, *Roorkee Treatise on Civil Engineering in India* (Roorkee: Thomason College Press, 1873), 1:133.

94 Scott-Moncrieff, "Sir James Browne and the Harnai Railway," 621.

95 Hudson diary, 25 January 1894.

96 Charles Cole wrote a technical paper on catch sidings in 1901 that compared the "gravity" design, consisting of an uphill track to slow wagons, and the sanded design, where the wagon's wheels ran through sand packed around the rails for use where an uphill slope was not available. C.J. Cole, "Note on the Experiments on Gravity Catch-Sidings and Sanded Kopcke-Sidings," BL MSS EUR F239/101).

97 W.J. Weightman, "Two Brave Deeds," *CHM* 13, no. 8, October 1940, 201–2. The second brave deed was an Indian storekeeper and his assistant calmly carrying explosives out of a burning magazine.

98 Hudson diary, 29 June 1895.

99 This route was closed by Pakistan Railways in 2006 when several bridges were deliberately destroyed. Restoration efforts were announced in 2016 and 95% of the work had been completed as of December 2019. "Railway Completes 95% Rehabilitation Work of Sibbi-Khost Section," Radio Pakistan, 4 December 2019, https://en.baaghitv.com/railway-completes-95-percent-rehabilitation-work-of-sibbi-khost-section/.

100 P.S.A. Berridge, "Abstract. Withdrawing Two 155-Foot Spans from Bridges on an Indian Frontier Railway in Baluchistan," *Journal of the Institution of Civil Engineers* 29 (November 1947): 63–71. Also see Berridge, *Couplings to the Khyber*, chap. 13. The supports of the ruined bridge and the galleried approach tunnel are visible on satellite images at 30°20'16.08"N, 67°29'28.47"E.

101 "Tunnelling the Malakand," *Civil & Military Gazette*, 29 April 1911, reprinted in *CHM* 8, no. 22, October 1911, 342–4.

102 John Benton, "Irrigation works in India," *Journal of the Royal Society of Arts* 61 (13 June 1913): 717–54. It is interesting that Benton still included the designation "F.C.H." – Fellow of Coopers Hill – after his name on this article.

103 "Malakand Tunnel: Irrigation in Upper Swat," *Times of India*, 5 May 1911, 10.

104 Sangster diary, 27 June 1913.

105 Sangster diary, 25 September 1913.

106 Sangster diary, 4 October 1913.

107 CH 1878–81, the third of his family to attend the college.

108 "A Famous Irrigation Work: Construction of the Upper Swat Canal," *Times of India*, 21 January 1926, 15.

109 "Maintenance of Bridges in Indian Railways: A Review," in *Twenty-Third Report of the Standing Committee on Railways, 2018–19* (New Delhi: Lok Sabha Secretariat, December 2018), 2.

110 Stephen Martin-Leake, "The Rupnarayan Bridge, Bengal-Nagpur Railway," *Minutes of Proceedings of the Institution of Civil Engineers* 151 (1903): 251–78. The bridge is at 22°26'16.5"N 87°53'02.9"E.

111 R.R. Gales, "The Hardinge Bridge over the Lower Ganges at Sara," *Minutes of Proceedings of the Institution of Civil Engineers* 205 (1918): 18–67. The bridge is at 24°04'04.5"N 89°01'47.0"E.

112 Hudson diary, 1 March 1890.

113 See Ann Clayton, *Martin-Leake Double VC* (London: Leo Cooper, 1994). It was the second-youngest brother, Arthur (1874–1953), who became the most well-known of the siblings for being the first of only three men to be twice awarded the Victoria Cross for gallantry (South Africa in 1902 and Belgium in 1914).

114 "Royal Indian Engineering College," *Times*, 17 August 1880, 8.

115 Astbury's Register of Coopers Hill students indicates that Leake left Coopers Hill after two years, in 1882. However, he was on the list of third-year students in residence in the 1882–3 RIEC Calendar, 163. Moreover the announcement of his death in the November 1941 *Coopers Hill Magazine* gave his years as 1880–3.

116 *Bengal-Nagpur Railway*, 11. Wynne's comment was not entirely accurate. Leake was indeed unplaced in first-year exams and performed very poorly in second-year mathematics, natural science, and Hindustani and Indian History. However, he placed near the middle in his second-year engineering exams in 1882.

117 *Bengal-Nagpur Railway*, 11.

118 Roland Silversmith, "The Great Professor," *Oracle* 5, no. 43, December 1878. Silversmith was presumably a student, although his name is absent from the college records.

119 "Major R.J. Woods," *Indian Engineering*, 11 June 1921, 324.

120 Medley, *India and Indian Engineering*, 64.

121 Martin-Leake, "Rupnarayan Bridge," 253.

122 In the absence of solid ground, a sufficiently deep foundation could be supported solely by the friction between it and the surrounding soil.

123 Medley, *India and Indian Engineering*, 68. During the construction of the Mand (Maand) River bridge east of Kharsia (completed 5 May 1890), Hudson wrote about the diver (who was improbably "once a black-faced minstrel in America and a most amusing fellow") working underwater to place bags of concrete where the water swirling around the piers of the bridge had undermined the foundations by scouring out the riverbed. Hudson diary, 12 April 1890.

124　P.T. Cautley, "On the Use of Wells, Etc. in Foundations as Practised by the Natives of the Northern Doab," *Journal of the Asiatic Society of Bengal* 8 (1839): 327–40.

125　In his 1903 testimony, Frederick Hebbert was of the view that many of the older British-built Indian bridges had failed because their foundations were too shallow. PP Cd. 2056 (1904), 73.

126　MacGeorge, *Ways and Works in India*, 328–9.

127　Hudson diary, 5 January 1900.

128　S. Martin-Leake, "Thirty Years Ago – and Now (Part 1)," *Bengal Nagpur Railway Magazine*, November 1930, 15–19 (Hertfordshire County Archives). Martin-Leake noted that the new girders were made almost entirely from Indian steel, unlike his day when "every girder used, even down to six-foot spans, came from England" ("Thirty Years Ago – and Now (Part 2)," *Bengal Nagpur Railway Magazine*, December 1930, 17–22).

129　As a result of India's longstanding debate about track gauge, the two approaches to the ends of the Hardinge Bridge were of different gauges.

130　The project also included stone river embankments to ensure that flooding did not spill around the bridge.

131　Gales, "Hardinge Bridge," 28.

132　"Sir Robert Gales," *Indian Engineering*, 20 December 1919, 337–9.

133　Nevil Shute, *Slide Rule* (London: Pan, 1968), 63. This quote has also been attributed to the Duke of Wellington.

134　PP Cd. 2056 (1904), 50.

135　For example: *1892 Public Works Department Code*, 1:81 and 306.

136　Newcombe, *Village, Town and Jungle Life*, 175–6.

137　Headrick, *Tentacles of Progress*, chap. 3; Kerr, *Building the Railways of the Raj*, chap. 2. Sweeney has examined the complex world of Indian railway financing after 1875, suggesting that extravagant practices persisted in private construction. Stuart Sweeney, *Financing India's Imperial Railways, 1875–1914* (London: Pickering & Chatto, 2011).

138　Johns, "Tunnels on the First Division," 258, 260.

139　Cole, "Tunnels on the Second Division," 273.

140　Martin-Leake, "Rupnarayan Bridge," 276.

141　"The Lower Bari Doab Canal: the Ravi River Level Crossing," *CHM* 8, no. 20, March 1911, 311–15.

142　*Report of the Indian Irrigation Commission*, 24. In common with many reports concerning India, this was expressed in Indian units as "two crores of rupees," where a crore represents 10,000,000 (written 1,00,00,000). Another common unit is the lakh, which represents 100,000 (written 1,00,000). Therefore 1 crore = 100 lakhs.

143　*Imperial Gazetteer of India*, 3:351.

144　Whitcombe, "Environmental Costs of Irrigation."

145 Todd, *Work Sport and Play*, 24–5.

146 "Taming the Frontier," *Times of India*, 16 April 1914, 6.

147 This refers to the command of the canal, that is the area of land at a lower eleva-
     tion than the canal and that can therefore be irrigated.

148 Benton, *Irrigation works in India*, 719.

149 Sidney Preston in Ross et al., "Discussion," 50.

150 "Sir John Benton, K.C.I.E.," *Indian Engineering*, 5 August 1922, 81–3. John Ben-
     ton (1850–1927) was considered to be a brilliant canal engineer for the scope
     and imagination of his large-scale schemes. He saw service in India and Burma,
     retired as a chief engineer in 1905, but was brought back within months to serve
     as inspector-general of irrigation for India. His final retirement was in 1912, after
     he had been made Knight Commander of the Order of the Indian Empire.

151 John Benton, "The Punjab Triple Canal System," *Minutes of Proceedings of the
     Institution of Civil Engineers* 201 (1916), 24–47. The awarding of the medal was
     recorded in Benton's Indian Engineering profile and his 1928 ICE obituary. That
     it was a *gold* medal comes from the *CHM* 12, no. 15, December 1927, 231–3.

152 The names of the *doabs* (meaning two waters) were derived from those of the riv-
     ers that bordered them, so the Bari Doab is between the Beas and the Ravi Rivers,
     while the Rechna Doab is between the Ravi and the Chenab Rivers.

153 *Report of the Indian Irrigation Commission*, 117.

154 Benton, "Triple Canal," 25.

155 Robert Grieg Kennedy (CH 1871–3), one of McLeod's three rough Scots, devel-
     oped "hydraulic diagrams" for designing silt-free canals, formulas for determin-
     ing the water distributed to the fields, and other practically applicable "original
     scientific advancements." Kennedy retired to Vancouver Island, where he died in
     1920 (*CHM* 11, no. 9, July 1920, 134–5).

156 "The Lower Bari Doab Canal: The Ravi River Level Crossing," *CHM* 8, no. 20,
     March 1911, 314.

157 Benton, "Triple Canal," 45.

158 For a detailed exploration of the social consequences of the Triple Canal Scheme
     and other irrigation works, see David Gilmartin, "Models of the Hydraulic Envi-
     ronment," in *Nature, Culture, Imperialism*, ed. David Arnold and Ramachandra
     Guha, 210–36 (Oxford University Press, 1995). See also Gilmartin, "Scientific
     Empire and Imperial Science."

159 S.C. Sharma, *Punjab: The Crucial Decade* (New Delhi: Nirmal Publishers, 1987),
     chap. 3 "Irrigation."

160 Gilmartin, "Models of the Hydraulic Environment," 211.

161 For a broader perspective, see Scott, *Seeing Like a State*. Scott argued that
     state-sponsored disasters occur when a combination of four factors are present –
     an administrative simplification of nature, technical over-confidence, an author-
     itarian state, and a powerless civil society – all of which were arguably present in
     India under British rule. This issue was not limited to India: David Blackbourn,

The Conquest of Nature: Water, Landscape and the Making of Modern Germany (New York: W.W. Norton, 2006).

162 Michael Lewis, "The Personal Equation: Political Economy and Social Technology on India's Canals 1850–1930," *Modern Asian Studies* 41, no. 5 (2007), 967–94.

163 This view of the profession brings it towards the concept of *métis*, the incorporation of practical experience and skilled detailed knowledge into technical solutions, outlined by Scott in part 4 of *Seeing Like a State*, entitled "The Missing Link."

164 Cole, *Notes on Permanent-Way Material*, preface.

165 Cuddy, "Royal Indian Engineering College," 219.

166 Rollo Appleyard, "The Closing of Coopers Hill College," reprinted from the *Electrical Review* (17 March 1905). In BL IOR MSS EUR F239/106.

8. Crisis, Diversification, and Closure: 1896–1906

1 Obituary for John Pennycuick, *Times*, 10 March 1911, 11. At his death, Pennycuick was the same age as Sir Alex Taylor when he stepped down as president of Coopers Hill.

2 *University of Madras, The Calendar for 1892–93*, xvi–xvii.

3 John Pennycuick, "The Diversion of the Periyar," *Minutes of Proceedings of the Institution of Civil Engineers* 128 (1897), 140–63.

4 Sandes, *Military Engineer in India*, 2:29.

5 Diver, *Unsung*, 171–83.

6 "Pennycuick Memorial, a Temple for Us, Say Farmers," *Times of India*, 15 January 2013, https://timesofindia.indiatimes.com/city/madurai/Pennycuick-memorial-a-temple-for-us-say-farmers/articleshow/18041254.cms. In January 2019, a statue of Pennycuick donated by the Chennai police chief was unveiled in the churchyard of St. Peter's church in Frimley, UK, where he is buried.

7 Diver, *Unsung*, 179.

8 "Colonel Pennycuick and the Chair of Engineering at Coopers Hill," *Indian Engineering*, 16 October 1897, 249.

9 Pennycuick obituary, *Times*, and *CHM* 2, no. 14, October 1899, 195.

10 "The Deaths of Three Engineers," *Engineer*, 29 April 1927, 457.

11 Reprinted in *CHM* 2, no. 14, October 1899, 195–6.

12 Pennycuick accepted an advisory position in Queensland, Australia, to review a planned scheme for reducing flooding of the Brisbane River. The haste of his departure from Coopers Hill is apparent from the date of his report, 27 November 1899.

13 PP Cd. 539 (1901), 82.

14 This quote from the testimony of F.E. Matthews, PP Cd. 539 (1901), 83.

15 PP Cd. 539 (1901), 63–4.

16 The copy of this photograph in the British Library is labelled "1904" on the back, but the *Coopers Hill Magazine* (5, no. 2, November 1903, 20) mentions enclosing

a staff photograph and listed the names. Mr. Edgar is identified in the image from the label of Aitchison's copy of this photograph. The original handwritten label identified the medical officer as Dr. Brockett, although on the basis of information in the college calendar and the magazine it was probably in fact Dr. Giffard.

17 Using data from PP Cd. 2055 (1904).

18 PP Cd. 2055 (1904), 4. Fees at Coopers Hill were several times higher than those at other institutions, and in 1903 Ottley believed that the college would have had no trouble reaching capacity if the fees could have been reduced, appointments or not.

19 "Prize Day at Cooper's Hill," *Indian Forester* 14 (1888), 401–8.

20 PP Cd. 490 (1901), 33.

21 See, for example, Prof. George Minchin's evidence in PP Cd. 2056 (1904), 31.

22 "Sir John Ottley," *CHM* 12, no. 22, March 1931, 347–8.

23 A later article in the *Times* reported a statement by Lord George Hamilton, secretary of state for India: "To fill the vacancy he investigated the claims of all the candidates and found that far the ablest candidate was Colonel Ottley." "Coopers Hill," *Times*, 13 February 1901, 11.

24 "Colonel Sir John Ottley, K.C.I.E.," *Indian Engineering*, 23 July 1921, 43–4.

25 *Notes by the President, Colonel J.W. Ottley, C.I.E.*, 10 February 1900, in PP Cd. 490 (1901), 19–21.

26 "Instruction in Canal and Railway Construction," addendum to *Notes by the President, Colonel J.W. Ottley, C.I.E.*, 21.

27 Report by the Board of Visitors, 24 July 1900, in PP Cd. 490 (1901), 40–4.

28 Herbert McLeod wrote that his removal from the college "was the beginning of the end" in a letter to John Ottley, 5 November 1904 ("Disposal of Assets in Library, Chapel, Forest Museum, & Physical Laboratory Presented to the College," BL IOR L/PWD/8/316).

29 Sir Horace G. Walpole was assistant under-secretary of state for India and clerk of the Council of India.

30 PP Cd. 490 (1901), 48–9.

31 "Coopers Hill," *Times*, 3 January 1901, 6.

32 PP Cd. 539 (1901), 51–2.

33 "Coopers Hill," *Times*, 3 January 1901, 6, also reproduced in PP Cd. 490 (1901), 51.

34 McLeod's diary, 30 December 1900 to 1 January 1901.

35 An editorial in *Nature* at the end of January summarized clearly the first month of the controversy ("The Royal Indian Engineering College," *Nature* 63, no. 1630 [24 January 1901]: 303–4).

36 Excluding the Forest Department, there were sixteen instructors at Coopers Hill before the dismissals.

37 "The Royal Indian Engineering College," *Nature* 63, no. 1628 (1901): 256.

38 J.A. Ewing, letter to the editor, *Times*, 19 January 1901, 7.

39 "The Royal Indian Engineering College," *Nature* 63, no. 1628 (1901): 256.

40 "Three Students," letter to the editor, *Times*, 28 January 1901, 15.

41 "The Senior Students," letter to the editor, *Times*, 30 January 1901, 6.

42 "D.Sc.," letter to the editor, *Times*, 14 January 1901, 9.

43 "Coopers Hill," *Times*, 13 February 1901, 11. One of the signatories was a former member of the Board of Visitors, Sir Douglas Fox, who said that he had received no notice of the "drastic changes" and was "astonished" at the seven dismissals.

44 John Ottley to Edmund Neel, secretary to the Public Works Department, 1 March 1901 (BL IOR MSS EUR F123/45, 137).

45 Namely PP Cd. 490 (1901).

46 "Cooper's Hill College," *Engineer* 91 (8 March 1901), 244.

47 *Times*, 13 February 1901, 11.

48 Secretary of state for India to W.A. Anson, 11 March 1901, in PP Cd. 539 (1901), 2. Anson's summary of the deputation's concerns and Hamilton's response were published in the *Times*, 13 March 1901, 7.

49 A. Godley to George Hamilton, 5 March 1901 (BL IOR MSS EUR F123/45, 119–20).

50 PP Cd. 539 (1901), 59.

51 A. Godley to George Hamilton, 1 March 1901 (BL IOR MSS EUR F123/45, 121–4). Cuddy, "Royal Indian Engineering College," chap. 6 contains an in-depth description of the events surrounding the 1901 dismissals and deputation.

52 Secretary of state for India to W.A. Anson, 11 March 1901.

53 PP Cd. 539 (1901). A detailed summary appeared in "Coopers Hill," *Times*, 2 April 1901, 10.

54 John Ottley to Charles Crosthwaite, 4 April 1901 (L IOR MSS EUR F123/45, 151–2).

55 PP Cd. 539 (1901), 11.

56 Some students appear to have started a campaign asking their parents to lobby their members of Parliament in support of the enquiry (John Ottley to Lord Hamilton, 1 March 1901, BL IOR MSS EUR F123/45, 138).

57 John Ottley to Arthur Godley, 11 dated April 1901 (BL IOR MSS EUR F123/45, 150).

58 Robert Harrison to George Minchin, 7 May 1901 (Royal Society archives, reference NLB/22/392).

59 Many of these relate to Minchin being absent from class due to sickness and are contained in BL IOR L/PWD/8/388.

60 T.A. Hearson, letter to the editor, *Engineer* 91 (15 March 1901), 274.

61 Because higher diploma students were exempt, those passing the ICE examinations were the lower-placed students at Coopers Hill.

62 PP Cd. 539 (1901), 12.

63 PP Cd. 831 (1901), 7–18.

64 PP Cd. 831 (1901), 8.

65 PP Cd. 831 (1901), 15.

66 PP Cd. 831 (1901), 15.

67 George Minchin to George Fitzgerald, 21 April (archive of the Royal Dublin Society, reference 10/128). Although the year is not given, the letter refers to the president retiring "at the end of July," which would make it 1896. Minchin also expressed his concern that Taylor "will try to get his son-in-law into the post" of professor of engineering construction on Reilly's retirement. The son-in-law in question was Richard Woods, who was then instructor in surveying and was later appointed assistant professor of engineering (see also chap. 7).

68 PP Cd. 831 (1901), 15.

69 Gay, "Science, Scientific Careers and Social Exchange in London."

70 McLeod diary, 20 December 1900.

71 McLeod diary, 22 December 1900.

72 McLeod diary, 4 January 1901.

73 McLeod diary, 29 December 1900.

74 McLeod diary, 14 March 1901.

75 The statement is appendix 5 of PP Cd. 539 (1901), 99–101.

76 McLeod to Ottley, 5 November 1904 ("Disposal of Assets in Library"). McLeod concluded his letter by observing, "I have not yet realised the 'blessing in disguise' that you promised me!"

77 Gay, "Questionable Project."

78 PP Cd. 831 (1901), 3–4.

79 By that time, universities at Cambridge, Glasgow, Dublin, Victoria, London, and Birmingham were offering degrees in engineering.

80 PP Cd. 831 (1901), 4.

81 PP Cd. 831 (1901), 5.

82 Statement by T. Shields, 14 March 1901, appendix 3 to PP Cd. 539 (1901), 91.

83 Electron beams and X-rays, respectively.

84 G.M. Minchin, "The Electrical Measurement of Starlight. Observations Made at the Observatory of Daramona House, Co. Westmeath, in April 1895. Preliminary Report," *Proceedings of the Royal Society of London* 58 (1895): 142–54.

85 Minchin called this a "photo-electric battery": George M. Minchin, "Experiments in Photoelectricity," *Philosophical Magazine* 31 (1891): 207–38.

86 G.M. Minchin, "An Account of Experiments in Photo-Electricity," *Telegraphic Journal and Electrical Review* 8 (1880): 309–12, 324–6, and 342–4. Minchin called this the "telephotograph" in a letter to his friend and collaborator George Fitzgerald, 19 April 1880. See also Oliver Lodge's obituary for Minchin, "O.J.L.," "George Minchin Minchin," *Proceedings of the Royal Society London*, ser. A, 92, no. 645 (2 October 1916): xlvi–l.

87 Charles Reynolds, memo, 2 August 1895, appendix 12, in PP Cd. 539 (1901), 131.

88 PP Cd. 2056, 31.

89 Thorp (CH 1887–9) was chief engineer to the tea company and constructed electrically powered aerial ropeways and a tramway to facilitate access to the steep hills of the plantation. Thorp returned to England in 1906 and established a consulting engineering company, but died just two years later, aged thirty-nine. R.F. Thorp, "Munaar Valley Electrical Power Scheme," *Minutes of Proceedings of the Institution of Civil Engineers* 169 (1907): 365–80; Thorp, "Kotagudi Aerial Ropeway and Connecting Roads in North Travancore," *Minutes of Proceedings of the Institution of Civil Engineers* 161 (1905): 332–43; "Richard Fenwick Thorp, 1868–1908," *Minutes of Proceedings of the Institution of Civil Engineers* 175 (1909): 328–9.

90 S.M. Rutnagur, ed., *Electricity in India, Being a History of the Tata Hydro-Electric Project with Notes on the Mill Industry in Bombay and the Progress of Electric Drive in Indian Factories* (Bombay: Indian Textile Journal, 1912); R.B. Joyner, "The Tata Hydro-Electric Power-Supply Works, Bombay," *Minutes of Proceedings of the Institution of Civil Engineers* 207 (1919): 29–62.

91 G.T. Barlow and J.W. Meares, *Hydro-Electric Survey of India: Preliminary Report on the Water Power Resources of India* (Calcutta: Superintendent Government Printing India, 1919). Barlow died of smallpox within a few weeks of completing the survey, aged fifty-four.

92 C.G. Barber, *History of the Cauvery-Mettur Project* (Madras: Government Press), 1940.

93 The system effectively records a current of $+i$ if key A is pressed, $-i$ if key B is pressed, and 0 if both or neither keys are pressed. Because the status of one's own key is known, the status of the other key can be inferred from the current, thereby allowing two messages to be transmitted at the same time.

94 Quadruplex signalling was similar to the duplex technique but used two magnitudes of currents as well as two directions.

95 "Postscript to Note on Telephones Revised to 1st January 1897," in Ernest Hudson, *File of Technical Instructions*, issued by H. Mallock, officiating director general of telegraphs, 4 October 1888 (Hudson Papers, CSAS).

96 With titles such as "Faults Commonly Found in Telephonic Apparatus," Hudson Papers, *File of Technical Instructions*.

97 *Report of the Telegraph Committee 1906–07*, 23.

98 Simpson started at Coopers Hill in the same year as Ernest Hudson and was appointed to the Telegraph Department in 1887.

99 *CHM* 4, no. 6, February 1903, 108.

100 Narrative and quotation from M.G. Simpson, "Wireless Telegraphy from the Andaman Islands to the Mainland of Burma," *Electrician*, 27 April 1906, 49–51.

101 *Summary of the Administration the Earl of Minto*, section on wireless telegraphy 78–82.

102 Geoffrey Rothe Clarke, "Postal and Telegraph Work in India," *Journal of the Royal Society of Arts* 71, no. 3680 (1 June 1923): 483–98.

103 "Sir Maurice Simpson," *Times*, 5 July 1954, 10. He later served as director-in-chief of the Indo-European Telegraph Department. John Parker was for some time in charge of the Port Bair Subdivision, where he was replaced by J.G.P. Cameron, author of *A Short History of the Royal Indian Engineering College*. Obituary for J.N. Parker, *CHM* 16, no. 1, 1958, 10.

104 PP Cd. 490 (1901), 28.

105 "Cooper's Hill College," *Engineer* (6 October 1871): 241.

106 Information from RIEC calendars for 1873–4, 1883–4, 1895–6, and 1902–3.

107 "The New Electro-Technical Laboratories at Coopers Hill," *CHM* 4, no. 4, November 1902, 62–4.

108 A.F.T.A., "Electric Light in College," *CHM* 5, no. 2, November 1903, 22.

109 G.R.N. Minchin, *Under My Bonnet* (London: G.T. Foulis & Sons, 1950), 14.

110 Frank McClean attended Coopers Hill from 1894 to 1897 and joined the Bombay PWD as an assistant engineer 3rd grade, stationed in Belgaum, although he soon resigned in 1902 (RIEC calendar for 1901–2 and "Sir Francis McClean," *Times*, 12 August 1955, 11). For a full biography of Frank McClean, see Philip Jarrett, *Frank McClean: Godfather to British Naval Aviation* (Barnsley, UK: Seaforth Publishing, 2011).

111 McClean also flew a floatplane along the Nile. *Sphere* on 10, 24, and 31 January, 28 February, and 25 April 1914.

112 Quotes from Hearson's testimony in Cd. 539 (1901), 44 and 45. Hearson was not alone in finding it difficult to work with Brightmore. Arthur Heath, assistant professor of engineering and another of the dismissed instructors, objected to being Brightmore's subordinate. In a letter to the India Office, 22 January 1900, Ottley wrote of Heath, "I cannot think that there is much probability of his working harmoniously with Dr. Brightmore in the future." PP Cd. 539 (1901), 126.

113 M.E. Grant Duff to the House of Commons, House of Commons Debate, 3 March 1871, vol. 204, cols. 1326–55. This was also quoted in Dadabhai Naoroji, *Poverty and Un-British Rule in India* (London: Swan Sonnenschein, 1901), 105. Naoroji's work was a spiritual predecessor of Shashi Tharoor's excoriating *Inglorious Empire*.

114 Naoroji, *Poverty and Un-British Rule*, 109.

115 The rise of the Indian engineering profession is described in Ramnath, *Birth of an Indian Profession*.

116 Ramnath, *Birth of an Indian Profession*, 127.

117 PP Cd. 539 (1901), 131.

118 C.M. Lane speech to the Coopers Hill Society Annual Dinner, 25 June 1936, in *CHM* 12, no. 34, August 1936, 548. Charles Lane (1882–1956) was an irrigation engineer, reaching chief engineer and secretary to Government, Bombay.

119 For example, see Hannam, "Utilitarianism and the Identity of the Indian Forest Service," 215.

120 *Government of India Public Works Department Code*, 1892, 1:45.

121 *Government of India Public Works Department Code*, 1907, 1:53.

122 In 1887, for example, fifteen engineers were appointed from Coopers Hill and six from the Royal Engineers. The nine appointments from Indian colleges consisted of four or five in alternate years from Roorkee, two or one from Sibpur College in Calcutta, one from Madras Civil Engineering College, and three from the Poona College of Science. *Report of the Public Service Commission 1886–87* (Calcutta: Superintendent of Government Printing, 1888), 122.

123 Mital, *History of the Thomason College of Engineering*, 114.

124 "Resolution 1413G, dated 13 December 1884," *Gazette of India*, 20 December 1884, 479–80. This also summarizes the background to the original Roorkee Resolution.

125 *Report of the Public Service Commission 1886–87*, 123–5.

126 R.N. Thakur, *The All India Services: A Study of Their Origin & Growth* (Patna: Bharati Bhawan, 1960), 133.

127 Arun Kumar, "Colonial Requirements and Engineering Education: The Public Works Department, 1847–1947," in *Technology and the Raj*, ed. Roy MacLeod and Deepak Kumar, 216–32 (New Delhi: Sage Publications, 1995). Also see Mital, *History of the Thomason College of Engineering*, 81.

128 Kumar, "Colonial Requirements and Engineering Education."

129 Thakur, *All India Services*, chap. 5.

130 Prof. McLeod's copy of the 1871 college prospectus (BL IOR MSS EUR F239/106).

131 In the "Prospectus" sections of RIEC calendars between 1880 and 1889.

132 John Ottley evidence in PP. Cd. 2056 (1904), 8.

133 In the absence of detailed demographic information, the numbers of students with non-European backgrounds were initially estimated simply on the basis of their names. In most cases, this information was verified using census records, the India Office List, and biographical information obtained for example from newspapers, obituaries, and the *Coopers Hill Magazine*. A register of students appointed to the Indian services was published annually in the RIEC calendar (until publication ceased after the 1902–3 issue) and in the various India Office Lists.

134 John Ottley evidence in Cd. 2056 (1904), 7.

135 John Ottley's draft response to an inquiry, 26 March 1901, from a prospective student at Central College, Bangalore, in the file "Natives of India." BL IOR L/PWD/8/243.

136 *CHM* 8, no. 17, July 1910, 272.

137 Obituary *CHM* 15, no. 4, June 1950, 60.

138 Phya Sarasastra, 8 August 1947, *CHM* 14, no. 3, 1948, 585–6.

139 Likewise, N. Sanid, the first Siamese to attend the college (1900–2), had adopted the title Phya Darubhand Bidaksha. He served in the Siam Forest Department under Ernest Hudson's friend W.P. Tottenham.

140 John Ottley to J. Algernon Brown, 18 January 1906, in "Report on Siamese Students." BL IOR L/PWD/8/247.

141 Kaustavmoni Boruah, "'Foreigners' in Assam and Assamese Middle Class," *Social Scientist* 8, no. 11 (June 1980): 44–57. Also, the India Office List for 1916.

142 "The Old Dogs," *CHM* 12, no. 20, January 1930, 308. India Office List for 1916.

143 Nariman's Post Office books are preserved in the Coopers Hill Society Papers in the British Library, "Nariman, Rustam Kai Khushro: Postcards, Dinner Menus, Railway Tickets and Post Office Savings books," BL MSS EUR F239/58.

144 The other was Faredun Cursetji Pavry (CH 1897–1900), who was awarded a C.I.E. (Companion of the Order of the Indian Empire) in 1930 as chief engineer of the North West Railway. He died in December 1943.

145 Feni is in the northern part of modern Bangladesh, near the border with India.

146 "The Feni Dak Bungalow Incident," *Amrita Bazar Patrika*, 6 March 1904, 7.

147 *Amrita Bazar Patrika*, 22 January 1914, 6.

148 *Royal Commission on the Public Services in India. Appendix to the Report of the Commissioners*, vol.17. *Minutes of Evidence Related to the Post Office of India and the Telegraph Department.* PP Cd. 7905. London: His Majesty's Stationery Office, 1915. Roy's evidence is on pp. 42–6.

149 The was a "chicken-and-egg" situation. With little recruitment of Indian engineers, there was little incentive either for government to invest in the colleges or for top candidates to apply. The poor performance of the resulting recruits justified the decision not to hire Indian-trained engineers, completing the cycle.

150 "Indian Officers of the Telegraph Department," *Leader* (Allahabad), 18 July 1913, 8.

151 "Long Distance Telephony," *Leader* (Allahabad), 17 November 1922, 7.

152 "Retirement of Sir Ganen Roy," *CHM* 12, no. 14, September 1927, 219; "Sir Ganendra Roy, Indian Posts and Telegraphs" *Times*, 16 January 1943, 6.

153 "Sir Ganendra Roy, Indian Posts and Telegraphs." Soorjocoomar Goodeve Chuckerbutty (1826–74) was a professor at the Calcutta Medical College and surgeon-major in the Bengal Army. On his conversion to Christianity, he adopted the name "Goodeve" in honour of his mentor and tutor Dr. H.H. Goodeve. Obituary, *British Medical Journal*, 17 October 1874, 511.

154 John Ottley evidence in PP Cd. 2056 (1904), 7.

155 PP Cd. 831 (1901), 17.

156 Royal Indian Engineering College: Financial Results, in PP Cd. 2055 (1904), 42.

157 PP Cd. 2055 (1904), 17.

158 A complementary view was expressed by the *Engineer*: "It ought to be clearly understood that Cooper's Hill was established not because engineers could not be got for India, but because they could not be got at the price which the Indian Government was willing to pay for their services." "Engineers for India," *Engineer*, 14 October 1881, 279.

159 Cuddy and Mansell, *Engineers for India*, 121–2.

160 *CHM* 8, no. 4, July 1907, 48.

161 George Chesney report to Council of India, 1 February 1872 (BL IOR L/PWD/8/7, 238–43).

162 George Chesney request for additional engineering teaching staff, December 7, 1874 (BL IOR L/PWD/8/7, 491). Ernest Hudson's copies of Reilly's lecture pamphlets show that the material was indeed theoretical.

163 From the Board of Visitors of the Royal Indian Engineering College to the Marquis of Hartington, secretary of state for India, 26 May 1882, 1 (BL IOR L/PWD/8/9, unnumbered).

164 PP Cd. 2056 (1904), 114.

165 PP Cd. 2056 (1904), 124.

166 PP Cd. 2056 (1904), 130. This is contrary to a modern assertion that Coopers Hill's camaraderie came at the expense of engineering proficiency. Christopher Hill, "Imperial Design: The Royal Indian Engineering College and Public Works in Colonial India," in *The British Empire and the Natural World*, ed. Deepak Kumar, Vinita Damodaran, and Rohan D'Souza, 71–85 (Oxford: Oxford University Press, 2011).

167 PP Cd. 2056 (1904), 101.

168 PP Cd. 2056 (1904), 64–6, 73–4. It has been implied that Coopers Hill engineers were both unaware and contemptuous of these issues. Hill, *South Asia*, 120–1. However, Indian techniques for building foundations on sand were adopted and enhanced by British engineers long before Coopers Hill was established.

169 Egerton, Hebbert, and Reynolds in PP Cd. 2056 (1904), 43, 65, 76–7, respectively. This is supported by the lengths of service demonstrated by engineer officers on the state railways in 1910; nearly one-third had been in India for ten to nineteen years and a quarter for twenty to twenty-nine years. Derbyshire, "Building of India's Railways," 177–215.

170 PP Cd. 2056 (1904), 29.

171 Eustace Kenyon to his mother, 7 June 1884 (CSAS). Kenyon also recommended the deferred system of payment for Coopers Hill whereby students' annual fees were reduced and later recovered by a deduction from their salary after they started work in India.

172 PP Cd. 2055 (1904), summary table of information received from British engineering schools, 46–7.

173 In 1903, G. Carey-Foster, principal of University College, London, estimated that no fewer than 400 students per year completed college training. PP Cd. 2055 (1904), 49. The upper number is estimated from testimony by Professors Unwin and Hudson-Beare in 1904, combined with table 1.2 in Anna Guagnini, "Worlds Apart: Academic Instruction and Professional Qualifications in the Training of Mechanical Engineers in England, 1850–1914," in *Education, Technology, and Industrial Performance in Europe, 1850–1939*, ed. Robert Fox and Anna Guagnini (Cambridge: Cambridge University Press, Editions de la Maison des Sciences de l'Homme, 1993), 33. Not all of these students would have completed diplomas or degrees.

174 PP Cd. 2056 (1904), 35.

175 The recognition by industry of university education was particularly slow in Britain. As late as the 1920s, a survey of 226 major engineering companies showed that only thirty-three (15 per cent) mentioned a degree being a qualification for entry. Austen Albu, "British Attitudes to Engineering Education: a Historical Perspective," in *Technical Innovation and British Economic Performance*, ed. Keith Pavitt, 67–87 (London: Palgrave Macmillan, 1980).

176 John Wolfe-Barry, PP Cd. 2056 (1904), 112–13.

177 *CHM* 5, no. 5, March 1904, 84. Lancelot Edward Becher (1882–1960) won many awards at Coopers Hill and was commissioned into the Royal Engineers, reaching a rank of lieutenant-colonel. He was awarded a Distinguished Service Order and mentioned four times in despatches (*CHM* 16, no. 3, 1960, 8).

178 *CHM* 7, no. 8, September 1906, 122; and *CHM* 8, no. 1, December 1906, 5.

179 Buchanan, "Rise of Scientific Engineering," 228.

180 PP Cd. 2056 (1904), 118–19.

181 PP Cd. 539 (1901), 132.

182 Denham, *Coopers Hill. A Poeme.*

183 *CHM* 12, no. 3, July 1924, 36. Francis Dempster (1858–1941) qualified for the Indian Telegraph Department after just six months at Coopers Hill. He was chosen to play rugby for England, but the president would not permit it. He saw service as a telegraph officer in several military campaigns and retired as director general of the department. He was both president and secretary of the Coopers Hill Society. *CHM* 13, no. 10, November 1941, 249–50.

184 *CHM* 8, no. 4, July 1907, 47.

185 W.H. Preece, PP Cd. 2056 (1904), 133.

186 *CHM*, 12, no. 1, January 1924, 10.

187 E.H. Stone, PP Cd. 2056 (1904), 86–7.

188 F.R. Upcott and J.G.H. Glass, PP Cd. 2056 (1904), 36 and 84, respectively. R.W. Egerton, a Coopers Hill graduate, naturally took the opposite view, PP Cd. 2056 (1904), 42.

189 PP Cd. 2056 (1904), 107.

190 PP Cd. 2055 (1904), 19.

191 In 1905 the PWD announced that candidates should have engineering qualifications from universities of Cambridge, London, St. Andrews, Glasgow, Edinburgh, Dublin, Ireland, and Wales, or an equivalent institution, as well as one year's practical experience. *CHM* 6, no. 7, June 1905, 98–9.

192 The government of India complained that "it is a matter of regret that our views regarding the continued existence of the College were not sought before the present enquiry was decided on." PP Cd. 2055 (1904), 65.

193 PP Cd. 2055 (1904), 62–7.

194 PP Cd. 2055 (1904), 66.

195 Cuddy, "Royal Indian Engineering College," 290.

196 *CHM* 5, no. 8, July 1904, 122. The only student with the initials H.I.B. was Herbert Ivor Bond (CH 1897–1900).

197 "Sir W. Schlich: Forestry in the Empire," *Times*, 1 October 1925, 16.

198 *CHM* 8, no. 3, May 1907, 31.

199 "Notes," *Times*, 1 February 1911, 16.

200 Moreover, Thomas Hearson argued that a fair accounting should also deduct the £2200 that had previously been spent on training foresters before Coopers Hill. Hearson, letter to the editor, *Engineer*, 15 March 1901.

201 The assistant postmaster-general (Captain Norton) replying to Mr. Gretton, House of Commons Hansard, 20 December 1912, vol. 45, col. 1863.

202 *Fifth Report of the Special Committee on Home Charges*, 29–35.

203 Appleyard, "Closing of Coopers Hill College." Rollo Appleyard (1867–1943) went on to a successful career in the material technology for telegraph cables, including the San Francisco–Honolulu cable. He and his assistant, Alfred Penfold, also developed the rubber-cored golf ball. Appleyard developed new technologies during the First World War before resuming his industrial career. "Commander Rollo Appleyard, O.B.E., J.P., R.N.V.R.," *Journal of the Institution of Electrical Engineers, Part I: General* 90, no. 36 (1943): 530.

204 A. Godley to Lord George Hamilton, dated 1 March 1901 (BL IOR MSS EUR F123/45, 121–4).

9. The Coopers Hill Society

1 "The Closing of Coopers Hill," *Times*, 29 October 1906, 15; "Closing of Coopers Hill College," *Engineer*, 2 November 1906, 454. A copy addressed to George Minchin with the president's compliments is also preserved in the British Library (IOR MSS EUR F239/106).

2 Henry Marsh, 30 June 1909, Coopers Hill Society Dinner. *CHM* 8, no. 13, August 1909, 200.

3 Frank Rawson, 9 July 1919, Coopers Hill Society dinner. *CHM* 11, no. 6, August 1919, 80.

4 "Prize Day at Coopers Hill," *Times*, 27 July 1906, 10. Brodrick's comment was a reference to a speech made half a century earlier by the president of Haileybury College when it was closed: "We are standing on the scaffold and the halter is around our necks, but though we are led to execution, yet we are not criminals; when Haileybury was wanted, Haileybury did its duty." This reference could not have made anyone in the audience feel better, and *Indian Engineering* later said that the remark "savoured … more of claptrap oratory than of sense." *CHM* 12, no. 4, November 1924, 51.

5 This section is based on a file of correspondence on the matter entitled "Disposal of Assets in Library, Chapel, Forest Museum, & Physical Laboratory presented to the College," located in the BL IOR L/PWD/8/316.

6 McLeod-Innes to Ottley, 14 May 1904, in *Disposal of Assets*. By the time Ottley was ready to return the books in 1906, McLeod-Innes was too infirm to want them back.

7   McLeod to Ottley, 5 November 1904, in *Disposal of Assets*.

8   Ottley to Lady Chesney, 19 February 1907, in *Disposal of Assets*.

9   Lady Chesney to Ottley, 20 February [1907], in *Disposal of Assets*.

10 Ultimately the kneelers were sent to Rev. Arthur Crawley-Boevey at St. Philips Church in Birmingham; he was the only ordained former student still practising. The cross was returned to Lady Taylor and the vases given to McLeod to pass on to the bishop of Southwark.

11 Details from *CHM* 8, no. 3, May 1907, 29. The memorial to the five staff consisted of a reredos, a tiled floor, and a plaque. The fates of these are not entirely clear. Ottley proposed that the reredos itself should be sold and the plaque should be installed in Egham Church "with an additional small brass plate explaining the circumstances of its removal." Ottley to Hicks, 28 March 1907, in *Disposal of Assets*. This is also what appears in the May 1907 draft Specifications of Work for the relocation. When the author visited Egham Church in 2018 the small explanatory plaque was found lying loose, and the original memorial plate was not found. The floor tiles were presumably left in situ at Coopers Hill.

12 *CHM* 8, no. 2, February 1907, 19.

13 L.S. Milford, *Haileybury College Past and Present* (London: T. Fisher Unwin, 1909), 271–2. See also "Ecclesiastical Intelligence," *Times*, 11 March 1907, 7. The silverware was stolen in 1996 but identified some years later at auction and re-turned once again to Haileybury Chapel (with thanks to Rev. Christopher Briggs, chaplain of Haileybury School, April 2019).

14 The correspondence related to the decision was put before Parliament in *Correspondence Relating to the Training of Forestry Students*, PP Cd. 2523 (1905).

15 Edward Beck, vice-chancellor, University of Cambridge, 20 March 1905 in PP Cd. 2523 (1905), 64.

16 "The Indian Forest Service," *CHM* 6, no. 5, March 1905, 68–70.

17 *CHM* 6, no. 8, July 1905, 113.

18 Alastair Howatson, *Mechanicks in the Universitie: A History of Engineering Science at Oxford* (Oxford: Department of Engineering Science, 2008), chap. 4. When your author was an engineering student at Oxford in the early 1980s, Howatson, like "Pomph" Reilly, would greet students coming late to his engineering classes with a "good morning." To avoid interrupting the class, Reilly simply pointed to the letters *G.M.* printed on the blackboard. Howatson, however, did it to draw students' attention to latecomers in the hope they would avoid the embarrassment next time. Like Reilly, Howatson was a popular and well-respected lecturer.

19 Nichol Finlayson Mackenzie (1875–8), letter 21 October 1908, in *CHM* 8, no. 10, November 1908, 155.

20 *CHM* 8, no. 1, December 1906, 1–2.

21 First appearing in the *CHM* 6, no. 6, May 1905, 82.

22 *CHM* 6, no. 8, July 1905, 114.

23 The idea of an independent Coopers Hill Club had in fact first been proposed by Maj.-Gen. Courtney in 1898. *CHM* 1, no. 8, October 1898, 111.

24 *CHM* 7, no. 2, November 1905, 17–18.

25 *CHM* 7, no. 3, December 1905, 35.

26 Coopers Hill Society Minute Book (1906–24), BL MSS EUR F.239/1, 1.

27 *CHM* 8, no. 1, December 1906, 4.

28 *CHM* 8, no. 16, May 1910, 245–6.

29 *CHM* 11, no. 4, January 1919, 58–9.

30 Quoted in his dinner speech by N. Pearce (CH 1898–1901), June 1929. *CHM* 12, no. 19, September 1929, 293–4.

31 *CHM* 10, no. 8, October 1914, 113–15.

32 *CHM* 10, no. 8, October 1914, 118–19.

33 *CHM* 11, no. 15, April 1922, 226.

34 Coopers Hill Society, Minutes of General Meetings, vol. 2, 1924–61. BL IOR MSS EUR F239/2.

35 Tufnell to the Editor, *CHM* 11, no. 15, April 1922, 239.

36 This section is based on information in the three issues of *Coopers Hill Magazine* published between September 1927 and May 1928. *CHM* 12, no. 14–16.

37 *CHM* 12, no. 14, September 1927, 215.

38 *CHM* 12, no. 16, May 1928, 245–6.

39 Newcombe, *Village, Town and Jungle Life*, 413.

40 Newcombe, *Village, Town and Jungle Life*, 412.

41 Alice Perrin, *The Anglo-Indians*, 6th ed. (London: Methuen, 1913), 196.

42 For discussions of the issues of retirement to England, see Gilmour, *British in India*, chap. 14; Allen, *Plain Tales from the Raj*, chap. 21; and Buettner, *Empire Families*, chap. 5.

43 Best, *Forest Life in India*, 279.

44 Perrin, *Anglo-Indians*, 204.

45 Melissa Edmundson, "Perrin [née Robinson], Alice (1867–1934), Novelist," 17 September 2005, *Oxford Dictionary of National Biography* (Oxford University Press, https://www.oxforddnb.com).

46 Perrin, *Anglo-Indians*, 183.

47 Names listed as "serving" or with addresses care of financial institutions or clubs were excluded from the numbers of retirees. The London Area was defined to be within the current M25, while Southern England refers to the counties along the south coast.

48 It is quite possible that people who retired abroad simply did not join the Coopers Hill Society.

49 Perrin, *Anglo-Indians*, 184.

50 Buettner, *Empire Families*, 219–38.

51 A Coopers Hill Colonial Settler, "New Zealand," *CHM* 11, no. 14, January 1922, 216–19, and 11, no. 15, April 1922, 230–3.

52 "Mente Manuque," "Nova Scotia for Proportional Pensioners," *CHM* 12, no. 5, February 1925, 67–70. The pseudonym chosen by this correspondent, *Mente Manuque*, was the motto of Coopers Hill, meaning "With mind and hand."

53 *CHM* 9, no. 2, June 1913, 21.

54 *CHM* 12, no. 5, February 1925, 67.

55 A Coopers Hill Colonial Settler, "New Zealand," 230.

56 Another Coopers Hill Colonial Settler, "Southern Rhodesia," *CHM* 12, no. 1, January 1924, 12–14; and 12, no. 2, April 1924, 29–30.

57 For simplicity, this section focuses on the former students' arguments rather than the government responses, and is based on PP 290 (1901), addressing "Memorials from the Officers of the Public Works Department Appointed from the Royal Indian Engineering College in the Years 1873–1878."

58 *CHM* 11, no. 11, January 1921, 161.

59 A. Piatt Andrew, "Indian Currency Problems of the Last Decade," *Quarterly Journal of Economics* 15, no. 4 (August 1901): 483–516.

60 To make things more confusing, there seemed to have been several versions of the prospectus distributed at different times in different ways.

61 "Return to an Address of the Honourable the House of Commons, dated 27 March 1903," PP 98 (1903).

62 For example, "The Pensions Scandal," *CHM* 11, no. 11, January 1921, 161–3.

63 *Oracle* 4, no. 34, December 1877, leading article.

64 Joseph Fayrer, *On Preservation of Health in India: A Lecture Addressed to the Royal Indian Engineering College at Cooper's Hill* (London: Kerby & Endean, 1880). When Ernest Hudson initially failed his medical examination in August 1888, he and his father visited Fayrer in his capacity as president of the India Office Medical Board.

65 "Engineers in India," *Fraser's Magazine* 18, n.s., no. 107 (November 1878): 565.

66 *Oracle* 5, no. 43, December 1878, leading article.

67 Newcombe, *Village, Town and Jungle Life*, 411.

68 "How Has Life Expectancy Changed over Time?," UK Office for National Statistics, 9 September 2015, http://visual.ons.gov.uk/, compiled using the UK decennial life tables.

69 Note that distributions for the general population show the age range of all male deaths in a single year. The Coopers Hill data, on the other hand, are a compilation of ages at death for men with dates of birth spanning ten years and dates of death over an eighty-eight-year period.

70 The high "31–40" value for 1896–1905 is consistent with deaths in the Great War.

71 Cameron Taylor, *General Sir Alex Taylor*, 2:289.

72 The magazine usually stated this just before making a political comment – e.g., *CHM* 12, no. 20, January 1930, 309, which proceeded to deplore leaving the future of India "at the mercy of men who know nothing of the intricacies of the problem of its Government."

73　Much was made at the time, and since, of the strong and ongoing support for Dyer's actions by the lieutenant-governor of the Punjab, Sir Michael O'Dwyer. In *India as I Knew It* (London: Constable, 1925), O'Dwyer gives a long justification of his own actions, and those of Dyer, in chapter 17, entitled "The Punjab Rebellion of 1919."

74　*Report of the Disorders Inquiry Committee 1919–1920* (Calcutta: Superintendent of Government Printing, India, 1920.

75　"General Dyer: The Amritsar Shooting," *Times*, 25 July 1927, 8. According to probate records, his estate was worth only half that amount when he died on 23 July 1927.

76　Presumably this was to avoid a conflict with their employer. Later the Government of India would prohibit military and civilian officials from contributing to the Dyer Fund. *Aberdeen Daily Journal*, 11 November 1920, 6.

77　Fowler was known in his student days as "The Baron" because of his "aristocratic appearance and distinguished mien." Despite his soubriquet, Fowler was the author of *Socialism Explained*, a short book published in 1920, the year he presided over the CH dinner. F.D. Fowler, *Socialism Explained* (London: Grant Richards, 1925).

78　Report on the AGM and the Dinner from *CHM* 11, no. 9, July 1920, 126–7, 129.

79　*CHM* 11, no. 16, August 1922, 243–7.

80　*CHM* 12, no. 7, September 1925, 100–4.

81　For a full discussion on the development of the Indian engineering profession after 1900, see Ramnath, *Birth of an Indian Profession*.

82　*CHM* 12, no. 19, September 1929, 293.

83　*CHM* 12, no. 20, January 1930, 309.

84　*CHM* 12, no. 22, March 1931, 343–4.

85　*CHM* 12, no. 22, March 1931, 343.

86　For example, C.M. Lane's speech to 1936 annual dinner, *CHM* 12, no. 34, August 1936, 547–8.

87　*CHM* 13, no. 2, October 1937, 26–8.

88　*CHM* 13, no. 2, October 1937, 27.

89　One letter from R.F. Stoney (CH 1894–7), 31 July 1947, Ootacamund, commented that it was too soon to tell whether there would be any discrimination against Europeans (*CHM* 14, no. 3, 1948, 587). According to the membership list of the Coopers Hill Society, Stoney was still living there in 1961.

90　R. MacGregor (CH 1900–3) chairing Coopers Hill Dinner 5 July 1939. *CHM* 13, no. 6, October 1939, 139.

91　*CHM* 15, no. 12, 1957, 205.

92　Beinart and Hughes, *Environment and Empire*, 132–47. William Willcocks, *Sixty Years in the East* (London: Blackwood, 1935), 33.

93　*CHM* 13, no. 16, October 1944, 424.

94　Newcombe, *Village, Town and Jungle Life*, 403.

95 Todd, *Work, Sport, and Play*, 54–60.

96 Stebbing, *Diary of a Sportsman Naturalist*, chaps. 21–3. Also Stebbing, *Forests of India*, 3:677–81.

97 Best, *Forest Life in India*, 275–313 (chaps. 13 and 14).

98 Best, *Forest Life in India*, 281.

99 Coopers Hill Society Annual Report 1925–6, Hudson Papers, CSAS. The report noted that in the fourteen years prior to 1920–1, revenues exceeded expenses by over £900, a sum that was added to the investments.

100 This version appeared in the *Coopers Hill Magazine* in October 1926, reporting on an article that had appeared earlier in *Indian Engineering*. In a letter published in *Coopers Hill Magazine* in 1946, Alfred Lines recalls the same *Indian Engineering* piece, and gives his version of the story. Lines' abbreviation of the Father of Coopers Hill to F.C.H. suggests a joke on "Fellow of Coopers Hill." By virtue of arriving in college the day before all the other students in 1871, Walter Bernard de Winton considered himself to be the father of Coopers Hill (*CHM* 13, no. 12, October 1942, 298).

101 *CHM* 12, no. 12, January 1927, 177.

102 Annual Report for 1936–7, in *CHM* 13, no. 1, May 1937. An affiliation between the society and the EIUS Club was contemplated but rejected as unworkable.

103 Society Annual Report in *CHM* 15, no. 7, 1952, 108.

104 *CHM* 15, no. 9, 1954, 145.

105 *CHM* 15, no. 10, 1955, 163–5.

106 *CHM* 10, no. 12, November 1915, 181–2.

107 "Indians Remove a British Statue," *Times*, 7 June 1957, 9.

108 "Removal of Foreigners' Statues," *Civic Affairs, Kanpur, India* 4, no. 11 (June 1957): 68; and "Statues of British Rulers," *Civic Affairs, Kanpur, India* 4, no. 12 (July 1957): 8.

109 A group of alumni from Shoreditch and Brunel operates a web site containing a history of the property and a space for posting reminiscences: The Runnymede Campus Archive, http://www.runnymedecampus.com. As part of the development of the site for retirement residences, the original mansion extensively remodelled and the block containing the student rooms, porter's lodge, and dining hall that was built for the college's opening in 1871 have been demolished except for the western facade.

110 *CHM* 16, no. 4, 1961, 6.

111 *CHM* 16, no. 4, 1961, 6.

112 *CHM* 13, no. 6, October 1939, 139.

113 Only 1 per cent indicated a different career, while the other 10 per cent did not provide any information.

114 *CHM* 12, no. 7, September 1925, 101.

115 *CHM* 12, no. 2, April 1924, 19.

116 HM Willmott (CH 1888–91) in *CHM* 13, no. 18, December 1945, 476.

117 *CHM* 12, no. 4, November 1924, 50–2.

118 *CHM* 12, no. 4, November 1924, 50–1.

119 Ernest Shadbolt, Coopers Hill Half-Yearly Dinner, December 1906, *CHM* 8, no. 1, December 1906, 4.

120 Cain, "Empire and the Languages of Character," 269.

121 A.R. Astbury, Register of Students Admitted to the Royal Indian Engineering College, 1966, in the archives of the Institution of Civil Engineers, London, barcode 1966ASTRSA.

10. Conclusions and Reflections

1 Dale H. Porter and Gloria C. Clifton, "Patronage, Professional Values, and Victorian Public Works: Engineering and Contracting the Thames Embankment," *Victorian Studies* 31, no. 3 (Spring 1988): 319–49; Casper Andersen, *British Engineers and Africa, 1875–1914* (London: Routledge, 2016).

2 Andersen, *British Engineers and Africa*, 70.

3 Andersen, *British Engineers and Africa*, 73.

4 John Ottley, evidence to the 1903 committee of enquiry, PP Cd. 2056 (1904), 12.

5 The question of whether non-resident students trained at other institutions could develop the necessary esprit de corps as engineers in India was important strand of the 1903 enquiry.

6 Taylor also worked on the construction of the Mushkaf-Bolan Railway.

7 *CHM* 11, no. 11, January 1921, 165.

8 Kipling's school friend George Beresford attended Coopers Hill, as mentioned in his quasi-fictional novel, *Stalky & Co.*

9 Andersen, *British Engineers and Africa*, 128–34.

10 Andersen, *British Engineers and Africa*, 90, and chap. 4.

11 Only 33 out of 226 engineering firms surveyed. Albu, "British Attitudes to Engineering Education," 69.

12 PP Cd. 2056 (1904) testimonies from George Minchin and A.M. Rendel, respectively.

13 *Engineer*, 6 October 1871, 241.

14 For example, David Goldberg and Mark Somerville, *A Whole New Engineer* (Douglas, MN: ThreeJoy Associates, 2014).

15 Annual degrees granted, from *Canadian Engineers for Tomorrow: Trends in Engineering Enrolment and Degrees Awarded in 2019* (Engineers Canada, 2019). The proportion of 2016 graduates who obtained their professional licence was approximately 40 per cent. *2021 National Membership Information* (Engineers Canada 2021), https://engineerscanada.ca.

16 The title of this section is the line in Alexander Pope's *Windsor Forest* prior to the more frequently quoted: "On Cooper's hill eternal wreaths shall grow, / While lasts the mountain, or while Thames shall flow."

17 Summary numbers from Cuddy and Mansell, *Engineers for India*, 121.

18 Rasoul Sorkhabi, "George Bernard Reynolds: A Forgotten Pioneer of Oil Dis-
    coveries in Persia and Venezuela," *Oil-Industry History* 11, no. 1 (2010): 157–72;
    "Obituary: The Hon. David Carnegie," *Geographical Journal* 17, no. 2 (February
    1901): 202–3; David W. Carnegie, *Spinifex and Sand* (London: Arthur Pearson,
    1898).

19 *Imperial Gazetteer of India*, 3:443.

20 Not to mention the miserable conditions those passengers endured on their jour-
    neys. Prasad, *Tracks of Change*, chap. 1.

21 *Imperial Gazetteer of India*, 3:386.

22 *Imperial Gazetteer of India*, 3:318.

23 Cameron, *Short History of the Royal Indian Engineering College*, 38.

24 Several Canadian engineering schools each graduate a similar number *annually*.

25 *CHM* 12, no. 32, October 1935, 499.

26 Wimbush, *Life in the Indian Forest Service*, 161–2.

27 "Deaths," *Times* (London), 3 January 1981, 22.

Bibliography

Archives

British Library

Coopers Hill Society
Collection area India Office Records and Private Papers (IOR), under the reference
MSS EUR F239, "Cooper's Hill Society: Papers Relating to the Royal Indian
Engineering College, Cooper's Hill, and the Careers of Some of Pupils." Materials
include *Coopers Hill Magazine*, minutes and annual reports of the Coopers Hill
Society, and a wide variety of photograph albums, technical works, and other
memorabilia donated by its members. A finding aid that includes papers and
photographs is available at http://hviewer.bl.uk/IamsHViewer/FindingAidHandler.
ashx?recordid=032-002290828, last accessed September 2021.
Coopers Hill Magazine
British Library IOR MSS EUR F239 contains almost all issues of *Coopers Hill
Magazine*. Most issues between 1897 and 1945 are also available in Hudson papers
at the University of Cambridge Centre of South Asian Studies, including the years
1897–9 and 1910–12 that are missing from the BL. The BL also contains issues of the
earlier college magazine, *Oracle* (1877–80). The magazines are cited here in the form
volume/issue, month year.
Calendar of the Royal Indian Engineering College
IOR/V/25/700/1-25 contain copies of the college calendar from 1873/4 to 1902/3.
India Office Records
Papers related to Coopers Hill are in BL IOR L/PWD/8/, including the Cooper's Hill
Papers IOR/L/PWD/8/7–9: 1868–1904, and individual files related to the operations
of the college.
Papers of Lord George Hamilton as Secretary of State for India 1895–1903
IOR MSS EUR F123/45 contains letters related to the 1901 remodelling of the Coopers
Hill curriculum.

Diary of William Sangster
In the William Peter Sangster papers, IOR MSS EUR F245.

Centre of South Asian Studies, University of Cambridge

https://www.s-asian.cam.ac.uk/archive/
Canning photographs
images of Coopers Hill and India.
Ernest Hudson papers and photographs
Diaries, field notebooks, Coopers Hill notes, *Coopers Hill Magazine,* annual reports of
   the Coopers Hill Society, photograph albums, Telegraph Department documents
Eustace Kenyon papers
Family letters
Osmaston papers and photographs
Diaries of Lionel Sherbrooke Osmaston and Selina Osmaston, letters, official papers,
   photographs
Anthony Wimbush papers
Typescript reminiscences "Life in the Indian Forest Service," 1964
John Peake Wildeblood papers and photographs
Copies of family letters and photographs

Other Archives

Admiralty records, UK National Archive
Thomas Keith Hudson (ADM 196/70/400, ADM 196/14/119), John Pasco
   (ADM/196/11/318, ADM/196/76/26, ADM/196/79/7)
Hertfordshire County Archives
Family Papers of the Martin-Leake family, Acc. 599, under DE/MI
Imperial College, London
Diaries of Herbert McLeod
Institution of Civil Engineers, London
A.R. Astbury, Register of Students Admitted to the Royal Indian Engineering College,
   1966, barcode 1966ASTRSA
Royal Dublin Society
Letters of George Francis Fitzgerald, https://digitalarchive.rds.ie/collections/show/1
Royal Society of London
Correspondence with George Minchin, reference NLB/22/392
Suffolk Archives
Walter J. Boyes, "Life in the XIIth Regiment: Fifty-Six Years Personal Reminiscences,"
   pages from *Suffolk Regimental Gazette,* nos. 203–10, 1907/8, archive reference GB
   554 Y1/32

Articles and Books

Abetti, G. "Patrick William Henderson." In *Memorie della Società Astronomia Italiana* 20 (1949): 345–6 (in Italian).

Adley, Charles. *The Story of the Telegraph in India*. London: Spon, 1866.

Albion, Robert Greenhalgh. *Forests and Sea Power: The Timber Problem of the Royal Navy 1652–1862*. Cambridge, MA: Harvard University Press, 1926.

Albu, Austen. "British Attitudes to Engineering Education: A Historical Perspective." In *Technical Innovation and British Economic Performance*, edited by Keith Pavitt, 67–87. London: Palgrave Macmillan, 1980.

Allen, Charles, ed. *Plain Tales from the Raj*. London: Futura, 1976.

Althaus, Julius. *On Sclerosis of the Spinal Cord*. New York: G.P. Putnam & Sons, 1885.

Andersen, Casper. *British Engineers and Africa, 1875–1914*. London: Routledge, 2016.

Andrew, A. Piatt. "Indian Currency Problems of the Last Decade." *Quarterly Journal of Economics* 15, no. 4 (August 1901): 483–516.

Appleyard, Rollo. "The Closing of Coopers Hill College." *Electrical Review* (17 March 1905). In BL IOR MSS EUR F239/106.

Appleyard, Rollo (Obituary). *Journal of the Institution of Electrical Engineers, Part I: General* 90, no. 36 (1943): 530.

Argyll. *His Grace the Duke of Argyll on Indian Public Works and Cooper's Hill College: A Reply with Correspondence between His Grace and the President of the Institution of Civil Engineers*. London: E. & F.N. Spon, 1882.

Arnold, David. *Science, Technology and Medicine in Colonial India*. The New Cambridge History of India. Cambridge: Cambridge University Press, 2000.

Arnold, David, and Ramachandra Guha, eds. *Nature, Culture, Imperialism*. Delhi: Oxford University Press, 1995.

"Arsenical Tissues." *British Medical Journal* 1 (23 June 1888): 1347–8.

Ayres, Peter G. *Harry Marshall Ward and the Fungal Thread of Death*. Minnesota: American Phytopathological Society, 2005.

Baden-Powell, B.H. *Forest Law*. London: Bradbury, Agnew, 1893.

Baden-Powell, B.H., and J. Sykes Gamble, eds. *Report of the Proceedings of the Forest Conference 1873–74*. Calcutta: Office of the Superintendent of Government Printing, 1874.

Baden-Powell, R.S.S. *Pig-Sticking or Hoghunting*. London: Harrison & Sons, 1889.

Bailey, F. "The Indian Forest School." *Transactions of the Scottish Arboricultural Society* 11, pt. 2 (1886): 155–61.

Baker, E.C. *Sir William Preece, F.R.S.: Victorian Engineer Extraordinary*. London: Hutchinson, 1976.

Baker, W.E., T.E. Dempster, and H. Yule. *The Prevalence of Organic Disease of the Spleen*. Calcutta: Office of the Superintendent of Government Printing, 1868.

Barber, C.G. *History of the Cauvery-Mettur Project*. Madras: Government Press, 1940.

Barlow, G.T., and J.W. Meares. *Hydro-Electric Survey of India: Preliminary Report on the Water Power Resources of India*. Calcutta: Superintendent Government Printing, 1919.

Barr, Pat. *The Memsahibs: The Women of Victorian India*. London: Secker & Warburg, 1976.

Barton, Gregory, and Brett Bennett. "Environmental Conservation and Deforestation in British India 1855–1947." *Itinerario* 32, no. 2 (2008): 83–104.

– "'There Is a Pleasure in the Pathless Woods': The Culture of Forestry in British India." *British Scholar* 3, no. 2 (September 2010): 219–34.

Bayley, Steuart C. "College Literature." In *Memorials of Old Haileybury College*, edited by Frederick Charles Danvers, Sir M. Monier-Williams, Sir Steuart Colvin Bayley, Percy Wigram, Brand Sapte, et al., 257–305. London: Archibald Constable, 1894.

Beckett, Ian F.W. *Riflemen Form: A Study of the Rifle Volunteer Movement 1859–1908*. Aldershot, UK: Ogilvy Trusts, 1982.

Beinart, William, and Lotte Hughes. *Environment and Empire*. Oxford: Oxford University Press, 2007.

Bell, Evans, and Frederick Tyrrell. *Public Works and the Public Service in India*. London: Trübner, 1871.

Bell, James Richard. "The First Section of the Kandahar Railway." *Minutes of Proceedings of the Institution of Civil Engineers* 61 (1880): 274–85.

The Bengal-Nagpur Railway: Its History and Development, Parts I and II, Reprinted from the January and February Issues of "The Indian Railway Gazette," ca. 1910, 12. DeGolyer Library, Southern Methodist University, http://digitalcollections.smu.edu/cdm/ref/collection/eaa/id/1511.

Benton, John. "Irrigation Works in India." *Journal of the Royal Society of Arts* 61 (13 June 1913): 717–54.

Berridge, P.S.A. "Abstract. Withdrawing Two 155-Foot Spans from Bridges on an Indian Frontier Railway in Baluchistan." *Journal of the Institution of Civil Engineers* (29 November 1947): 63–71.

– *Couplings to the Khyber*. Newton Abbot, UK: David & Charles, 1969.

Best, James W. *Forest Life in India*. London: John Murray, 1935.

Bhatt, Vikram. *Resorts of the Raj*. Middletown, NJ: Grantha, 1998.

Black, John. "A More Vigorous Prosecution of Public Works: The Reforms of the Indian Public Works Department." *Accounting History* 6, no. 2 (2001): 91–119.

Black, John, and John Richard Edwards. "Accounting Careers Traversing the Separate Spheres of Business and Government in Victorian Britain." *Accounting History* 21, no. 2–3 (2016): 306–28.

Blackbourn, David. *The Conquest of Nature: Water, Landscape and the Making of Modern Germany*. New York: W.W. Norton, 2006.

Bladin, Peter F. "Julius Althaus (1833–1900): Neurologist and Cultural Polymath; Founder of Maida Vale Hospital." *Journal of Clinical Neuroscience* 15 (2008): 495–501.

Boruah, Kaustavmoni. "'Foreigners' in Assam and Assamese Middle Class." *Social Scientist* 8, no. 11 (June 1980): 44–57.

Boyes, Walter J. "Life in the XIIth Regiment: Fifty-Six Years Personal Reminiscences." *Suffolk Regimental Gazette* 1907/8, nos. 203–10.

Bramwell, F.J., and H. Truman Wood. "Education in Forestry." *Journal of the Society of Arts* 30 (14 July 1882): 879.

Brander, Michael. *The Victorian Gentleman*. London: Gordon Cremonesi, 1975.

Broadfoot, W. "Addiscombe: The East India Company's Military College." *Blackwood's Edinburgh Magazine*, May 1893, 647–57.

Browne, James (Obituary). *Minutes of Proceedings of the Institution of Civil Engineers* 125 (January 1896): 428–30.

Buchanan, R.A. *The Engineers: The History of the Engineering Profession in Britain 1750–1914*. London: Jessica Kingsley Publishers, 1989.

– "Gentlemen Engineers: The Making of a Profession." *Victorian Studies* 26, no. 4 (1983): 407–29.

– "The Rise of Scientific Engineering in Britain." *British Journal for the History of Science* 18, no. 2 (July 1985): 218–33.

Buck, Edward J. *Simla Past & Present*. Calcutta: Thacker, Spink, 1904.

Buettner, Elizabeth. *Empire Families*. Oxford: Oxford University Press, 2004.

Cain, Peter J. "Character and Imperialism: The British Financial Administration of Egypt, 1878–1914." *Journal of Imperial and Commonwealth History* 34, no. 2 (June 2006): 177–200.

– "Empire and the Languages of Character and Virtue in Later Victorian and Edwardian Britain." *Modern Intellectual History* 4, no. 2 (2007): 249–73.

Calvert, Monte A. *The Mechanical Engineer in America*. Baltimore, MD: Johns Hopkins University Press, 1967.

Cameron Taylor, A. *General Sir Alex Taylor C.G.B., R.E.: His Times, His Friends, and His Work*. London: Williams & Norgate, 1913.

Cameron, J.G.P. *A Short History of the Royal Indian Engineering College*. [Richmond]: Coopers Hill Society, 1960.

Cannadine, David. *Ornamentalism: How the British Saw Their Empire*. Oxford: Oxford University Press, 2001.

Canning, Frederick. "Uttar Pradesh 1903–1937." In *100 Years of Indian Forestry*, 1:37–40. Dehra Dun: Forest Research Institute, 1961.

Carnegie, David (Obituary). *Geographical Journal* 17, no. 2 (February 1901): 202–3.

Carnegie, David W. *Spinifex and Sand*. London: Arthur Pearson, 1898.

Carroll, Lewis. *The Hunting of the Snark*. London: Macmillan, 1876.

Cautley, P.T. "On the Use of Wells, Etc. in Foundations as Practised by the Natives of the Northern Doab." *Journal of the Asiatic Society of Bengal* 8 (1839): 327–40.

Center, Dr. "A Note on Reh or Alkali Soils and Saline Well Waters." *Indian Forester* 7 (1882): 61–83.

Chaudhury, Deep Kanta Lahiri. "First Virtual Community and the Telegraph General Strike of 1908." In "Uncovering Labour in Information Revolutions, 1750–2000," supplement 11, *International Review of Social History* 48 (2003): 45–71.

– *Telegraphic Imperialism: Crisis and Panic in the Indian Empire c. 1830.* New York: Palgrave Macmillan, 2010.

Chesney, George. *Indian Polity,* 2nd ed. London: Longmans, Green, 1870.

– *Memorandum on the Employment of the Corps of Royal Engineers in India.* London: Spottiswoode, 1868.

Chota Mem. *The English Bride in India,* 2nd ed. Madras: Higginbotham, 1909.

Chuckerbutty, Soorjocoomar Goodeve (Obituary). *British Medical Journal* (17 October 1874): 511.

Clarke, Geoffrey Rothe. "Postal and Telegraph Work in India." *Journal of the Royal Society of Arts* 71, no. 3680 (1 June 1923): 483–98.

Clarke, George Sydenham. *Practical Geometry and Engineering Drawing.* London: Spon, 1875.

Clarke, I.F. "Before and after 'The Battle of Dorking.'" *Science Fiction Studies* 24, no. 1 (1997): 33–46.

Clayton, Ann. *Martin-Leake Double VC.* London: Leo Cooper, 1994.

Cleghorn, Hugh. "Address Delivered at the Nineteenth Annual Meeting." *Transactions of the Scottish Arboricultural Society* 7 (1875): 1–9.

– *Forests and Gardens of South India.* London: W.H. Allen, 1861.

Cole, Charles John. "The Tunnels on the Second Division of the Mushkaf-Bolan Railway, India." *Minutes of Proceedings of the Institution of Civil Engineers* 128 (1897): 265–77.

Cole, W.H. *Notes on Permanent-Way Material, Platelaying and Points and Crossings.* London: Spon, 1890.

Collen, Edwin. "The Volunteer Force of India: Its Present and Future." United Service Institution of India prize essay, 1883.

– "The Volunteer Forces of India." *Empire Review* 5, no. 27 (April 1903): 221–32.

Cuddy, Brendan. "Royal Indian Engineering College, Cooper's Hill (1871–1906): A Case Study of State Involvement in Professional Civil Engineering Education." PhD diss., London University, 1980.

Cuddy, Brendan, and Tony Mansell. "Engineers for India: The Royal Indian Engineering College at Cooper's Hill." *History of Education* 23, no. 1 (1994): 107–23.

Curzon of Kedleston. *The Romanes Lecture 1907: Frontiers,* 2nd ed. Oxford: Clarendon, 1908.

D'Arcy, W.E. *Preparation of Forest Working-Plans in India.* Calcutta: Office of the Superintendent of Government Printing, India, 1895.

Das, Pallavi. "Colonialism and the Environment in India: Railways and Deforestation in 19th Century Punjab." *Journal of Asian and African Studies* 46, no. 1 (2011): 38–53.

– *Colonialism, Development, and the Environment: Railways and Deforestation in British India, 1860–1884.* New York: Palgrave Macmillan, 2015.

– "Hugh Cleghorn and Forest Conservancy in India." *Environment and History* 11, no. 1 (2005): 55–82.

- "Railway Fuel and Its Impact on the Forests in Colonial India: The Case of the Punjab, 1860–1884." *Modern Asian Studies* 47, no. 4 (2013): 1283–1309.

Dasgupta, Subrata. *Awakening.* Noida: Random House India, 2010.

Davis, Mike. *Late Victorian Holocausts : El Niño Famines and the Making of the Third World.* London: Verso, 2002.

The Declaration of Students of the Natural and Physical Sciences. London: Simpkin, Marshall, 1865.

Denham, John. *Coopers Hill: A Poeme.* 2nd ed. with additions. London: Humphrey Moseley, 1650.

Dennis, Graham. *Englefield Green in Pictures.* Englefield Green Village Association, 1994.

Dennis, Graham, and Richard Williams. *The Englefield Green Picture Book.* Egham-by-Runnymede Historical Society, 2000.

Derbyshire, Ian. "The Building of India's Railways." In *Technology and the Raj,* edited by Roy MacLeod and Deepak Kumar, 177–215. New Delhi: Sage Publications, 1995.

Dewey, C.J. "The Education of a Ruling Caste: The Indian Civil Service in the Era of Competitive Examination." *English Historical Review* 88, no. 347 (1973): 262–85.

"Dietrich Brandis, the Founder of Forestry in India." *Indian Forester* 10, no. 8 (August 1884): 343–57.

Diver, Maud. *The Englishwoman in India.* London: William Blackwood & Sons, 1909.

- *The Unsung.* London: William Blackwood & Sons, 1945.

Douglas, John Christie. *Manual of Telegraph Construction.* London: Griffin, 1877.

Duncan, Jeannette. *Simple Adventures of a Memsahib.* New York: D. Appleton, 1893.

Duthie, John Lowe. "Pressure from Within: The 'Forward' Group in the India Office during Gladstone's First Ministry." *Journal of Asian History* 15, no. 1 (1981): 36–72.

Eagles, Thomas Henry (Obituary). *Minutes of Proceedings of the Institution of Civil Engineers* 110 (1892): 392–3.

Edgcome, William Henry (Obituary). *Minutes of Proceedings of the Institution of Civil Engineers* 168 (1907): 364.

"Edmund du Cane Smithe (1853–1908)," *Minutes of Proceedings of the Institution of Civil Engineers* 178 (1909): 369.

Edmundson, Melissa. "Perrin [née Robinson], Alice (1867–1934), Novelist." *Oxford Dictionary of National Biography,* 17 September 2005.

Elrington, Stephen N. *Election of Fellows at Trinity College: Report of the Proceedings at a Visitation.* Dublin: Hodges, Foster, 1871.

Elsmie, G.R. *Thirty-Five Years in the Punjab 1858–1893.* Edinburgh: David Douglas, 1908.

Fayrer, Joseph. *On Preservation of Health in India: A Lecture Addressed to the Royal Indian Engineering College at Cooper's Hill.* London: Kerby & Endean, 1880.

Fisher, W.R. "The Forestry Branch at Coopers Hill." *Indian Forester* 31 (December 1905): 679–86.

- "Practical Forest Training Grounds for the Coopers Hill College." *Indian Forester* 18 (1892): 12–18.

"Football: The Killed and Wounded." *British Medical Journal* 1, no. 1214 (5 April 1884): 689.

"The Forestry Department of Edinburgh University." *Nature* 106 (1921): 706–7.

"The Forestry School at Cooper's Hill." *Nature* 37, no. 962 (1888): 529–31.

Foster, G.C. "The Intended Engineering College." *Nature*, 18 August 1870, 316–17.

Fowler, F.D. *Socialism Explained*. London: Grant Richards, 1925.

"Frederick Benbow Hebbert (1854–1905)." *Minutes of Proceedings of the Institution of Civil Engineers* 166 (1906): 387–8.

G.H.F.N. "Sir Ronald Ross (1857–1932)." *Obituary Notices of Fellows of the Royal Society* 1, no. 2 (December 1933): 108–15.

Gadgil, Madhav, and Ramachandra Guha. *This Fissured Land: An Ecological History of India*. Oxford: Oxford University Press, 2012.

Gales, R.R. "The Hardinge Bridge over the Lower Ganges at Sara." *Minutes of Proceedings of the Institution of Civil Engineers* 205 (1918): 18–67.

Gay, Hannah. "A Questionable Project: Herbert McLeod and the Making of the Fourth Series of the Royal Society Catalogue of Scientific Papers, 1901–25." *Annals of Science* 70, no. 2 (2013): 149–74.

- "Science and Opportunity in London, 1871–85: The Diary of Herbert McLeod." *History of Science* 41 (2003): 427–58.

- "Science, Scientific Careers and Social Exchanges in London: The Diary of Herbert McLeod, 1885–1900." *History of Science* 46 (2008): 457–96.

Gilmartin, David. "Models of the Hydraulic Environment." In *Nature, Culture, Imperialism*, edited by David Arnold and Ramachandra Guha, 210–36. Oxford: Oxford University Press, 1995.

- "Scientific Empire and Imperial Science: Colonialism and Irrigation Technology in the Indus Basin." *Journal of Asian Studies* 53, no. 4 (November 1994): 1127–49.

Gilmour, David. *The British in India: Three Centuries of Ambition and Experience*. UK: Allen Lane, 2018.

- *The Ruling Caste: Imperial Lives in the Victorian Raj*. New York: Farrar, Straus and Giroux, 2006.

Gilmour, Robin. *The Idea of the Gentleman in the Victorian Novel*. London: George Allen & Unwin, 1981.

Glass, James George Henry (Obituary). *Journal of the Royal Society of Arts* 59, no. 3049 (28 April 1911): 592.

Goldberg, David, and Mark Somerville. *A Whole New Engineer*. Douglas, MI: ThreeJoy Associates, 2014.

Gracey, T. *Administration Report of the Railways in India for 1896–97*. London: Her Majesty's Stationery Office, 1897.

[Gregory, Richard A.] R.A.G.? "Prof. G.M. Minchin, F.R.S." *Nature* 93, no. 2318 (2 April 1914): 115–16.

Guagnini, Anna. "Worlds Apart: Academic Instruction and Professional Qualifications in the Training of Mechanical Engineers in England, 1850–1914." In *Education, Technology, and Industrial Performance in Europe, 1850–1939*, edited by Robert Fox and Anna Guagnini, 16–41. Cambridge: Cambridge University Press Editions de la Maison des Sciences de l'Homme, 1993.

Guggisberg, F.G. *The Shop*. London: Cassell, 1900.

Guha, Ramachandra. "An Early Environmental Debate: The Making of the 1878 Forest Act." *Indian Economic and Social History Review* 27, no. 1 (1990): 65–84.

– "Forestry in British and Post-British India: A Historical Analysis." *Economic and Political Weekly* 18, no. 45/46 (5–12 November 1983): 1940–7.

Guha, Ramachandra, and Madhav Gadgil. "State Forestry and Social Conflict in British India." *Past & Present* 123 (May 1989): 141–77.

A Handbook for Travellers in India and Ceylon. London: John Murray, 1892.

A Handbook for Travellers in India, Burma and Ceylon. London: John Murray, 1911.

A Handbook for Travellers in India, Burma and Ceylon. London: John Murray, 1919.

Handley, Mrs. M.A. *Roughing It in Southern India*. London: Edward Arnold, 1911.

Hannam, Kevin. "Utilitarianism and the Identity of the Indian Forest Service." *Environment and History* 6, no. 2 (May 2000): 205–28.

Harper, Brian. "Civil Engineering: A New Profession for Gentlemen in Nineteenth-Century Britain?" *Icon* 2 (1996): 59–82.

Hay, Ian. Preface to *Tiger Tiger* by W. Hogarth Todd. London: Heath Cranton, 1927, 8.

Headrick, Daniel. *The Tentacles of Progress: Technology Transfer in the Age of Imperialism 1850–1940*. New York: Oxford University Press, 1988.

Hele Shaw, H.S. "The Royal Indian Engineering College." *Engineer* 58 (21 November 1884): 392.

Hill, Christopher. "Imperial Design: The Royal Indian Engineering College and Public Works in Colonial India." In *The British Empire and the Natural World*, edited by Deepak Kumar, Vinita Damodaran, and Rohan D'Souza, 71–85. Oxford: Oxford University Press, 2011.

– *South Asia: An Environmental History*. Santa Barbara, CA: ABC-CLIO, 2008.

Hirose, Shin. "Two Classes of British Engineers: An Analysis of Their Education and Training, 1880s–1930s." *Technology and Culture* 51, no. 2 (2010): 388–402.

Holmes, G.D. "History of Forestry and Forest Management." *Philosophical Transactions of the Royal Society of London. Series B, Biological Sciences* 271, no. 911 (10 July 1975): 69–80.

Horne, W.O. *Work and Sport in the Old I.C.S.* Edinburgh: Blackwood, 1928.

Howatson, Alastair. *Mechanicks in the Universitie: A History of Engineering Science at Oxford*. Oxford: Department of Engineering Science, 2008.

Jarrett, Philip. *Frank McClean: Godfather to British Naval Aviation*. Barnsley, UK: Seaforth Publishing, 2011.

Jenkin, Fleeming. *A Lecture on the Education of Civil and Mechanical Engineers in Great Britain and Abroad*. Edinburgh: Edmonston & Douglas, 1868.

Jepsen, Thomas C. *My Sisters Telegraphic: Women in the Telegraph Office, 1846–1950*. Athens: Ohio University Press, 2000.

Johns, William Arthur. "The Tunnels on the First Division of the Mushkaf-Bolan Railway, India." *Minutes of Proceedings of the Institution of Civil Engineers* 128 (1897): 257–64.

Joshi, Gopa. "Forests and Forest Policy in India." *Social Scientist* 11, no. 1 (January 1983): 43–52.

Joyner, R.B. "The Tata Hydro-Electric Power-Supply Works, Bombay." *Minutes of Proceedings of the Institution of Civil Engineers* 207 (1919): 29–62.

Kerr, Ian J. "Strachey, Ralph (1868–1923)." *Oxford Dictionary of National Biography*, 23 September 2004.

Kerr, Ian. *Building the Railways of the Raj.* Delhi: Oxford University Press, 1995.

– *Engines of Change: The Railroads That Made India.* Westport, CT: Praeger, 2007.

Kincaid, Dennis. *British Social Life in India.* London: George Routledge & Sons, 1938.

Kipling, Rudyard. *Many Inventions.* New York: Appleton, 1893.

Kirkwood, Patrick M. "The Impact of Fiction on Public Debate in Late Victorian Britain: The Battle of Dorking and the 'Lost Career' of Sir George Tomkyns Chesney." *Graduate History Review* 4, no. 1 (Fall 2012): 1–16.

Knowles, Elizabeth, ed. *Oxford Dictionary of Quotations*, 7th ed. Oxford: Oxford University Press. 2009.

Kumar, Deepak, Vinita Damodaran, and Rohan D'Souza, eds. *The British Empire and the Natural World: Environmental Encounters in South Asia.* Oxford: Oxford University Press, 2011.

Kumar, Arun. "Colonial Requirements and Engineering Education: The Public Works Department, 1847–1947." In *Technology and the Raj,* edited by Roy MacLeod and Deepak Kumar, 216–32. New Delhi: Sage Publications, 1995.

Lalvani, Kartar. *The Making of India.* London: Bloomsbury, 2016.

Laski, Harold. "The Danger of Being a Gentleman: Reflections on the Ruling Class in Britain." In *The Danger of Being a Gentleman and Other Essays*, chap. 1. London: Allen and Unwin, 1939.

Lawrence, Rosamund. *Indian Embers.* Oxford: George Ronald, 1949.

Layton, Edwin T. *The Revolt of the Engineers: Social Responsibility and the American Engineering Profession.* Baltimore, MD: Johns Hopkins University Press, 1986.

Lecky, W.E.H. *The Empire: Its Value and Its Growth.* London: Longmans, Green, 1893.

Ledzion, Mary MacDonald. *Forest Families.* Putney, London: British Association for Cemeteries in South Asia, 1991.

Leinster-Mackay, Donald P. "The English Private School 1830–1914, with Special Reference to the Private Proprietary School." PhD diss., Durham University, 1971.

Lewis, Michael. "The Personal Equation: Political Economy and Social Technology on India's Canals 1850–1930." *Modern Asian Studies* 41, no. 5 (2007), 967–94.

"Lionel S. Osmaston." *Commonwealth Forestry Review* 48, no. 2 (1969): 97–8.

Lodge, Alfred (Obituary). *Nature* 938, no. 3561 (29 January 1938): 191.

Lodge, Oliver J. "George Minchin Minchin." *Proceedings of the Royal Society London*, ser. A, 92, no. 645 (2 October 1916): xlvi–l.

Luke, P.V. "Early History of the Telegraph in India." *Journal of the Institution of Electrical Engineers* 20 (1891): 102–22.

Lundgreen, Peter. "Engineering Education in Europe and the U.S.A., 1750–1930: The Rise to Dominance of School Culture and the Engineering Professions." *Annals of Science* 47, no.1 (1990): 33–75.

Lunney, Linde, Enda Leaney, and Patricia M. Byrne. "Minchin, George Minchin." *Dictionary of Irish Biography*, rev. March 2013, https://www.dib.ie/.

MacGeorge, George Walter. *Ways and Works in India*. Westminster: Archibald Constable, 1894.

MacLeod, Roy, and Deepak Kumar, eds. *Technology and the Raj*. New Delhi: Sage, 1995.

MacMillan, Margaret. *Women of the Raj*. Thames & Hudson, 1988.

Mammen, E., M.S. Tomar, and N. Parmeswaran. "A Salute to William Schlich, Forestry Pioneer." *Journal of the Forestry Commission* 34 (1965): 170–5.

March, E.W., and A.E. Osmaston. "Coopers Hill." *Commonwealth Forestry Review* 50, no. 3 (September 1971): 243–6.

Martin-Leake, S. "Early Recollections of the B.-N. R." *Bengal-Nagpur Railway Magazine*, July 1929, 47–51.

– "Early Recollections of the B.-N. R. (continued)." *Bengal-Nagpur Railway Magazine*, August 1929, 29–34.

– "The Rupnarayan Bridge, Bengal-Nagpur Railway." *Minutes of Proceedings of the Institution of Civil Engineers* 151 (1903): 251–78.

– "Thirty Years Ago – and Now (Part 1)." *Bengal Nagpur Railway Magazine*, November 1930, 15–19.

– "Thirty Years Ago – and Now (Part 2)." *Bengal Nagpur Railway Magazine*, December 1930, 17–22.

Matin, A. Michael. "Scrutinizing the Battle of Dorking: The Royal United Service Institution and the Mid-Victorian Invasion Controversy." *Victorian Literature and Culture* 39 (2011): 385–407.

Matthew, H.C.G. "Campbell, George Douglas, Eighth Duke of Argyll in the Peerage of Scotland, and First Duke of Argyll in the Peerage of the United Kingdom (1823–1900)." *Oxford Dictionary of National Biography*, 23 September 2004.

McLeod, Herbert. "Apparatus for Measurement of Low Pressures of Gas." *Proceedings of the Physical Society* 1 (1874): 30–4.

Medley, Julius George. *India and Indian Engineering*. London: Spon, 1873.

– , ed. *Roorkee Treatise on Civil Engineering in India*. Vol. 2. Roorkee: Thomason College Press, 1867.

Medley, J.G., and A.M. Lang. *Roorkee Treatise on Civil Engineering in India*. Vol. 1. Roorkee: Thomason College Press, 1873.

Milburn, William. *Oriental Commerce*. London: Black, Parry, 1813.

Milford, L.S. *Haileybury College Past and Present*. London: T. Fisher Unwin, 1909.

Minchin, G.R.N. *Under My Bonnet*. London: G.T. Foulis & Sons, 1950.

Minchin, George M. "An Account of Experiments in Photo-Electricity." *Telegraphic Journal and Electrical Review* 8 (1880): 309–12, 324–6, and 342–4.

– "The Electrical Measurement of Starlight. Observations Made at the Observatory of Daramona House, Co. Westmeath, in April 1895. Preliminary Report." *Proceedings of the Royal Society of London* 58 (1895): 142–54.

– "Experiments in Photoelectricity." *Philosophical Magazine* 31 (1891): 207–38.

Mital, K.V. *History of the Thomason College of Engineering (1847–1949)*. Roorkee: Indian Institute of Technology, 2008.

Mitcham, Carl. "A Historico-Ethical Perspective on Engineering Education: From Use and Convenience to Policy Engagement." *Engineering Studies* 1, no. 1 (2009): 35–53.

Mitchell, P. Chalmers. "Zoological Gardens and the Preservation of Fauna." *Science* n.s. 36, no. 925 (20 September 1912): 353–65.

Mitchell, Sally. *Daily Life in Victorian England*. Westport, CT: ABC-CLIO, 2008.

Moir, Martin. *A General Guide to the India Office Records*. London: British Library, 1988.

Moyle, George. "The Platelaying of the Jacobabad or Broad Gauge Section of the Kandahar Railway." *Minutes of Proceedings of the Institution of Civil Engineers* 61 (1880): 286–94.

Mulhern, Kirsteen Mairi. "The Intellectual Duke: George Douglas Campbell, 8th Duke of Argyll, 1823–1900." PhD diss., University of Edinburgh, 2006.

Naoroji, Dadabhai. *Poverty and Un-British Rule in India*. London: Swan Sonnenschein, 1901.

Newcombe, Alfred C. *Village, Town and Jungle Life in India*. Edinburgh: Blackwood & Sons, 1905.

O'Callaghan, Francis Langford (Obituary). *Minutes of Proceedings of the Institution of Civil Engineers* 179 (1910): 364–5.

O'Dwyer, Michael. *India as I Knew It*. London: Constable, 1925.

O'Shaughnessy, William. "Memoranda Relative to Experiments on the Communication of Telegraphic Signals by Induced Electricity." *Journal of the Asiatic Society of Bengal* 8, no. 93 (1839): 714–31.

Osmaston, B.B., Henry Osmaston, and B.E. Smythies. *Wild Life and Adventures in Indian Forests*. Ulverston, UK: Henry Osmaston, 1999.

Osmaston, Henry. "The Osmaston Family, Foresters & Imperial Servants." *Commonwealth Forestry Review* 68, no. 1 (March 1989): 77–87.

Osmaston, L.S. "A Man-Eating Panther." *Journal of the Bombay Natural History Society* 15 (1903): 135–8.

Osmaston, Lionel S. (Obituary). *Commonwealth Forestry Review* 48, no. 2 (1969): 97–8.

Oxford Dictionary of National Biography. Oxford University Press, 2004, https://www.oxforddnb.com.

Parsons, J.H. *The History of the Third Burmese War; 1890–91 Period VI. The Winter Campaign of 1890–91*. Simla: Government Central Printing Office, 1893.

Partridge, Eric, and Jacqueline Simpson. *A Dictionary of Historical Slang*. London: Penguin, 1972.

Patterson, Steven. *The Cult of Imperial Honor in British India*. New York: Palgrave Macmillan, 2009.

Pearson, G.F. "The Teaching of Forestry." *Journal of the Society of Arts* 30 (3 March 1882): 422–33.

Pennycuick, John. "The Diversion of the Periyar." *Minutes of Proceedings of the Institution of Civil Engineers* 128 (1897): 140–63.

Perrin, Alice. *The Anglo-Indians*, 6th ed. London: Methuen, 1913.

Porter, Dale H., and Gloria C. Clifton. "Patronage, Professional Values, and Victorian Public Works: Engineering and Contracting the Thames Embankment." *Victorian Studies* 31, no. 3 (Spring 1988): 319–49.

Porter, Mrs. Gerald. *Annals of a Publishing House: John Blackwood*. Edinburgh: William Blackwood & Sons, 1898.

Prasad, Ritika. *Tracks of Change: Railways and Everyday Life in Colonial India*. Delhi: Cambridge University Press, 2015.

Preece, W.H., and J. Sivewright. *Telegraphy*. London: Longmans, Green, 1876.

Procida, Mary A. "Good Sports and Right Sorts: Guns, Gender, and Imperialism in British India." *Journal of British Studies* 40, no. 4 (October 2001): 454–88.

– *Married to the Empire*. Manchester: Manchester University Press, 2014.

Pycroft, James. *Oxford Memories: A Retrospect after Fifty Years*. London: Richard Bentley & Son, 1886.

Ramnath, Aparajith. *Birth of an Indian Profession: Engineers, India, and the State, 1900–47*. Delhi: Oxford University Press, 2017.

Ramsay, James. "The Mushkaf-Bolan Railway, Baluchistan, India." *Minutes of Proceedings of the Institution of Civil Engineers* 128 (1897): 232–56.

Reilly, Callcott (Obituary). *Minutes of Proceedings of the Institution of Civil Engineers* 142 (1900): 376–9.

Reilly, Callcott. "On the Longitudinal Stress of the Wrought Iron Plate Girder." *Civil Engineer and Architect's Journal* 23 (September 1860): 261–4.

– "On Uniform Stress in Girder Work, Illustrated by Reference to Two Bridges Recently Built." *Minutes of Proceedings of the Institution of Civil Engineers* 24 (1865): 391–425.

– "Studies of Iron Girder Bridges, Recently Executed, Illustrating Some Applications of the Modern Theory of the Elastic Resistance of Materials." *Minutes of Proceedings of the Institution of Civil Engineers* 29 (1870): 403–500.

Ribbentrop, B. *Forestry in British India*. Calcutta: Office of the Superintendent of Government Printing, India, 1900.

"Richard Fenwick Thorp, 1868–1908," *Minutes of Proceedings of the Institution of Civil Engineers* 175 (1909): 328–9.

Roberts, Frederick Sleigh Roberts, *Forty-One Years in India, from Subaltern to Commander-in-Chief*. London: Macmillan, 1901.

Rogers, Charles Gilbert. *A Manual of Forest Engineering for India*, 3 vols. Calcutta: Office of the Superintendent of Government Printing, India, 1900, 1901, and 1902.

Ross, A., J.S. Beresford, S. Preston, H. Marsh, E. Benedict, L.W. Dane, and J. Benton. "Discussion: Punjab Triple Canal System." *Minutes of Proceedings of the Institution of Civil Engineers* 201 (January 1916): 48–62.

Ross, Ronald. *Memoirs, with a Full Account of the Great Malaria Problem and Its Solution.* London: Murray, 1923.

Russell, J. Scott. *Systematic Technical Education for the English People.* London: Bradbury, Evans, 1869.

Rutnagur, S.M., ed. *Electricity in India, Being a History of the Tata Hydro-Electric Project with Notes on the Mill Industry in Bombay and the Progress of Electric Drive in Indian Factories.* Bombay: Indian Textile Journal, 1912.

Sandes, E.W.C. *The Military Engineer in India*, 2 vols. Chatham: Institution of Royal Engineers, 1933 and 1935.

Sarmah, Dipak. *Forestry in Karnataka.* Chennai: Notion, 2019.

Schlich, William. "Forestry Education." *Indian Forester* 24 (1898): 228–44.

Scoones, W. Baptiste. "The Civil Service of India." *Macmillan's Magazine* 30, May–October 1874, 365–76.

Scott-Moncrieff, G.K. "The Frontier Railways of India." *Professional Papers of the Corps of Royal Engineers* 11 (1885): 213–56.

– "Sir James Browne and the Harnai Railway." *Blackwood's Magazine* 177 (May 1905): 608–21.

Scott, James C. *Seeing Like a State: How Certain Schemes to Improve the Human Condition Have Failed.* New Haven, CT: Yale University Press, 2005.

Sharma, S.C. *Punjab: The Crucial Decade.* New Delhi: Nirmal Publishers, 1987.

Shute, Nevil. *Slide Rule.* London: Pan, 1968.

Simms, Frederick Walter, and D. Kinnear Clark. *Practical Tunnelling*, 4th ed. London: Crosby Lockwood & Son, 1896.

Simpson, M.G. "Wireless Telegraphy from the Andaman Islands to the Mainland of Burma." *Electrician*, 27 April 1906, 49–51.

Sinha, S.N., ed. *Mutiny Telegrams.* [Lucknow]: Department of Cultural Affairs, U.P., 1988.

Sorkhabi, Rasoul. "George Bernard Reynolds: A Forgotten Pioneer of Oil Discoveries in Persia and Venezuela." *Oil-Industry History* 11, no. 1 (2010): 157–72.

Stearn, Roger T. "Chesney, Sir George Tomkyns (1830–1895)." *Oxford Dictionary of National Biography*, 23 September 2004.

– "General Sir George Chesney." *Journal of the Society for Army Historical Research* 75, no. 302 (Summer 1997): 106–18.

Stebbing, E.P. *The Diary of a Sportsman Naturalist in India.* London: John Lane, The Bodley Head, 1920.

– *The Forests of India*, 3 vols. London: John Lane, The Bodley Head, 1922–6.

– *Jungle By-Ways in India.* London: John Lane, The Bodley Head, 1911.

Steel, Flora Annie, and Grace Gardiner. *Complete Indian Housekeeper & Cook.* 7th ed. London: William Heinemann, 1909.

Stephen, Leslie. *Sir Leslie Stephen's Mausoleum Book.* Oxford: Clarendon, 1977.

Sweeney, Stuart. *Financing India's Imperial Railways, 1875–1914.* London: Pickering & Chatto, 2011.

Sydenham of Combe. *My Working Life.* London: J. Murray, 1927.

Thakur, R.N. *The All India Services: A Study of Their Origin & Growth.* Patna: Bharati Bhawan, 1960.

Thana, S. Abdul. "Forest Policy and Ecological Change." In *Environmental Encounters in South Asia,* edited by Deepak Kumar, Vinita Damodaran, and Rohan D'Souza, 262–80. Oxford: Oxford University Press, 2011.

Tharoor, Shashi. *Inglorious Empire: What the British Did to India.* London: Hurst, 2016.

Thiselton-Dyer, William. "Harry Marshall Ward 1854–1906." In *Makers of British Botany,* edited by F.W. Oliver, 261–79. Cambridge: Cambridge University Press, 1913.

Thomas, Wilson. "Forest Department to Pay Homage to Hugo Wood." The Hindu, 12 December 2017. https://www.thehindu.com/todays-paper/tp-national/tp-tamilnadu/forest-department-to-pay-homage-to-hugo-wood/article21457571.ece.

Thorp, R.F. "Kotagudi Aerial Ropeway and Connecting Roads in North Travancore." *Minutes of Proceedings of the Institution of Civil Engineers* 161 (1905): 332–43.

– "Munaar Valley Electrical Power Scheme." *Minutes of Proceedings of the Institution of Civil Engineers* 169 (1907): 365–80.

Tian, Hanqin, Kamaljit Banger, Tao Bo, and Vinay K. Dadhwal. "History of Land Use in India during 1880–2010: Large-scale Land Transformations Reconstructed from Satellite Data and Historical Archives." *Global and Planetary Change* 121 (2014): 78–88.

Todd, W. Hogarth. *Tiger Tiger.* London: Heath Cranton, 1927.

– *Work Sport and Play.* London: Heath Cranton, 1928.

Tomes, Jason. "Clarke, George Sydenham, Baron Sydenham of Combe (1848–1933)." *Oxford Dictionary of National Biography,* 23 September 2004, https://www.oxforddnb.com.

Tremper, Ellen. "'The Earth of Our Earliest Life': Mr. Carmichael in *To the Lighthouse.*" *Journal of Modern Literature* 19, no. 1 (Summer 1994): 163–71.

Troup, R.S. "Sir William Schlich, K.C.I.E., F.R.S." *Nature* 116, no. 2921 (25 October 1925): 617–18.

– *The Work of the Forest Department in India.* Calcutta: Superintendent Government Printing, 1917.

Troup, R.S., rev. Andrew Grout. "Schlich, Sir William Philipp Daniel (1840–1925)." *Oxford Dictionary of National Biography,* 23 September 2004. *Oxford Dictionary of National Biography,* https://www.oxforddnb.com.

Vibart, H.M. *Addiscombe: Its Heroes and Men of Note.* London: Archibald Constable, 1894.

Walker, E.G. *The Life and Work of William Cawthorne Unwin.* London: Unwin Memorial Committee, 1938.

Weightman, Walter James. "The Khojak Rope-Inclines." *Minutes of Proceedings of the Institution of Civil Engineers* 112 (1893): 310–20.

Weil, Benjamin. "Conservation, Exploitation, and Cultural Change in the Indian Forest Service, 1875–1927." *Environmental History* 11, no. 2 (April 2006): 319–43.

Wells, H.G. *Experiment in Autobiography Discoveries and Conclusions of a Very Ordinary Brain (since 1866).* London: Gollancz, 1934.

Wenzlhuemer, Roland. *Connecting the Nineteenth-Century World: The Telegraph and Globalization.* Cambridge: Cambridge University Press, 2013.

Whitcombe, Elizabeth. "The Environmental Costs of Irrigation in British India: Waterlogging, Salinity, Malaria." In *Nature, Culture, Imperialism*, edited by David Arnold, and Ramachandra Guha, 237–59. Delhi: Oxford University Press, 1995.

Whorton, James C. *The Arsenic Century.* Oxford: Oxford University Press, 2010.

Willcocks, William. *Sixty Years in the East.* London: Blackwood, 1935.

Williams, W. *Manual of Telegraphy.* London: Longmans, Green, 1885.

Wilson, J.S. "William Cawthorne Unwin 1838–1933." *Proceedings of the Royal Society* 1, no. 3 (December 1934): 167–78.

With the Cooper's Hill Volunteer Company at Aldershot by One of the Awkward Squad. Beccles: Caxton, 1890[?].

Yule, Henry, and A.C. Burnell. *Hobson-Jobson: A Glossary of Colloquial Anglo-Indian Words and Phrases.* London: John Murray, 1886.

Official Publications

Account of Roorkee College Established for the Instruction of Civil Engineers, with a Scheme for Its Enlargement. Agra: Secundra Orphan, 1851.

Administration Report of the Baluchistan Agency for 1895–96. Calcutta: Office of the Superintendent of Government Printing India, 1896.

Administration Reports of the Forest Department in the Bombay Presidency Including Sind for the Year 1891–92. Bombay: Government Central Press, 1893.

Administration Reports of the Forest Department in the Bombay Presidency Including Sind for the Year 1893–94. Bombay: Government Central Press, 1895.

Administration Reports of the Forest Department in the Bombay Presidency Including Sind for the Year 1894–95. Bombay: Government Central Press, 1896.

Administration Reports of the Forest Department in the Bombay Presidency Including Sind for the Year 1896–97. Bombay: Government Central Press, 1898.

Canadian Engineers for Tomorrow: Trends in Engineering Enrolment and Degrees Awarded in 2019. Engineers Canada, 2019.

"Cooper's Hill College: Return of the Working of the New Scheme." UK parliamentary paper 166, 1884.

"Correspondence in Reference to the Establishment of an Engineering College at Cooper's Hill." UK parliamentary paper 115, 1871.

"Correspondence Relating to the Remodelling of the Studies and the Retirement of Certain of the Professors and Lecturers." UK parliamentary paper Cd. 490, 1901.

Education and Status of Civil Engineers in the United Kingdom and in Foreign Countries. London: Institution of Civil Engineers, 1870.

"Fifth Report of the Special Committee on Home Charges, 26 October 1888." In "East India Home Charges." UK parliamentary paper 327, 1893.

"Further Papers in Regard to the Royal Indian Engineering College." UK parliamentary paper Cd. 831, 1901.

"Further Papers Relating to the Indian Civil Engineering College." UK parliamentary paper 148, 1871.

General Report on Public Instruction in the North Western Provinces of the Bengal Presidency for 1843–44. Agra: Agra Ukhbar, 1845.

Government of India circular no. 22-F, reprinted in *Punjab Record* 30, 40–4. Lahore: Civil and Military Gazette, 1896.

Government of India Public Works Department Code, vol. 1. Calcutta: Office of the Superintendent of Government Printing, India, 1892.

Government of India Public Works Department Code. Vol. 1, *General Regulations.* Calcutta: Superintendent of Government Printing, India, 1908.

Hansard. London: British Parliament.

History of Services of the Officers of the Indian Telegraph Department Corrected to 1 July 1907. Calcutta: Government of India Department of Commerce and Industry, 1907. Available in Hudson papers, Centre of South Asian Studies, University of Cambridge.

Imperial Gazetteer of India. Vol. 3, *Economic.* Oxford: Clarendon, 1908.

India List and India Office List for 1898, 1902, and 1905. London: Harrison & Sons, in those years.

India List Civil and Military, July 1888. London: Allen, 1888.

India Office List 1893, 1916, 1928. London: Harrison, in those years.

Indian Civil Engineering College Calendar, 1873–4 and 1874–5. London: Allen, in those years.

Indian Civil Engineering College, Coopers Hill. Syllabus of the Course of Study. London: Wm. H. Allen, 1871.

"Indian Volunteer Corps List Corrected to 31 December 1912," in *The Quarterly Army List for the Quarter Ending 31st December 1913* (London: His Majesty's Stationery Office, 1914), 2: 1904–17.

India Office and Burma Office List 1945. London: His Majesty's Stationery Office, 1945.

"Maintenance of Bridges in Indian Railways: A Review." In *Twenty-Third Report of the Standing Committee on Railways, 2018–19.* New Delhi: Lok Sabha Secretariat, December 2018.

"Minutes of the Evidence Taken Before the Committee Appointed in 1903 to Enquire into the Expediency of Maintaining the Royal Indian Engineering College." UK parliamentary paper Cd. 2056, 1904.

"National Forest Policy (1988): New Direction in Forest Management. Resolution No. 3A/86-FP Ministry of Environment and Forests." *Social Change* 33, nos. 2 & 3 (2003): 192–203.

National Membership Information. Engineers Canada, 2021.

Quarterly Army List of Her Majesty's British and Indian Forces on the Bengal Establishment. Calcutta: Military Orphan, 1859.

"Remodelling of Course of Instruction and Retirement of Certain of the Professors and Lecturers. Report of the Board of Visitors (dated 25th March 1901), and Minutes of Evidence Taken before Them." UK parliamentary paper Cd. 539, 1901.

Report of the Indian Irrigation Commission 1901–1903, Part I: General. UK Parliamentary Paper Cd. 1851, 1903.

Report of the Public Service Commission 1886–87. Calcutta: Superintendent of Government Printing, India, 1888.

Report of the Telegraph Committee 1906–07. Calcutta: Superintendent of Government Printing, India, 1907.

Report of the Disorders Inquiry Committee 1919–1920. Calcutta: Superintendent of Government Printing, India, 1920.

"Reports and Correspondence Relating to the Expediency of Maintaining the Royal Indian Engineering College, and Other Matters." UK parliamentary paper Cd. 2055, 1904.

Royal Commission on the Public Services in India. Appendix to the Report of the Commissioners, vol. 17. *Minutes of Evidence Related to the Post Office of India and the Telegraph Department.* UK parliamentary paper Cd. 7905. London: His Majesty's Stationery Office, 1915.

Royal Indian Engineering College Calendars, 1875–6 to 1902–3. Published in London, first by Allen, then by Harrison & Sons.

Summary of the Administration of the Earl of Minto, Viceroy and Governor-General of India in the Department of Commerce and Industry, November 1905 to July 1910. Simla: Government Central Branch Press, 1910.

Thomason Civil Engineering College Calendar, 1871–72. Roorkee: Thomason Civil Engineering College Press, 1871.

University of Madras, The Calendar for 1892–3. 1892.

Periodicals

Amrita Bazar Patrika (Calcutta, India)
Builder (London, England)
Civic Affairs (Kanpur, India)
Daily Graphic (London, England)
Engineer (London, England)
Engineering (London, England)
Fraser's Magazine (London, England)

Indian Engineering (Calcutta, India)*
Jackson's Oxford Journal (Oxford, England)
Lincoln, Rutland and Stamford Mercury (Stamford, England)
Madras Mail (Madras, India)
Morning Chronicle (London, England)
Nature (London, England)
Overland Ceylon Observer (Sri Lanka)
Pall Mall Gazette (London, England)
Pioneer (Allahabad, India)
Spectator (London, England)
Times (London, England)
Tribune (Lahore, Pakistan)

---

* Profiles of Coopers Hill men from Indian Engineering are collected in BL MSS EUR F239/108.

Index

Italicized page numbers represent figures and tables.

Milton Keynes UK
Ingram Content Group UK Ltd.
UKHW011254210424
441408UK00003B/42/J